Mobile Telecommunications in a High-Speed World

To Mair, Peter's much-loved first grandchild.
And in memory of Jason's much-loved grandmother, Joan.

Mobile Telecommunications in a High-Speed World

Industry Structure, Strategic Behaviour and Socio-Economic Impact

PETER CURWEN
Visiting Professor of Mobile Telecommunications, University of Strathclyde, UK

JASON WHALLEY
Reader, University of Strathclyde, UK

Routledge
Taylor & Francis Group
LONDON AND NEW YORK

First published in paperback 2024

First published 2010 by Gower Publishing

Published 2016 by Routledge
4 Park Square, Milton Park, Abingdon, Oxon OX14 4RN

and by Routledge
605 Third Avenue, New York, NY 10158

Routledge is an imprint of the Taylor & Francis Group, an informa business

Gower Applied Business Research
Our programme provides leaders, practitioners, scholars and researchers with thought provoking, cutting edge books that combine conceptual insights, interdisciplinary rigour and practical relevance in key areas of business and management.

British Library Cataloguing in Publication Data
Curwen, Peter J.
Mobile telecommunications in a high-speed world : industry structure, strategic behaviour and socio-economic impact.
1. Data transmission equipment industry.
2. Telecommunication equipment industry. 3. Online information services industry.
I. Title II. Whalley, Jason.
384.5'34-dc22

Library of Congress Cataloging-in-Publication Data
Curwen, Peter J.
Mobile telecommunications in a high-speed world : industry structure, strategic behaviour and socio-economic impact / Peter Curwen and Jason Whalley.
p. cm.
Includes index.
ISBN 978-1-4094-0361-6 (hbk) -- ISBN 978-1-4094-0362-3 (ebook) 1. Cell phone services industry. 2. Wireless telecommunication systems. 3. Telecommunication. I. Whalley, Jason. II. Title.
HE9713.C87 2010

ISBN: 978-1-4094-0361-6 (hbk)
ISBN: 978-1-03-292408-3 (pbk)
ISBN: 978-1-315-59569-6 (ebk)

DOI: 10.4324/9781315595696

Contents

List of Figures

List of Tables

List of Abbreviations

1G	first generation
2G	second generation
3G	third generation
3GPP	Third Generation Partnership Project
3GPP2	Third Generation Partnership Project 2
4G	Fourth generation
ANSI	American National Standards Institute
API	application programming interface
ARPU	average revenue per user
AS	assignment
ATM	automatic teller machine
AU	auction
AWS	advanced wireless service
BC	beauty contest
CDG	CDMA Development Group
CDMA	code division multiple access
CEO	chief executive officer
CEPT	Conference of European Post and Telecommunication Administrations
CMA	cellular market area
DL	downlink
DoT	Department of Telecommunications
DS	direct sequence/spread
DS-FDD	direct sequence frequency division duplex
EA	economic area
E-DCH	enhanced dedicated channel
EDGE	enhanced data (rates) for GSM evolution
EDGE+	evolved EDGE/EDGE evolution
E-GSM	enhanced global system for mobile
ERC	European Radiocommunications Committee
ETSI	European Telecommunications Standards Institute
EU	European Union
EUL	enhanced uplink
E-UTRAN	evolved UMTS terrestrial radio access network
EV-DO	evolution-data optimized
EV-DV	evolution-data and voice
FCC	Federal Communications Commission
FCFS	first-come first-served
FDD	frequency division duplex
FDMA	frequency division multiple access
GHz	Gigahertz

GNI	gross national income
GPRS	general packet radio service
Group 3G	Orange, Telefónica, Sonera
GSM	global system for mobile
H3GA	Hutchison 3G Australia
HDR	high data rate
HK	Hong Kong
HSDPA	high-speed downlink packet access
HSPA	high-speed packet access
HSUPA	high-speed uplink packet access
ICT	information, computing and technology
iDEN	integrated digital enhanced network
IEEE	Institute of Electrical and Electronics Engineers
IMS	IP multimedia subsystem
IMT-2000	International Mobile Telecommunications-2000
IP	Internet protocol
IPS	Internet service provider
ITFS	instructional television fixed service
ITU	International Telecommunication Union
JIL	Joint Innovation Lab
JMCS	Japanese mobile cellular system
kbps	kilobits per second
kHz	kilohertz
LBC	loosely backward compatible
LI	licensing
LTE	long-term evolution
MANs	metropolitan area networks
MAP	mobile application pack
M&A	merger and acquisition
Mbps	megabits per second
MC	multi-carrier
MC-FDD	multi-carrier frequency division duplex
MDS	multipoint distribution service
MHz	megahertz
MIMO	multiple input multiple output
MNP	mobile number portability
MSCs	mobile switching centres
MTR	mobile termination rate
MVNO	mobile virtual network operator
NGMN	next generation mobile network
NGNs	next generation networks
NSN	Nokia Siemens Networks
OECD	Organisation for Economic Co-operation and Development
Ofcom	Office of Communications
OFDM	orthogonal frequency division multiplexing
OFDMA	orthogonal frequency division multiple access
OS	operating system

PCN	personal communications network
PCS	personal communications services
PDC	personal digital cellular
PHS	personal handy-phone system
POP	per head of population
R&D	research and development
RA1	regional allocation 1
RA2	regional allocation 2
REAG	regional economic area group
RIM	Research In Motion
RSA	rural services/statistical area
RTT	radio transmission technology
SABA	simultaneous ascending-bid auction
SAE	system architecture evolution
SARI	Sustainable Access in Rural India
SBC	strictly backward compatible
SC-FDMA	single carrier frequency division multiple access
SDK	software development kit
SDMA	space division multiple access
SEC	Securities and Exchange Commission
SIM	subscriber identity module
SMA	spectrum management authority
SMS	short message/messaging service
TC	time code
TDD	time division duplex
TDMA	time division multiple access
TD-SCDMA	time division synchronous code division multiple access
TE	tender
TMT	technology, media & telecommunications
UAS	unified access services
UHF	ultra-high frequency
UL	uplink
UAE	United Arab Emirates
UMB	ultra mobile broadband
UMTS	universal mobile telecommunications system
UTRA	universal terrestrial radio access
UTRAN	UMTS terrestrial radio access network
UWB	ultra-wideband
VHA	Vodafone Hutchison Australia
WAP	wireless application protocol
W-CDMA	wideband code division multiple access
WiMAX	worldwide interoperability for microwave access
W-LAN	wireless local area network
WLL	wireless in the local loop
WMAN	wireless metropolitan area network
WRC	World Radiocommunications Conference

Preface

Ten years ago, so-called 3G (third generation) mobile telecommunications was a hot topic as the first batch of licences had recently been issued. At the time, a particular debate was going on about the merits of auctions versus beauty contests as a method for assigning licences. But then, for a while, nothing much happened. More licences were issued, a few network roll-outs were set in hand and a lot of people began to argue that it had all been a horribly expensive mistake. It is still unclear how much money is going to be made or lost in the quest to provide high-speed data services, but over the past three years or so there have been massive developments in terms of the number of licence issues and launches worldwide, in the speed of data transfer, in the range of services that can be accessed, in the range of devices that can be used to access them and in the strategies of operators, equipment vendors and software writers. This book sets out to tell the story of these developments.

The core research which underpins this book was intended to be a fairly straightforward affair in the sense that it involved real-world events, namely the award of high-speed data licences and the launch of the corresponding networks. As one of the authors had been collecting this information on a regular, if sporadic, basis for many years, it was already clear that there existed conflicting versions of these events, and that many details were, for whatever reason, being withheld from the public domain.

Nevertheless, when the research was formally set in hand in September 2008, with the assistance (for three months) of a full-time student, on the Masters in Communications degree at Strathclyde University, who was writing a dissertation on the topic, it rapidly became evident that the situation was far worse than the authors could possibly have imagined. The existing public domain databases agreed on remarkably little, and most of that was in conflict with statements published on the Internet.

Even after an initial database had finally been created by mid-2009, information would suddenly appear on the Internet that appeared to conflict with it, necessitating further research to verify the truth. That this should be happening after nine months of intensive work may seem unlikely, so one example, dating from as late as September 2009, will suffice to indicate the problems that needed to be addressed.

At that point, some information was published that threw into doubt the agreed database entry for a two-operator country in the Middle East. The problem did not appear to relate to the new entrant because it had been awarded a high-speed (3G) licence at the same time as its basic (2G) licence, but rather to the incumbent. So the authors wrote to the incumbent asking it to verify the award date and conditions of its own high-speed licence. The reply, asking us to consult the regulator, was not merely unhelpful but rather odd. Did the operator really not know the details of its licence?

So we wrote to a contact we had at the regulator who sent us three documents. These made clear, *inter alia*, that the incumbent was entitled to apply for a high-speed licence on the same terms and conditions as the new entrant once the latter was awarded its licence, *but there was no reference to when the incumbent obtained the licence* (which the incumbent, as noted, would not tell us directly). Furthermore, one document gave a date for the licence award to the new entrant *seven months later than the date on which that operator had clearly stated in public that it had been awarded the licence*.

Now the latter issue could be rationalized by arguing that being awarded a licence is not the same thing as the licence being handed over. This indicates one of the reasons for the above discussion, namely the need to devise a series of definitions so as to match up the various entries in the database. These definitions, of which there are a significant number, are clearly set out in the relevant chapters. Nevertheless, readers are given due forewarning that because these definitions are unique to this book – which has the obvious benefit that it now contains the only internally compatible databases on high-speed licences and launches in the public domain – they are of necessity incompatible with such limited alternative databases as can be found elsewhere.

Overall, the authors have been taken aback by the unwillingness or inability (though one might wonder how that could be) of operators and regulators to disclose what are surely matters of fact that should be in the public domain. But the most significant revelation – in reality, perhaps, a confirmation of a conclusion drawn from previous research – has been the astonishing degree of misinformation published on the Internet. Crudely expressed, a common process appears to be that one website publishes a story, which is then picked up by others which repeat it, often with facts or, more commonly, percentages and other numbers, carelessly altered. But one common thread is that many of those who place entries on the Internet have only a passing understanding of the topics on which they comment. A classic example relates to a company called Turkcell. For a considerable period, the authors (who are somewhat unhappy to admit it) were unaware that there are two companies that use this name – one a holding company the other a mobile operator. The reason for this unawareness on the part of the authors is that no document they had seen – among hundreds – had ever used anything other than the word 'Turkcell' – or the full name of the operator, often when referring to the holding company – and it was only when devising the schematic for another operator's assets that they realized that there had in fact to be two Turkcells if the ownership percentages were to tally.

Another issue worthy of comment is the apparently huge degree of freedom that companies have in determining what they put in their annual reports and submissions to regulatory bodies. For example, most annual reports cite ownership percentages for directly held subsidiaries, and, usually in footnotes elsewhere in the report, the basis on which they consolidate such assets. But even if this is noted – and it is easy to miss – there is rarely any reference in the footnotes to the actual percentage owned. Often, the report will simply say that since the subsidiary is consolidated, all of its subscribers are claimed to be owned by the parent. Furthermore, it is rare for a report to include any reference to indirectly held assets, especially if they are held by a minority-owned subsidiary.

However, the purpose of this Preface is not simply to carry on about the problems of doing quantitative research in the mobile communications industry. The problems exist and they have been addressed. The rest of the book is an attempt to tie together a series of related topics that add flesh to the bare bones of which operators have been licensed to provide high-speed data services and whether their networks have been launched. Other issues that immediately spring to mind include, among others:

- How has auction theory contributed to, and in turn been affected by, the issue of high-speed licences?
- How has the issue of these licences affected the industry structure?
- Which equipment vendors are involved and how have they fared during the past decade?
- What mobile platforms are involved?
- How have mobile devices evolved in recent times?
- What role is being played by the main software developers?
- What is the relevance of diffusion standards and value chains?
- What effect has the licensing of progressively more sophisticated mobile networks had upon emerging economies?

The book attempts to answer these and other questions although, of necessity, the discussion must sometimes be circumscribed by the need to keep the book within a pre-ordained length. The most obvious factor in this respect is that there are hundreds of operators and hundreds of countries, and these must be discussed either selectively or on some kind of aggregated basis.

It was also decided that the book's content should adhere to certain principles. Primary among these was that the databases should be as up to date as humanly possible, with a large proportion of them being based upon 31 December 2009. Bearing in mind that the book was delivered to the publishers in January 2010, we believe that this is almost unprecedented among academic texts.

Secondly, the book is intended to be accessible to a wide audience encompassing academics, those who work in the telecommunications industry, students on relevant courses and possibly even the general public. Partly for this reason there are no equations. There are, no doubt, certain elements in the academic community who will (not for the first time) invoke the criticism that a book without equations is simply reportage. However, we have read a very large number of books and articles on mobile communications and we have become increasingly convinced that, since the databases being used are seriously flawed, the equations are themselves often misleading: To claim a cause and effect relationship based on faulty data does not 'prove' anything.

Thirdly, clarity is a keynote of the book. For example, some of the material is necessarily presented in a historical context and it is essential to keep the reader's attention while the narrative unfolds. For this reason, a relatively light touch has been used in the style of writing employed

Turning now to the content, we have tried to do several things. In the first place, we have set out to report the results of some original research. Secondly, we have set out to summarize a vast amount of material available in the public domain, having sifted through it carefully to distinguish truth from fiction. Thirdly, we have produced a number of up-to-date databases. Fourthly, we have produced some case studies not available elsewhere. Overall, we have set out to produce a far-reaching overview of how mobile telecommunications has evolved during the past decade with the emphasis upon things that can be characterized as 'high-speed' compared to what went before.

Professor Peter Curwen
Dr Jason Whalley
15 January 2010

Acknowledgements

We would firstly wish to thank Carlos Rebellon, a student on the Masters in Communications Management course at Strathclyde University, for his assistance in tracing information on high-speed data licences and launches.

This book has been written in its entirety by the authors, but certain sections have appeared in a somewhat different form in previous publications. We would accordingly like to express our appreciation to Emerald Publishing for its willingness to let the authors reproduce work that has appeared in its journals. In this case, it applies to material in a number of articles published by the authors in the journal *info*.

CHAPTER 1

High-Speed Data Transmission

Introduction

The mobile – alternatively, wireless or cellular radio – communications market encompasses a variety of radio telephone services that are provided by connecting together two or more mobile devices or terminals via a base station. For the purposes of the discussion below, these devices will for the most part be taken to be handsets since they dominate current usage, although they are constantly evolving both in form and function (see Chapter 7). Suffice it to say for now that whatever form a device may take – including data cards and modems in computers – it is currently already capable of incorporating some or most of the following: blogs, calculator, calendar, camera with flash, clock, diary, digital music player, digital purse, email facility, games, instant messaging, modem, satellite tracking, spreadsheet, television, text messaging, torch, video camera and word processor. This list is incomplete, and becomes ever more incomplete with each passing day. The device is also useful for conveying voice, which remains its primary function in its guise as a handset. In practice, the owner of a modern 'converged' handset can do without, for example, a watch, alarm clock, camera and music player, and there is growing evidence that the loss of a handset is regarded as more distressing than that of a wallet or credit card.

Mobile communications have a long and complex history (Anderson, Daim and Kim, 2008; Curwen and Whalley 2008: Chapter 2; Fuentelsaz, Maicas and Polo 2008; Hausman 2002). Throughout the world, the standard digital technology is currently known as second-generation or 2G. The original, analogue, technology (1G), where each user has exclusive use of a specific frequency, is still to be found, but is being phased out wherever possible and the spectrum transferred to 2G networks which, because they are digital, use the spectrum much more efficiently.

In terms of data transfer speeds, 2G broadly – greater precision is not possible since even the various trade associations fail to agree on this – provides a maximum downlink of between 10 kbps (kilobits per second) and 30 kbps, and 2.5G a maximum lying between 100 kbps and 200 kbps. Such speeds clearly do not represent what is meant by a high-speed data network. However, an awkward aspect is that technology providers prefer to define technologies in terms of their maximum theoretical speed – that is, achievable in a laboratory with a single user – whereas retail customers are more concerned with what minimum speed can reasonably be delivered. The reality is that the typical speeds enjoyed by retail customers fall well short of the claimed maxima. In order to make sense of this, a technological framework is needed.

A Technological Framework

Table 1.1 provides such a framework. It has been specifically developed for the purposes of this book and is wholly original. As is evident, it involves the categorization of mobile technologies into a series of generations. Originally, as indicated above, this simply formed the sequence '1, 2, 3', but even the addition of a '.5' division is no longer sufficient. As also indicated above, the second obvious method for segregating technologies is according to the speed of data transfer that they can handle, but whereas there were originally minor, if any, overlaps, such overlaps are now the norm because of the tendency for equipment vendors to cite a maximum speed rather than the (generally much lower) average speed delivered to subscribers.

Table 1.1 also indicates how technology has progressed using a rather primitive regional breakdown (a more sophisticated method is utilized in subsequent chapters). As shown, certain regions of the world have developed either along one technological path (Western Europe), along competing pathways (for example, North and Latin America) or – but only in the case of individual countries – developed unique pathways that have run alongside those followed elsewhere.

The other crucial factor to bear in mind is that each pathway can either progress using the spectrum bands allocated at the 2G stage, or by introducing new spectrum bands primarily in order to avoid congestion due to the build-up of data traffic. Licences always need to be issued at the 2G stage, but there is strictly no need to issue further licences provided the technological pathway remains within the same spectrum band – that is, the technology advances by overlaying the existing network primarily through the use of new software (but occasionally of pieces of hardware). However, if a new spectrum band needs to be utilized, it becomes necessary to introduce a new set of licences or to amend the existing licences to encompass the expanded spectrum.

Spectral efficiency varies according to the spectrum band utilized, and hence a mobile network has to be reconfigured somewhat to make best use of any new spectrum that is introduced. However, the essential configuration of a mobile network (Curwen and Whalley 2008: Chapter 2) is not affected. Mobile networks currently use packet-switching which divides up data into individual packets of a specified size and in a specified format before sending them along the network. The correct sequence for delivery and the address for delivery are determined via control information attached to the packets before they are sent. As a packet-switched network is permanently available to all data from any source, the data inevitably get jumbled up together and may take a variety of routes from any one source to any one destination. However, the control instructions should ensure that the data is reassembled into the correct order before delivery. Since this may be subject to delays, it is advisable to keep voice data – which are time-critical – separate from non-voice data.

It is evident that this system must recognize the identity of individual subscribers, and this is normally achieved via the subscriber identity module (SIM) which comes as a small card to be inserted in a mobile device. It contains a unique international mobile subscriber identity used to identify the subscriber when they are logging into the system.

Table 1.1 Technological framework

Region/country	2G	2.5G	2.75G	2.75G+	3G	3.5G	3.75G	3.75G+	4G
Western Europe	GSM	GPRS	EDGE	EDGE+	W-CDMA (UMTS)	HSDPA	HSPA	HSPA+	LTE
North America Latin America Asia-Pacific Eastern Europe/CIS Middle East Africa	CDMA GSM iDEN	cdma2000 1x RTT GPRS	1xRTT EDGE	 EDGE+	cdma2000 1xEV-DO Rel. 0 W-CDMA	EV-DO Rev. A HSDPA	Rev. B HSPA	HSPA+	UMB LTE
China	CDMA GSM	cdma2000 1x RTT GPRS	1xRTT EDGE EDGE	 EDGE+ EDGE+	cdma2000 1xEV-DO Rel. 0 W-CDMA TD-SCDMA	EV-DO Rev. A HSDPA HSDPA	Rev. B HSPA HSPA	Rev. B HSPA+ HSPA+	LTE LTE TD-LTE
Japan	CDMA PDC	cdma2000 1x RTT i-mode	1xRTT EDGE	 EDGE+	cdma2000 1xEV-DO Rel. 0 W-CDMA	EV-DO Rev. A HSDPA	Rev. B HSPA	Rev. B HSPA+	LTE LTE

Source: Compiled by authors

Notes: The main abbreviations are as follows:

CDMA = code division multiple access, also known as IS-95 or cdmaOne.
cdma2000 1xRTT (radio transmission technology) is also known as IS-2000
cdma2000 1xEV-DO (evolution-data optimized)
EDGE = enhanced data (rates) for GSM evolution
EDGE+ is otherwise known as evolved EDGE or EDGE evolution
GSM = global system for mobile, first launched by Radiolinja on 1 July 1991
GPRS = general packet radio service.
HSDPA = high-speed downlink packet access
HSPA = high-speed packet access involving upgrades to HSDPA and use of HSUPA
HSPA+ is otherwise known an evolved HSPA, HSPA evolution or Internet HSPA (I-HSPA)
HSUPA = high-speed uplink packet access
iDEN = integrated digital (sometimes dispatch) enhanced network
JMCS = Japanese mobile cellular system
LTE = long-term evolution
PDC = personal digital cellular
TD-SCDMA = time division synchronous code division multiple access
TDMA = time division multiple access
UMB = ultra mobile broadband, otherwise known as EV-DO Rev. C
W-CDMA = wideband code division multiple access

As shown in Table 1.1, global system for mobile (GSM) is the common base digital system in Western Europe. It uses the time division multiple access (TDMA) standard which allows each call to run along a channel within the available bandwidth and assigns it a fraction of the time available within that channel. In contrast, code division multiple access (CDMA) assigns a special electronic code to each signal and hence allows the entire frequency band to be occupied simultaneously. This is claimed to provide greater capacity, better sound quality, lower power consumption and a decreased potential for fraudulent use.

Most of the 2G licences in Western Europe were issued between 1992 and 1996, and utilized spectrum in the 900 MHz GSM band. In the 900 MHz band, the uplink is 890–915 MHz and the downlink is 935–960 MHz. The 25 MHz band is divided into 124 carrier frequency channels, each spaced 200 kHz apart, with eight speech channels per frequency channel. The term GSM is also used generically in relation to spectrum in the 1800 MHz band (1710–1785 and 1805–1880 MHz) otherwise known as a personal communications network (PCN). PCN spectrum was typically licensed in Western Europe after 1997, either to existing GSM licensees to ameliorate spectrum shortages as mobile telephony surged in popularity or to permit the entry of new operators. A crucial factor was that the administrative arm of the European Union, the European Commission, enforced the exclusive usage of these two spectrum bands via the auspices of the European Telecommunications Standards Institute (ETSI).

Elsewhere in the world a variety of spectrum bands are in use for 2G. Overall, however, by far the most popular technology is GSM, although outside Europe it is generally used in alternative spectrum bands – typically the 800 MHz band (824–849 and 869–894 MHz) and/or the 1900 MHz band (1850–1910 and 1930–1990 MHz) known in the USA as personal communications services (PCS). The use of GSM is fostered by the factor of international 'roaming' – that is, the ability of a GSM handset owner to take it to another country where it can still be used both internally and to make international calls to other countries where GSM is in use (although a multi-band handset may be required).

Given the use of GSM throughout Europe, as well as its partial use in the USA, there was accordingly an obvious advantage for countries switching from analogue to digital to adopt it. However, some countries adopted one of the alternative standards to GSM before it became so well-established. In particular, the government in the USA made no attempt to enforce a national standard, and hence three different technologies were adopted (with (integrated digital enhanced network) iDEN also attaining a modest popularity in Latin America, but in essence being ignored elsewhere). In addition to GSM, there was widespread use of both CDMA, transmitting in the 450 MHz, 800/850 MHz, 1700 MHz and 1900 MHz bands, and TDMA.

Third-generation mobile telephony (3G) is essentially an extension of 2G capabilities, although the developmental process has in practice required an intermediate stage, known commonly as 2.5G, to be introduced. The wideband code division multiple access (W-CDMA) version of 3G, incorporating both terrestrial and satellite components, is known in Western Europe as the universal mobile telecommunications system (UMTS), and that term has increasingly become used in preference to W-CDMA. UMTS has been described by ETSI as 'personalised globally accessible high quality mobile communications services', but it is more commonly described as 'mobile interactive multimedia services'. The official European Commission version is that UMTS is an advance on GSM capable of combining the use of terrestrial and satellite components (Council of the European Communities,

1998, 1999). UMTS did not strictly represent a major technological breakthrough because general packet radio service (GPRS) already provided many services that were not originally expected to evolve until a later stage of development.

2.5G

For the purposes of this book, 2.5G does not qualify as a high-speed data service and is accordingly not discussed in detail. The GPRS which emerged from GSM was slow to establish itself, partly because of the somewhat adverse reaction to the wireless application protocol (WAP) in Western Europe. The GPRS infrastructure and handsets supported 'always-on' data transmission at up to 21 kbps per channel over a maximum of eight channels – that is, a theoretical maximum of around 168 kbps combining several downlink and uplink channels. Typically, a device accommodated four downlink and two uplink channels, with the bandwidth divided up between data and voice according to the strategy of the operator, delivering in the range 60–80 kbps with bursts up to 115 kbps. cdma2000 1xRTT, the equivalent development along the CDMA pathway, tended to deliver somewhat higher data speeds and hence was touted as a superior technology – it was not unknown for CDMA advocates to identify it misleadingly as 3G. In Japan, a proprietary version of 2.5G known as i-mode was developed by NTT DoCoMo with considerable success (Peltokorpi, Nonaka and Kodama 2007), but this success has yet to be emulated on the international stage (Fransman 2002).

2.75G

Although cdma2000 1xRTT is clearly not 3G, it can reasonably be included in the 2.75G category where its equivalent along the GSM pathway is not GPRS but enhanced data [rates] for GSM evolution (EDGE). EDGE offers very good spectral efficiency up to 100 kbps – until recently its transmission speed in typical circumstances. In essence, the advantage of EDGE is that it offers improvements upon GPRS by utilizing more error-tolerant modulation and coding schemes. Each timeslot is theoretically capable of providing 59.3 kbps, and hence bundling eight timeslots provides a theoretical maximum of 473.6 kbps. Until the initial delays with UMTS became severe, few expected EDGE to be introduced other than very occasionally. The reason for this is that EDGE needs its own handsets and possibly new hardware in base stations, although the latest GSM/GPRS networks need only changes to radio planning and operational parameters via software upgrades. It should also be borne in mind that EDGE cells are smaller than GSM cells. Because EDGE was developed as a means for TDMA operators to provide more sophisticated data services, it was thought that it would not be introduced in Europe other than, perhaps, by operators that either chose not to bid for, or failed to acquire, UMTS licences. The early adopters, accordingly, were not European.

By early 2006, when EDGE provided up to 240 kbps downlink (DL) and uplink (UL), it had become evident that EDGE would increasingly be seen as a means for covering relatively rural areas that were expensive to cover with 3G and which could be left uncovered for some considerable time because the maximum 3G coverage requirements were, typically, around 70 per cent of the population. At this point Nokia announced that it would launch a faster version of EDGE, known as EDGE+ (EDGE evolution or evolved EDGE), with work proceeding to produce a 3GPP[1] standard for EDGE+ via Release 7 (3G Americas 2006; Wikipedia 2009a). In principle, this would use two simultaneous radio channels compared to the one channel used for EDGE at the time, and hence would double the data download speed. However, it had the potential to double this again because it used an enhanced modulation and coding scheme.

As ever, there was a delay until compatible hardware could be made available on the requisite scale, so a wide-scale launch did not occur until 2009. At this point, the introduction of Dual Carrier enabled a roughly doubled downlink speed. So-called EGPRS-2 will provide up to 1.2 Mbps (DL) and 474 kbps (UL) in 2010 with a theoretical maximum of 1.9 Mbps (DL) and 947 kbps (UL).

The Origins of 3G

In late 1997, the International Telecommunication Union (ITU) published a report – *Framework for Modularity and Radio Commonality within IMT-2000* – which advocated the introduction of a standardized system within the context of International Mobile Telecommunications-2000 (IMT-2000) by the year 2002 (Commission of the European Communities 1997). The idea was that both carriers and equipment manufacturers would be provided with a set of common building blocks with which any third-generation system had to be made compatible. This was a critical aspect since it signified that there was not to be a unique standard, but rather a 'family of standards' which would allow all existing networks, with their attendant historic heavy investment, to be modified rather than discarded.

The key requirement placed upon a 3G network by the ITU was its data transfer speed which would, in practice, apply to the downlink:

- 2 Mbps in fixed or in-building environments;
- 384 kbps in pedestrian or urban environments;
- 144 kbps in wide-area mobile environments;
- variable data rates via satellite.

It is immediately apparent that EDGE+, for example, is now technically qualified to be called 3G, and in the light of subsequent developments a target of under 400 kbps while on the move appears to have been conservative. Nevertheless, at the time – a mere decade ago – it was a serious challenge.

1 3GPP is an international standardization initiative concerned with GSM and its derivatives that was originally founded by ETSI and similar bodies elsewhere in the world. 3GPP periodically produces a complete current list of its specifications in a sequence of Releases. Much of the history can be traced via publications of 3G Americas, notably *The Evolution of UMTS – 3GPP Release 5 and Beyond*, a White Paper first published in June 2004 and updated in November 2004, and *The Evolution of UMTS/HSDPA – 3GPP Release 6 and Beyond*, a White Paper published in July 2005. 3GPP2 is the equivalent of 3GPP applied to derivatives of CDMA (see www.3GPP2.org). 3GPP does not publish a complete list of its specifications, but announces each one as it is approved.

The optimum spectrum for a 3G network did not really differ from that chosen for 2G, and indeed 3G could utilize any spectrum lying between 400 MHz and 3 GHz. However, international roaming between networks was a major objective, and hence it had already been determined, as a result of a decision made in 1992 by the World Radiocommunications Conference (WRC),[2] that the common frequencies in all third-generation systems would be restricted to 1885–2025 MHz and 2110–2200 MHz. Not surprisingly, some commentators did question whether the technical advantages derived from a move to new spectrum would be greater than the drawbacks arising from the need to design, build and test new networks from scratch, build additional base stations and deal with the need to migrate progressively from 2G networks using lower frequencies.

It is important to note, first, that the designated spectrum was not necessarily unoccupied; and, secondly, that even if it was (as in the then European Union) it might prove to be inadequate in the face of an explosion of data. In June 1997, the European Radiocommunications Committee (ERC) decided to divide up the WRC frequency bands, allocating 1900–1980 MHz, 2010–2025 MHz and 2110–2170 MHz to terrestrial 3G, as well as 1980–2010 MHz (earth-to-space) and 2170–2200 MHz (space-to-earth), to satellite applications which use a space division multiple access (SDMA) interface, but recognized that this would probably be insufficient and that additional spectrum would almost certainly need to be freed or reallocated by 2005 within the 900, 1800 and 1900 MHz bands used most commonly by 2G systems (Commission of the European Communities 1998a).

In a further move, the ITU divided the new system into two parts based respectively upon the core network architecture and the radio (air) interface – linking base station and handset and known as the universal terrestrial radio access (UTRA) – and specified common components in each case. The telecoms industry was then left to work out how best to meet the requirements of each. So far as the core architecture was concerned, the choice was fairly limited in practice given the predominance of GSM in Europe, and cdmaOne and TDMA in the USA – the two respective core architectures are generally referred to as GSM-MAP (mobile application pack) and ANSI (American National Standards Institute)-41. The ITU wanted everything to be settled by the end of 1999 so that services could commence in 2002 (*The Economist* 1998).

Europe's leading mobile handset manufacturers announced that they would be backing GSM-MAP for the core architecture. For the radio interface the choice lay in essence between wideband CDMA (W-CDMA – now called 3GSM by the GSM Association) and a hybrid wideband TDMA/CDMA (W-TD/CDMA). Whereas Japan opted for W-CDMA throughout, the European (UMTS) approach to the radio interface contained elements of W-TD/CDMA. Work was set in hand to bring these into close compatibility.

The issue of compatibility with standards in the USA presented the most awkward problems because Lucent Technologies and Qualcomm wanted to use cdma2000. The ITU agreed to accept cdma2000 1xEV-DO as equivalent to 3G in August 2001. In its original Release 0 variant, 1xEV-DO was capable of delivering downlink at up to 2.4 Mbps while Revision A, approved by the 3GPP in April 2004, upgraded the downlink to a theoretical maximum of 3.1 Mbps with a maximum of 1.8 Mbps for the uplink. Although

2 The WRC meets every two or three years – it met, for example, in 1995, 1997, 2000 and 2003. Its conferences are organized by the International Telecommunication Union, a specialized agency of the UN – see www.itu.int/ITU-R/conferences/wrc. The WRC's job is to review and, where necessary, revise the Radio Regulations – the international treaty obligations governing the global use of the radio spectrum and of satellite orbits.

Revision A only came on stream towards the end of 2006, it was already determined that Revision B would in due course upgrade the downlink to a maximum of 4.9 Mbps (Wikipedia 2009b).

RADIO INTERFACE MODES

The present position is that there are three main radio interface modes within IMT-2000, known respectively as IMT DS (direct sequence or spread), IMT MC (multi-carrier) and IMT TC (time code). These modes are known more commonly as direct sequence frequency division duplex (DS-FDD – a type of CDMA (W-CDMA) supported by GSM carriers which uses paired (separated uplink and downlink) spectrum and is sometimes referred to as UTRA FDD); multi-carrier frequency division duplex (MC-FDD – which uses paired spectrum and is based on cdma2000); and Time Division Duplex (TDD – which comes in a version known as UTRA TDD or one harmonized with the new TD-SCDMA (time division synchronous code division multiple access) standard, both of which use unpaired spectrum for the provision of both uplink and downlink).

It is to be noted that FDD/W-CDMA uses a channel spacing of 5 MHz as compared to 1.25 MHz for cdma2000 1xRTT, and that an operator normally needs either three or four channels to provide an acceptable high-speed service. The standard allocation of spectrum is accordingly 1920–1980 MHz paired with 2110–2170 MHz – that is, 60 MHz paired – normally providing either 15 MHz paired (four licences) or 20 MHz paired (three licences) per licensee. TDD/SCDMA simply provides 5 MHz of spectrum within the 1900–1920 and 2010–2025 MHz bands. However, it is unusual in practice for seven lots of TDD to be allocated, and TDD spectrum can be paired up, if so desired, just as it is possible to get by with a minimum of 10 MHz of paired spectrum (enabling six licences to be issued).

WRC 2000

At the World Radiocommunications Conference 2000 (WRC-2000), governments of 150 countries again addressed the issue of spectrum allocation in the hope of agreeing a single global range for further expansion after 2005, in essence consisting of 160 MHz in one continuous band in the 2.5 GHz (2,500 MHz) band. However, the best that could be agreed was a choice of three frequency bands, namely 806–960, 1710–1885 and 2500–2690 MHz – the third band options being heavily favoured by European organizations because it was the only one not currently occupied by 2G. In addition, the bands 1525–1544, 1545–1559, 1610–1625.5, 1626.5–1645.5, 1646.5–1660.5 and 2483.5–2500 MHz could be used for satellite transmission, as well as 2500–2520 and 2670–2690 MHz should the need arise. For its part, the European Commission issued, on 9 March 2001, a fourth Mandate to the Conference of European Post and Telecommunication Administrations (CEPT) to harmonize frequency usage in the additional bands identified in 2000. This resulted in a Report and an ECC Decision in line with the scope and tasks given under Mandate 4. The key points included in the adopted EEC Decision, ECC/DEC/(02)06 of 15 November 2002, were:

- designation of the whole 2500–2690 MHz band to IMT-2000/UMTS systems;
- making the whole so-called expansion band available for use by 1 January 2008, subject to market demand and national licensing schemes;
- designation of the frequency band 2520–2670 MHz for terrestrial use.

For its part, the USA refused to commit itself to any of the original spectrum bands because much of all this bandwidth was already occupied (Curwen 2005). To ameliorate matters somewhat, the WRC also accepted that 3G networks could be built using existing 2G spectrum. In certain cases the spectrum could be in the 450 MHz band (CDMA-450) which would enable large, mainly rural, areas to be covered without building large numbers of base stations. This is because whereas a GSM cell has a maximum 37 km radius, a CDMA-450 cell can 'breathe' and hence cover large but empty territories, a characteristic shared, in practice, by any variant of W-CDMA.[3]

Towards the end of 2005, there was much discussion as to whether the 900 MHz band, already in heavy use for GSM, could also provide 3G services. The potential use of the GSM band was generally thought to be limited to rural areas where GSM usage was less heavy, but in any event the regulatory and spectrum-management issues were far from full resolution – for example, the EU GSM Directive reserved the 900 MHz band exclusively for 2G services – so the UMTS900 band was not going to be used before 2007 at the earliest, not least because of the need to develop the requisite handsets.

There was pressure for the GSM Directive to be amended – the European Commission officially proposed this in July 2007. Meanwhile, at the beginning of 2007, the CEPT adopted a decision to permit the use of UMTS900 throughout the EU, subject to authorization by national regulators. The first countries to take advantage of this opportunity were Finland, Portugal and France where, due to the existence of only three operators, GSM spectrum was relatively plentiful. In March 2007, the Portuguese regulator decided to open up the 900 MHz band for 3G services without imposing any conditions on technical aspects. What was claimed to be the world's first network was launched by Elisa in November 2007 and covering two municipalities.

After holding up proceedings for some time, the European Parliament eventually relented, and on 24 March 2009 the required amendment to the GSM Directive was approved (Global Insight 2009b; Global mobile Suppliers Association 2009a) with a formal authorization date of October. By May 2009, 10 networks had been launched and a number of non-EU countries, such as Australia, Indonesia and New Zealand, had also authorized refarming of the GSM band (Global mobile Suppliers Association 2009b). However, the biggest difficulty that remains is how to appease those new entrants who only have 3G spectrum and who are not going to be handed GSM spectrum by existing licensees on a voluntary basis (Global mobile Suppliers Association 2007; Parker 2009).

As can be seen, the issue of spectrum use for 3G has become increasingly complicated over time, albeit largely in relation to the GSM pathway. A full list of UMTS frequency bands can be found in Wikipedia (2009e).

---

3 The obvious justification for exploring the use of CDMA-450 (often CDMA450) is that one base station provides roughly the same coverage as 25 UMTS base stations, and a CDMA-450 network is roughly 30 per cent cheaper than a CDMA-800 network and up to 70 per cent cheaper than a CDMA-1900 network. However, the absence of a facility to roam out of the 450 MHz band presents a problem. As analogue networks are shut down, the vacated 400–450 MHz spectrum band becomes available for other purposes. Because the typical spectrum available is less than the 5 MHz paired required by W-CDMA, the 450 MHz band can in general only be developed along the cdma2000 route.

cdma2000 1xEV-DO

Because of the complexity involved in shifting from GSM to W-CDMA, the previous discussion has necessarily concentrated upon this version of 3G. However, it is also necessary to clarify the situation with respect to the cdma2000 1x family of technologies. The initial design of 1xEV-DO was created by Qualcomm in 1999 to meet IMT-2000 requirements for a stationary communications downlink that would operate at over 2 Mbps. Originally called high data rate (HDR), it evolved into evolution-date optimized (EV-DO) after ratification by the ITU and was allocated the designation IS-856. Although DO initially stood for 'data only', this was altered to 'data optimized', probably because it had more positive connotations. In order to function alongside a voice connection, a further radio channel of 1.25 MHz had to be added (Wikipedia 2009b).

It is of interest that a variant of cdma2000 1xEV-DO, known as cdma2000 1xEV-DV (which combines circuit-switched voice and packet data in the same radio channel), has effectively died the death. Initially, operators with existing voice networks favoured EV-DV because it was backwards-compatible without the need for an overlay, while those without such a network and design engineers favoured evolution data optimized (EV-DO) because it did not need to be backwards compatible and hence could be designed more flexibly within a fully-internet protocol (IP)-enabled network, doing away with the likes of mobile switching centres (MSCs) and hence cutting costs. In the event, EV-DV equipment was slow to come to market and, in the meantime, operators were able to meet their objectives with 1xEV-DO. Hence, they were inclined to resort to 1xEV-DV only if they expected to run short of spectrum, which was not the norm. When LG Telecom of South Korea struggled to roll out a network using its 3G licence, EV-DV was consigned to the technological dustbin.

cdma2000 1xEV-DO, like other technologies, is undergoing progressive development which could in principle be termed 3.5G, 3.75G and so forth. Revision (Rev.) A came on stream in 2006 – Motorola unveiled its prototype solution in November 2005 – and had the potential to provide 3.1 Mbps downstream and up to 1.8 Mbps upstream which made it roughly equivalent to high-speed downlink packet access (HSDPA). At the same time, Revision (Rev.) B was nearing completion within standards body Third Generation Partnership Project 2 (3GPP2). Rev. B was designed to enable dynamic scalable bandwidth capability, potentially enabling operators to combine up to fifteen 1.25 MHz channels, although initially two or three channels would probably be used to provide a maximum of 14.7 Mbps, comfortably in excess of the maximum possible with Rev. A. It is treated in Table 1.1 as roughly equivalent to high-speed packet access (HSPA) although it certainly had the potential to achieve equivalence with HSPA+. Huawei launched the first commercially ready Rev. B platform in June 2009 with a maximum of 9.3 Mbps (DL) and 5.4 Mbps (UL). Significantly, it was capable of a smooth migration to long-term evolution (LTE) (Cellular-news 2009a). However, only Japan's KDDI and China Telecom had made a commitment to Rev. B by mid-2009 (Gabriel 2009a).

Rev. C was already under development. In July 2006, 3GPP2 selected from among 11 different proposals one that was loosely backward compatible (LBC), using orthogonal frequency division multiple access (OFDMA) (IEC 2005) on the forward link and a combination of CDMA and OFDMA on the reverse link. The strictly backward compatible (SBC) approach favoured by many, and which enabled compatibility between legacy and new handsets, was rejected. In December 2006, the CDMA Development Group announced that Rev. C would be known as ultra mobile broadband (UMB) and claimed that it would operate at up to 288 Mbps in a 20 MHz bandwidth. A standard was published in September 2007 with a view to commercial availability in 2009 (Cellular-news 2007a), although no operator had committed to its use by the end of 2007 (Cellular-news 2007b). As a consequence of the widespread commitment by 1xEV-DO operators to move towards the adoption of LTE during 2008 (see below), UMB was effectively abandoned in November 2008, but there are those who believe that it is merely hibernating pending the appearance of compatible handsets (Cellular-news 2009b).

The Search for New Spectrum

At this juncture it is necessary to pause in order to consider the most important issue which arises once it becomes possible to transfer data at speeds in excess of 20 Mbps. Spectrum, as noted, is scarce in the bands that are regarded as optimal for mobile communications. Once 20 Mbps or thereabouts is achieved, these bands will quickly fill to capacity (Cellular-news 2009d), generating a demand for new spectrum.[4] But in which bands?

In practice, although the use of IMT-2000 frequency bands for mobile worldwide interoperability for microwave access (WiMAX) is particularly contentious, the World Radiocommunication Conference held in October/November 2007 (WRC-07) needed to grapple more broadly with how to find additional spectrum wherever it could for what used to be referred to in the ITU as 'systems beyond IMT-2000' – that is, beyond 3G – but which are now commonly known as IMT-Advanced (Office of Communications 2007) although officially everything – whether IMT-2000 or IMT-Advanced, is henceforth referred to as IMT. The ITU Radiocommunication Sector Working Party 8F is heavily responsible for progressing this project. The full list of bands under consideration in 2007 is set out in Table 1.2.

According to Report ITU-R M.2078 produced by WP8F, IMT would require 1,280 MHz of spectrum (as compared to the 580 MHz already identified for IMT-2000) by the year 2020. Hence, in effect, a further 700 MHz of spectrum was being sought in the bands identified in Table 1.2. In this respect, it is worthy of note that the ITU's Radio Regulations contain a table of frequency allocations whereby frequency bands are allocated to services such as fixed-wire or mobile. A 'primary' allocation means that protection from interference by other 'secondary' services can be sought. Furthermore, certain frequency bands are identified as suitable for particular uses such as IMT. Although this provides no specific benefits, it does tend to mean in practice that throughout the world the same bands are favoured for specific uses and hence much increases the probability of agreement about such uses at the WRC. So far as Table 1.2 is concerned, it includes many instances where no 'primary' allocation yet existed, so such an allocation would need to be assigned if there was sufficient backing at WRC-07.

4 According to Unwired Insight (2009), several factors will account for a substantial increase in demand, primarily:

- the migration of 2G subscribers to 3G networks;
- an upsurge in the availability of data cards and dongles;
- the increasing use of smartphones;
- flat-rate priced broadband bundles;
- the introduction of ever-more usage-intensive services;
- the improved performance of networks;
- a switch from fixed-wire to mobile connectivity.

Table 1.2 Candidate bands for WRC-07

Candidate band	Anticipated use	Support	Available
410 to 430 MHz	Coverage extension of existing IMT-2000 networks	Very little	No
450 to 470 MHz	Coverage extension of existing IMT-2000 networks	Limited	No
470 to 862 MHz	Coverage extension and possible IMT-Advanced	Yes, but in 2011	>112 MHz
2300 to 2400 MHz	IMT-Advanced	For mobile WiMAX	No
2700 to 2900 MHz	IMT-Advanced	Very little	No
3400 to 3600 MHz	IMT-Advanced	Considerable	Partly
3600 to 3800 MHz	IMT-Advanced	Considerable	Partly
3800 to 4200 MHz	IMT-Advanced	Limited	Very little
4400 to 4990 MHz	IMT-Advanced	Very little	No

Source: Adapted from Ofcom (2007: 2)

Getting agreement on a worldwide basis is no mean feat – it can be problematic enough at the level of each regional grouping. For example, within the EU there was no unanimity in the run-up to WRC-07. The full range of preferences is set out in Ofcom (2007) Annex 6. This shows that 18 of the 29 European countries listed favoured use of the 3400–3600 MHz and 3600–4200 MHz bands but disagreed on almost everything else. So far as the use of the 470–862 MHz band was concerned, the main disagreement was whether this band should be considered at WRC-07 or WRC-11 in 2011.

WRC-07 concluded in mid-December 2007. The frequency bands identified for use by IMT were:

- 450–470 MHz
- 698–862 MHz in Region 2 (the Americas) and nine countries in Region 3 (Asia)
- 790–862 MHz in Regions 1 (Europe, Africa, Middle East, former USSR, Mongolia) and 3
- 2.3–2.4 GHz
- 3.4–3.6 GHz (on voluntary basis).

3.5G

The easiest way to describe high-speed packet access (HSPA) is by recourse to its data transfer capability. HSPA has gone through several upgrades. Initially it provided up to 1.8 Mbps (HSDPA), but rapidly upgraded to 3.6 Mbps (HSDPA) and 384 kbps (high-speed uplink packet access (HSUPA)) followed by up to 7.2 Mbps (HSDPA) and 1.45 Mbps (HSUPA). The next upgrade provided up to 14.4 Mbps (HSDPA) and 5.7 Mbps (HSUPA). HSPA+ is currently being introduced for the latest upgrade which can provide up to 28.0 Mbps (HSDPA) and 11.5 Mbps (HSUPA). At the end of July 2009, HSDPA at 7.2 Mbps was declared to be the official baseline at a time when half of all active networks had achieved that data transfer rate.

In effect, HSPA does for W-CDMA what EDGE does for GSM (Wikipedia 2009c). In that respect it can be referred to as 3.5G, but it has evolved, as noted previously, to provide what is effectively 3.75G and, in its HSPA+ variant, 3.75G+. High-speed downlink (sometimes data) packet Access (HSDPA) technology, which is a standardized feature in 3GPP Release 5, adds a new downlink channel dedicated to carrying data with additional intelligence added to the network to ensure an efficient allocation of channel capacity between users within a cell. Speed is not the only virtue of HSDPA since it also doubles network capacity and much improves quality of service. Nevertheless, although HSDPA is in essence an overlay and hence requires no more than a software upgrade, this is true only for operators that installed 3G infrastructure with built-in HSDPA capability (essentially since 2002). Earlier 3G networks have needed some hardware additions such as channel cards and traffic processing units. Furthermore, hardware additions have been needed at customer premises including additional handset memory.

3GPP Release 5 merely specified that the uplink associated with HSDPA should operate at a minimum of 64 kbps. 3GPP Release 6, finalized in April 2005, also contained specifications for HSUPA, otherwise known as the enhanced dedicated channel (E-DCH) or enhanced uplink (EUL). This permits symmetrical two-way high-speed data communications. It was first demonstrated by Ericsson and 3 Scandinavia in May 2005. HSPA+ is defined in 3GPP Release 7 and 8 and began to be launched during 2009, although it will not become reasonably commonplace until well into 2010. At that time, for example, Huawei intends to launch HSPA+ providing 56 Mbps downstream (3g.co.uk 2009b).

In principle, HSPA+ can operate on existing GSM spectrum, but in many cases operators intend to use spectrum in the 2.5/2.6 GHz bands as and when it is made available. Thus, for example, SmarTone in Hong Kong intends to launch on 1800 MHz in 2010, whereas CSL, Hutchison (3), PCCW and Peoples' intend to use their newly acquired 2.6 GHz spectrum. The first launch on 900 MHz (using enhanced global system for mobile (E-GSM) spectrum) was made by new entrant Aero2 in Poland in November 2009.

4G

The ITU working definition for 4G is variously reported as a downlink speed of at least 100 Mbps when fully mobile or (according to the ITU which is operating via Working Party 8F) 1 Gbps with limited or local mobility, although it remains unclear for now exactly how this is to be measured. An alternative definition which would take account of the fact that, in practice, nothing like 100 Mbps is going to be achieved, no matter what the technical possibilities are, would concentrate upon the existence of an all-IP network capable of delivering in excess of the current target set for HSPA+ – that is, more than, say, 28 Mbps downstream. However, 100 Mbps is certainly going to be achievable under highly controlled circumstances and, indeed, it was first achieved in September 2006 by Samsung in a bus being driven around Jeju Island in South Korea at speeds of up to 38 mph.

There are those who will doubtless wonder why anyone wants a mobile network to operate at 100 Mbps in the first place, especially if it requires investment in new infrastructure. For now, there is no clear answer, although advocates tend to argue that there are so many activities being transferred to mobile devices that by the time higher speeds are available there will be a clear demand for them. It may also be argued that once one national operator introduces 4G to gain a competitive advantage, the others will be forced to respond as they have done in respect of HSPA.

LONG-TERM EVOLUTION

The approach based upon a 100 Mbps downlink is also reflected in 3GPP Release 8 (which specifies 50 Mbps on the uplink) although this is not necessarily referred to as 4G even though it is certainly a significant step beyond HSPA (3G Americas 2007c). Networks achieving these kinds of speeds are based upon packet-switched technology combined with a new orthogonal frequency division multiplexing (OFDM) air interface – OFDMA for the downlink combined with Single Carrier FDMA (SC-FDMA for the uplink) – and multiple input multiple output antenna (MIMO) technology (Global mobile Suppliers Association 2009c). MIMO means that it is more problematic to step up from HSPA to what is now generally called LTE than from W-CDMA to HSPA.

In March 2006, T-Mobile, Orange, KPN and Sprint Nextel set out their joint vision in a White Paper entitled *Next-Generation Mobile Networks. Beyond HSPA & EV-DO*. This led to the formation of the NGMN Forum in June to establish performance targets and deployment scenarios. China Mobile (HK) and DoCoMo and others joined the Forum with a view to creating standards for LTE by the end of 2008 and conducting operator trials during 2009. 3GPP were expected to include the E-UTRAN (evolved UTRAN (UMTS terrestrial access network)) standard and the standards for the system architecture evolution (SAE) which addresses the core network supporting the radio access network as part of Release 8 (Wikipedia 2009f). This timetable placed LTE roughly one year behind WiMAX in terms of development. DoCoMo was keen to pre-empt the discussion, and to this end issued its proposed pre-standard version of LTE called Super 3G (3g.co.uk 2009a). However, other members of the Forum were not keen on this arrangement. As a result, a LTE/SAE Trial Initiative was set up in May 2007 to demonstrate the potential of LTE and SAE through joint tests, including radio transmission performance, early interoperability, field trials and full customer trials. The Initiative was expected to last for up to two years. As part of this, Alcatel-Lucent and LG Electronics announced the completion of test calls in November 2007 while Nokia Siemens Networks (NSN) completed the first multi-user field trial in December 2007, following up with the first call using a commercial base station and fully compliant (with 3GPP Rel. 8) software in September 2009.

Finland was the first European country specifically to allocate new spectrum for 4G to incumbents in the GSM band without the need for new licences (Global Insight 2009a), but it also assigned licensed spectrum in the 2.5–2.69 GHz band in November 2009, with frequency division duplex (FDD) blocks suitable for LTE awarded to the 2G incumbents and a TDD block suitable for WiMAX awarded to Pirkanmaan Verkko (CommunicationsDirect 2009). Applications for new licences in Japan – unusually referred to as 3.9G – were invited with a deadline of May 2009. These were awarded to DoCoMo, KDDI and Softbank Mobile (each receiving 10 MHz in the 1.5 GHz band), as well as eMobile (10 MHz in the 1.7 GHz band) which has committed to the launch of 3.9G LTE in 2010 and 4G LTE in 2012. A wide bandwidth of between 10 MHz and 20 MHz is preferable because it allows for higher speeds.

LTE is expected to primarily utilize existing GSM spectrum[5] together with the 2.5/2.69 GHz band, although Verizon (and to a lesser extent AT&T) intends to use the 700 MHz spectrum won in the course of Auction 73 (Federal Communications Commission 2007; Middleton 2009). During 2010, launches are also expected from the likes of DoCoMo (Wood 2009), Vodafone, TeliaSonera (TelecomWeb 2009) and Telecom Italia (Global mobile Suppliers Association 2008; Wikipedia 2009f). Sweden is so far the only country in the world where the three largest operators have committed to a launch of LTE during 2010 with Tele2 and Telenor intending to share a network, but networks are to be restricted to the main urban areas for several years. TeliaSonera launched in Stockholm and Oslo in December 2009 (Wood 2009). It is of interest that, despite being forced to deploy TD-SCDMA, China Mobile has also committed to launch LTE in the guise of TD-LTE in 2011 (Haslam 2009; In-Stat 2009). However, many operators are taking a wait-and-see approach to roll-outs, and Telstra, for example, intends to use it only in major cities.

In October 2009, the ITU received six proposals for IMT-Advanced that would, potentially, be sanctioned in October 2010. These can be referred to as 'true' 4G – as against, say, 3.9G – in that they have to meet the specifications laid out in 3GPP LTE Release 10 (Telecompaper 2009a). However, irrespective of which of these is sanctioned, there will remain the outstanding controversy of whether or not voice should be included in the initial LTE roll-outs. Some operators have indicated that they see LTE as a data-only play with the possibility of a voice add-on at a later date, whereas others argue that without voice LTE will struggle to become established. The problem here is that the technology for LTE with voice is not fully developed and hence there may be technology fragmentation (Cellular-news 2009e) despite the formation of the One Voice Initiative (Telecompaper 2009b). Assuming an initial launch on a data-only basis using a dongle, that then raises the issue of whether the higher data-transfer speed compared to HSPA+, paralleled by higher prices, will attract customers – it is of interest that in Sweden the fourth-largest operator, 3, was the first to launch HSPA+ and will not be launching LTE.

WiMAX

Another way forward for 4G is encapsulated in the Worldwide Interoperability for Microwave Access (WiMAX) Forum which is behind the Institute of Electrical and Electronic Engineers' (IEEE) 802.16-2004 *fixed* standard first published in April 2002 (WiMAX Forum 2009). This is a data-only standard – not strictly a technology as such – that primarily covers the 2 GHz to 11 GHz band, making it suitable for connection to Wi-Fi hotspots, and the 56 GHz to 66 GHz band. The intervening bandwidth tends to be already occupied and hence of limited practical use. The *mobile* variant of WiMAX is known as 802.16e-2005 and was approved in December 2005.

With a peak range of at least 50 kilometres – and hence suitable for metropolitan area networks (MANs) – and a peak shared data rate of 70 Mbps, WiMAX certainly appears to offer an attractive prospect for areas currently not covered at all, or poorly covered, by other types of network. However, this speed reflects the optimum channel size, power and best-case modulation, and takes no account of the network overheads (Wikipedia

5 Finland was the first European country to allocate additional spectrum (in the 1800 MHz band) specifically for 4G (to TeliaSonera, Elisa and DNA) (Global Insight 2009a).

2009d). In fact, roughly thirty per cent of the bandwidth is needed for error correction and encoding, and what is left – roughly 50 Mbps – covers both directions of data transfer given that WiMAX is a TDD system. Hence, with equal directional flows, each could carry only 25 Mbps, and a further 10 per cent is needed for other purposes. What is left – 20 Mbps downstream and upstream – is accordingly rather less than the original amount claimed, and given that WiMAX is a point-to-multipoint network, the result of sharing the available spectrum among all those logged on at any one time means that a realistic downstream speed is currently 3–5 Mbps combined with 1 Mbps upstream.

In June 2007, the International Telecommunication Union (ITU) gave its initial approval to plans to include a specific subset of WiMAX known as orthogonal frequency division multiple access time division duplexing (OFDMA TDD WMAN) as a terrestrial radio interface under IMT-2000, meaning that spectrum would be reserved for this on an international basis (Global Insight 2007a). Final approval was forthcoming in October 2007 with the result that WiMAX could henceforth be rolled out in the 2.5–2.69 GHz band and the 1.9 GHz 3G band. It is of interest that in gaining acceptance as a 3G technology, WiMAX appeared to have given up its pretensions to be treated as part of the 4G family of technologies despite the high theoretical speeds of which it is capable (Global Insight 2007b).

The WiMAX Forum has sanctioned licensing in the 2.3 GHz, 2.5 GHz and 3.5 GHz bands. Unfortunately, there is no agreement on a single spectrum band to be used on a worldwide basis – the USA favours 2.5 GHz, Asia-Pacific countries favour 2.3 GHz and 2.5 GHz and Europe and Latin America favour 3.4–3.6 GHz.

But does it have a future as a competitor for LTE? Currently, it is arguable that only the USA is seeing the roll-out of a network that might be seen in this light. When Sprint took over Nextel, it ended up with the great bulk of the spectrum in the 2.5 GHz band in the largest 100 markets in the USA. Given the existence of other users in the band, it appeared unlikely that Sprint Nextel could use it to roll out any kind of nationwide service, but the Federal Communications Commission (FCC) imposed a condition for sanctioning the takeover of Nextel in the form of a requirement that the merged company offer services in the 2.5 GHz band to at least 15 million citizens by 2009 and to 15 million more by 2011. Technically, Sprint Nextel did not need to launch nationwide to achieve its FCC-mandated targets, but it appeared to be determined to achieve this, although details were hazy until August 2006 when Sprint Nextel opted for mobile WiMAX using 802.16e-2005 technology (Harrowell 2006).

In March 2007, AT&T sold its entire 2.5 GHz spectrum to Clearwire in order to satisfy one of the conditions attached to its purchase of BellSouth. This meant that, in effect, Clearwire and Sprint Nextel had the 2.5 GHz field to themselves. In July 2007, Sprint Nextel and Clearwire announced that they intended to build a joint nationwide WiMAX network branded as Xohm. Each would build out its own network and these would then be linked via a roaming agreement. By January 2008, Sprint Nextel was said to be desperately seeking a partner in Xohm. In May 2008 came the announcement that the new partners would in fact be Comcast, Time Warner Cable, Bright House Networks, Intel and Google. They would get a combined 22 per cent stake in a new network branded as Clear, with Clearwire itself holding a further 27 per cent and Sprint Nextel the residual 51 per cent. The deal was sanctioned by the FCC in November 2008 and closed in December (PriMetrica 2008). Sprint Nextel is now providing services over the Clear network acting as a mobile virtual network operator (MVNO)-branded as 'Sprint 4G'. The initial launch was in Baltimore at the end of September 2008 (Unstrung 2009).

Hence mobile WiMAX has a first-mover advantage over LTE which is due on stream in 2010, and if it is a success it may entice others outside the USA to follow where the future of WiMAX has been adversely affected by the decision of some major equipment vendors to concentrate upon LTE (Cellular-news 2009c). So far, the best prospects rest in India, where state-owned BSNL launched in December 2009, and Indonesia (Gabriel 2009b). Despite these developments, it is fair to conclude, overall, that since mobile WiMAX has developed out of fixed WiMAX, it will remain of limited interest to mobile operators and that, given its typical downstream speed, these will regard it as more equivalent to HSPA than to LTE.

Conclusion

3G is a fact of life and 3.5G is almost omnipresent in developed countries. For reasons of capacity constraints and interoperator competition, not to mention the unstoppable driving force of ever-improving technology, some parts of the world have reached, and occasionally passed, the 3.75G mark. During 2010, 4G will become a reality.

However, precisely who will benefit is not yet clear-cut. Consumers will acquire super-fast connectivity, although it is difficult to say at what point they will cease to notice any improvement, but they will need new (expensive, unless heavily subsidized) handsets. The iPhone case study (see Chapter 7) suggests that they will mostly be happy to pay up to acquire the hardware, but it is worth bearing in mind that LTE will initially appear only in respect of laptop computers, with handsets unlikely to arrive until 2011 (Penton Media 2009). Operators, for their part, will be obliged to invest huge sums even though they are not obliged to pay for costly new licences as was often the case for 3G. To justify this they need consumers to download lots of data. That data volumes are on a rising trend is not at issue, but operators have got to find the best pricing model and contend with increasingly fierce competition in saturated markets. Not all will survive. Indeed, not all have survived the transition to 3G.

References

Anderson, T., Daim, T. and Kim, J. 2008. Forecasting wireless technologies with DEA. *Technovation*, 28(9), 602–614.

Ballon, P. 2007. Changing business models for Europe's telecommunications industry: The impact of alternative wireless technologies. *Telematics and Informatics*, 24(3), 192–205.

CDG. 2007. Maximizing network value: Capitalizing on the next wave of mobile broadband data applications. Available at: http://www.cdg.org (accessed 1 November 2007).

Cellular-news. 2007a. Ultra mobile broadband (umb) specification is published. 24 September 2007. Available at http://www.cellular-news.com (accessed 25 September 2007).

----- 2007b. Poor outlook for Umb, but Qualcomm's future still secure. 30 December. Available at http://www.cellular-news.com (accessed 3 January 2008).

----- 2009a. Huawei launches CDMA2000 1EV-DO Revision B solution (Online 4 June). Available at: http://www.cellular-news.com (accessed: 5 June 2009).

----- 2009b. UWB – Is the egg finally about to hatch? 9 June. Available at http://www.cellular-news.com (accessed 10 June 2009).

----- 2009c. Decline in commercial WiMAX offer to drive migration to LTE in Latin America. 5 August. Available at http://www.cellular-news.com (accessed 6 August 2009).

----- 2009d. Voice over LTE's future is far from loud and clear. 22 October. Available at http://www.cellular-news.com (accessed 29 October 2009).

----- 2009e. Network operators to face 3G network capacity crisis in 2010. 28 October. Available at http://www.cellular-news.com (accessed 23 October 2009).

Clark, D. 2006. New WiFi devices get faster as gear improves. 13 June. Available at http://www.totaltele.com (accessed 14 June 2006).

Commission of the European Communities. 1997. *Communication from the Commission to the Council, the European Parliament, the Economic and Social Committee and the Committee of the Regions: Strategy and policy orientations with regard to the further development of mobile and wireless communications (UMTS).* COM(97) 513 final. Brussels: CEC.

----- 1998a. *Proposal for a European Parliament and Council Decision on the co-ordinated introduction of mobile and wireless communications (UMTS) in the Community.* COM(1998) 58 final. Brussels: CEC.

----- 1998b. Amended proposal for a European Parliament and Council Decision on the co-ordinated introduction of mobile and wireless communications in the Community. *Official Journal of the European Communities*, C276/4.

CommunicationsDirect. 2009. Elisa, TeliaSonera and DNA win radio spectrum for 4G services. 24 November. Available at http://communicationsdirectnews.com (accessed 25 November 2009).

Council of the European Communities. 1998. Common position (EC) No 52/98 adopted by the Council on 24 September 1998 with a view to adopting European Parliament and Council Decision No/98/EC, of ... on the co-ordinated introduction of a third-generation mobile and wireless communications system (UMTS) in the Community. *Official Journal of the European Communities*, C333/56. Annex 1.

----- 1999. Decision No 128/1999/EC of the European Parliament and of the Council of 14 December 1998 on the co-ordinated introduction of a third-generation mobile and wireless communications system (UMTS) in the Community. *Official Journal of the European Communities*, L17/1. Annex 1.

Curwen, P. 2005. The Prospects for 3G in the USA: The view from Europe. *info*, 7(3), 16–32.

----- 2006. Fixed-mobile convergence. *info*, 8(3), 3–11.

Curwen, P. and Whalley, J. 2008. *The Internationalisation of Mobile Telecommunications: Strategic Challenges in a Global Market.* Cheltenham, UK: Edward Elgar

Federal Communications Commission. 2001. *Final report. Spectrum study of the 2500–2690 MHz band.* Available at: www.fcc.gov/3G (accessed 12 November 2001).

----- 2007. Auction of 700 MHz band licences. FCC Public Notice DA 07-4514 of 2 November.

Fransman, M. 2002. Explaining the success of NTT DoCoMo's i-mode wireless internet service. A paper available from M.Fransman@ed.ac.uk.

Fuentelsaz, L., Maicas, J. and Polo, Y. (2008) The evolution of mobile communications in Europe: The transition from the second to the third generation. *Telecommunications Policy*, 32(6), 436–49.

Gabriel, C. 2009a. China Telecom second telco to commit to EV-DO rev B. 4 June. Available at http://www.rethink-wireless.com (accessed 5 June 2009).

Gabriel, C. 2009b. BSNL goes live with first Mobile WiMAX network in India. 14 December. Available at http://www.rethink-wireless.com (accessed 15 December 2009).

Global Insight. 2007a. WiMAX gains initial approval as IMT-2000 3G standard. 5 June. Available at https://www.communicationsdirectnews.com (accessed 6 June 2007).

----- 2007b. WiMAX as part of 3G – a boost or a step back? 22 October. Available at https://communicationsdirectnews.com (accessed 23 October 2007).

----- 2009a. Finland allocates new 4G frequencies to TeliaSonera, Elisa, DNA. 24 April. Available at https://communicationsdirectnews.com (accessed 25 April 2009).

----- 2009b. EU approves release of 900-Mhz GSM spectrum for mobile broadband. 28 July. Available at https://communicationsdirectnews.com (accessed 29 July 2009).

Global mobile Suppliers Association. 2007. GSM/3G market/technology update. Available at http://www.gsacom.com (accessed 2 December 2007).

----- 2008. Evolution in radio access networks. Available at http://www.gsacom.com (accessed 10 July 2008).

----- 2009a. UMTS 900 regulatory aspects. 28 May. Available at http://www.gsacom.com (accessed 10 August 2009).

----- 2009b. UMTS 900 commercialization. 28 May. Available at http://www.gsacom.com (accessed 10 August 2009).

----- 2009c. GSM3G market update. 28 May. Available at http://www.gsacom.com (accessed 10 August 2009).

Harrowell, A. 2006. Sprint Nextel comes over all WiMAX. 10 August. Available at http://www.telecoms.com (accessed11 August 2006).

Haslam, C. 2009. China Mobile fast-tracks TD-LTE. 26 June. Available at http://www.lightreading.com (accessed 27 June 2009).

Hausman, J. 2002. Mobile telephone, in M. Cave, S. Majumbdar and I. Vogelsang (eds) *Handbook of Telecommunications*, Volume 1. Amsterdam: Elsevier.

In-Stat. 2009. LTE the only road to wireless 4G in China. 28 July. Available at https://communicationsdirectnews.com (accessed 29 July 2009).

International Engineering Consortium (IEC). 2005. OFDM for Mobile data communications. 20 November. Available at http://ww.iec.org (accessed 21 November 2005).

Kattan, F. 2004. Accelerating MMS adoption and usage. 31 March. Available at https://telephonyonline.com (accessed 3 April 2004).

Middleton, J. 2009. Verizon makes LTE data call. 17 August. Available at http://www.telecoms.com (accessed 18 August 2009).

Morris, A. 2006. Nokia sees commercial LTE solution from 2009. 5 October. Available at http://www.totaltele.com (accessed 6 October 2006).

Mossberg, W. 2005. New WiFi products expand range, speed of networks. 6 January. Available at http://www.totaltele.com (accessed 7 January 2005).

Office of Communications. 2007. *WRC-07 agenda item 1.4: Consultation on candidate bands under consideration at WRC-07 for IMT*. London: Ofcom. 27 February. See also *Statement* of 7 June 2007.

Parker, A. 2009. Vodafone and O2 prevail in airwaves row. 13 May 2009. Available at http://www.ft.com (accessed 12 August 2009).

Pelkmans, J. 2001. The GSM standard: Explaining a success story. *Journal of European Public Policy*, 8, 432–53.

Peltokorpi, V., Nonaka, I., and Kodama, M. 2007. NTT DoCoMo's launch of I-mode in the Japanese mobile phone market: A knowledge creation perspective. *Journal of Management Studies*, 44(1), 50–72.

Penton Media. (2009) LTE handsets in 2010 a longshot despite 4G iPhone hopes. 13 May. Available at https://communicationsdirectnews.com (accessed 14 May).

PriMetrica. 2008. FCC approves wireless mergers and the use of white spaces. 5 November. Available at http://www.telegeography.com (accessed 6 November 2008).

Rao, B. and Parikh, M. 2003. Wireless broadband drivers and their social implications. *Technology in Society*, 25(4), 477–89.

Sehier, P., Gabriagues, J-M. and Urie, A. 2001. Standardisation of 3G mobile systems. *Alcatel Telecommunications Review*, 1st quarter, 11–18.

Srivastava, L. 2001. 3G mobile policy: The case of Japan. *info*, 3(6), 459–84.

Tanner, J. 2002. The other 3G. 26 September. Available at http://www.telecomasia.net (accessed 27 September 2002).

Telecompaper. 2009a. ITU receives six 4G broadband technology submissions. 22 October. Available at http://www.telecompaper.com (accessed 23 October 2009).

----- 2009b. Mobile industry agrees IMS standard for LTE intro. 4 November. Available at http://www.telecompaper.com (accessed 5 November 2009).

TelecomWeb. 2009. Telenor/Tele2 deal sparks Swedish race to 4G. 17 April. Available at https://communicationsdirectnews.com (accessed 18 April 2009).

The Economist. 1998. *Public Network Europe 1998 Yearbook*. London: The Economist.

Tilak, J. 2006. Camera phones to account for 87 per cent of handsets by 2010. 19 January. Available at http://www.dmeurope.com (accessed 20 January 2006).

Turner, D. 2007. DoCoMo loses two foreign mobile carriers. 18 July. Available at http://www.ft.com (accessed 19 July 2007).

Unstrung. 2009. Sprint's summer of WiMax. 17 August. Available at http://unstrung.com (accessed 18 August 2009).

Unwired Insight. 2009. *3G Networks Will Evolve, But Will They Cope?* White Paper. September 2009. Huntingdon: Unwired Insight.

Wikipedia. 2009a. Enhanced data rates for GSM evolution. Available at http://en.wikipedia.org/wiki/EDGE (accessed 1 June 2009).

----- 2009b. Evolution-Data Optimized. Available at http://en.wikipedia.org/wiki/EVDO (accessed 1 June 2009).

----- 2009c. High speed packet access. Available at http://en.wikipedia.org/wiki/High_Speed_Packet_Access (accessed 1 June 2009).

----- 2009d. WiMAX. Available at http://en.wikipedia.org/wiki/Wimax (accessed 1 June 2009).

----- 2009e. UMTS Frequency bands. Available at http://en.wikipedia.org/wiki/UMTS_frequency_bands (accessed 1 June 2009).

----- 2009f. 3GPP Long Term Evolution. Available at http://en.wikipedia.org/wiki/3GPP_Long_Term_Evolution (accessed 1 June 2009).

WiMAX Forum. (2009). WiMAX Forum overview. Available at http://www.wimaxforum.org/about (accessed 1 June 2009).

Wood, N. 2009. O2 in UK LTE trial. 15 December. Available at http://www.totaltele.com (accessed 16 December 2009).

3G Americas. 2006. Mobile Broadband: The global evolution of UMTS/HSPA Release 7 and beyond. 3 July. Available at http://www.3gamericas.com (accessed 17 August 20060.

----- 2007. UMTS Evolution from 3GPP Release 7 to Release 8 HSPA and SAE/LTE. 23 July. Available at www.3gamericas.com (accessed 13 October 2007).

3G.co.uk. 2009a. NTT DoCoMo's Super 3G (LTE). 22 July. Available at http://www.3g.co.uk (accessed 23 July 2009).

----- 2009b. HSPA+ solution to double 3G download speeds. 21 September. Available at http://www.3g.co.uk (accessed 23 September 2009).

CHAPTER

2 *Spectrum Issues*

Introduction

Historically, radio spectrum in general has been viewed as a scarce natural resource which, for reasons of efficiency and equity, has needed to be rationed by governments and/or international agencies. However, demand for spectrum has tended to concentrate upon particular wavelengths which have become increasingly valuable, while others have tended to be treated in a somewhat profligate manner because, at the time of their assignment, there did not appear to be any valuable use to which they could be put. Although technological advances have made it possible to reduce interference between wavelengths (Webb 2009), this permitting more efficient use – and hence expanding the supply – of suitable spectrum, the recent rapid development of mobile telecommunications has nevertheless transformed the balance between the supply and demand of suitable spectrum and forced governments to reappraise how this should be assigned and, in particular, whether to involve markets in the process.

It is customary to assign spectrum to specific parties via a system of licences issued by governments/regulators – the alternative term 'allocation' is used below to refer only to the specifying of permitted services under the terms of a licence. The objectives behind such assignments are generally agreed to be making spectrum available to those parties best able to make use of it (in the eyes of the licensing body), promotion of competitive markets and raising revenue for the licensing body. Secondary objectives include promoting the rapid deployment of new technologies. Another way of approaching this is to distinguish between economic efficiency (maximizing the total benefits net of costs of licensed services), equity (fairness) and revenue-raising aspects.

In principle, an auction can be expected to assign licences to those who value them most highly. This is because, unlike the main alternative, the bureaucratic process known as a beauty contest – which is also referred to as a 'comparative hearing' in the USA and 'comparative bidding' in European Commission documents – whereby licences are awarded by comparing applicants' proposals against a set of pre-determined criteria and selecting the closest matches, an auction works via market prices.[1] Combining a beauty contest with spectrum trading (see below) may produce a similar outcome, but this is a relatively long-winded process and involves significant transaction costs. Spectrum trading can, in any event, be used to improve upon an auction by permitting licence winners to exchange new and existing licences in such a way as to bring together licences in adjacent areas. In addition, the use of prices allows the auctioneer to calculate more

1 See, for example, McMillan (1995) and Noam (1997 and 1998). Professor Noam argues that whereas auctions as a method of assignment should be seen in a positive light, this does not make them the best way to deal with spectrum. He advocates the use of 'open spectrum access' whereby, using digital equipment for measurement purposes, users 'pay an access fee that is continuously and automatically determined by the demand and supply conditions at the time'. See also Pritchard (1997) and Bulow and Klemperer (1996).

accurately what spectrum is worth for future purposes. It may be counter-argued that auctions plus secondary trading allow operators to collect sufficiently large swathes of spectrum so as to create excessive market power, but there is no obvious reason why the auction rules should not contain clauses designed to prevent this outcome, such as spectrum caps, and secondary trading can be made subject to regulatory oversight.

On the face of it, auctions score well in terms of equity provided they take place in the public domain, whereas beauty contests invariably take place behind closed doors (and hence can be politicized) and there is bound to be disagreement about exactly what the criteria for success should be. Not that auctions are themselves necessarily entirely non-political since there may be a pre-qualifying stage designed to weed out 'undesirable' bidders or restrictive conditions may be placed upon bidders to oblige them to, for example, create consortia encompassing domestic companies. However, the bottom line question is always as follows: Why should a group of people who do not rely upon market-based activities for their living be entrusted to choose the best candidate for a licence?

By and large, an auction produces a good deal more revenue for the awarding authority than a beauty contest, even though the latter normally incorporates annual fees. However, in principle, a lump-sum payment can be incorporated into a beauty contest (as in the case of France), but if that is done then the beauty contest begins to look rather like a form of auction.

Beauty Contests

A few comments are in order at this point, although the pros and cons of beauty contests are also discussed alongside those of auctions in what follows. In essence, a beauty contest is more concerned with quality than with price. Thus, the regulator's primary task is to specify the qualitative criteria that will be applied to the licence applicants, which criteria will typically include geographical coverage, speed of roll-out and services to be provided. However, price does come into the equation because a fee, either upfront, annual or both will have to be specified. Furthermore, in (the rare) cases where the upfront fee runs into hundreds of millions of dollars, it may be argued, as noted, that there is little operational difference between a beauty contest and a sealed-bid auction.

An immediate problem is evident in that applicants have discretion over the numbers that they insert in their proposals, and the regulator can never be fully certain either that these numbers are correct or that the applicant actually intends to deliver them once the licence is awarded – the problem of asymmetric information. This, in turn, suggests that it is preferable to keep upfront fees at a modest level (to attract applicants) while setting out a series of (hopefully) ascending annual payments in line with the financial success of the licensees (Cartelier 2003). A further problem may arise in the context of regulatory capture, whereby the applicants may be able to influence significantly the regulator's decision. Finally, it is probable that a beauty contest will be a lengthy affair compared to an auction.

Auction Methodology

A rulebook for a licence auction needs to address a wide range of issues including eligibility to bid, the right to withdraw, the right to pass on a round of bidding, the size of bid increments, the number of rounds per day and so forth (OECD 2001). The simplest approach is to utilize a single-round sealed-bid Dutch auction which requires each bidder to make a single 'first-and-final' bid hidden from other bidders, and for winners to pay the sums they bid (Cartelier 2003). This not only prevents collusive behaviour during the auction (though not prior to it), but also induces new entrants to make bids in the hope that incumbents will try to get away with too low a bid (Bajari and Yeo 2009). For their part, incumbents must pitch what are effectively their opening bids at a much higher level than they would have chosen had they been able to make subsequent adjustments. Whether the outcome will be efficient if, say, a new entrant outbids an incumbent is, unfortunately, dependent upon the exact circumstances of individual auctions.

This is because it must be assumed that the incumbent would have tried to outbid the new entrant had it had the opportunity, and hence that the bidder with the highest potential valuation did not get the licence. On the other hand, it may be desirable for some high-speed licences to be awarded to operators without 2G licences if only to keep licensees on their toes. The precise set of conditions under which a licence award can be said to be efficient is ultimately a technical matter for economists to resolve, and is one where there is some disagreement (Börgers and Dustmann 2002; Grimm, Riedel and Wolfsetter 2002; Klemperer 2002a; van Damme 2002a).

In contrast, a crucial feature of a simultaneous ascending-bid auction (SABA) is that it permits bidders to switch their bids from one licence to another as the auction proceeds, after taking account of the previous behaviour of other bidders. Hence, they do not need to make opening bids above the established minimum.

A standard version of a SABA has evolved whereby prospective bidders become eligible to bid by paying an entrance fee;[2] a 'standing high bid' is established at the end of each round of bidding and bidding recommences at a specified minimum percentage above that amount; an eligible bid consists of the standing high bid plus at least the specified minimum increment; bidders are deemed to be 'active' if they have either tabled the standing high bid or made a subsequent eligible bid; and all valid bids are made known to all participants.[3] In addition, bidders may or may not lose the right to re-enter the bidding for a batch of spectrum in respect of which they have previously ceased to be active,[4] be able to rectify mistakes and/or be able to withdraw bids.

One virtue of a sealed-bid auction is that it is over within minutes whereas a SABA can potentially drag on for months. There is much to be said for restricting minimum bid increments to modest amounts in terms of keeping bidders interested, and in terms of allowing them to take advice and to seek to raise more funds, and whereas much larger

---

2 Occasionally, a simplified form of beauty contest is initially employed to check that, *inter alia*, prospective bidders are financially sound. A non-refundable entry fee is commonly used to discourage those hoping to get licences at the minimum price but who would otherwise drop out immediately the list of applicants demonstrated that this would not happen.

3 But sometimes only the standing high bid is made known to all. It may be noted that although a SABA is much more transparent than a sealed-bid auction, only individual bidders ever know how much higher than their actual winning bids they would have been willing to go if pressed.

4 Deeming bidders to have lost the right to bid for a batch of spectrum if, say, they have remained inactive for a full day (covering multiple rounds of bidding) is intended to discourage 'wait-and-see' behaviour.

increments would certainly bring the bidding process to an earlier closure, they might equally bring it to a halt well before most bidders have reached anywhere near their maximum potential bids.

What this suggests is that while there are certain general conditions that are likely to bring about a satisfactory outcome to an auction, much also depends upon the particular circumstances in question – one size does not necessarily fit all and, clearly, bidding rules cannot be expected to bring about the precise outcomes to be expected in a free market (Melody 2001). Hence, for example, countries may choose to ensure that a new entrant is licensed (as in the UK), that entrepreneurial companies are prioritized (as in the USA) or that a minority group is guaranteed a licence (as in New Zealand). In addition, licences may be geographically restricted or a cap may be fixed for the total amount of spectrum available to one licensee (as in Australia). It should, however, be noted that these kinds of restrictions are not unique to auctions, but may equally well be applied to beauty contests.

A second issue is whether licences should specify whether additional spectrum will be made available at a later date. This is rarely addressed when licences are first put up for sale, but if some licences are not awarded it obviously becomes a major issue, especially in relation to any differences in the amounts to be paid on different occasions. A third issue arises from the fact that, understandably, governments have wanted to discourage potential bidders from staying away from auctions in the hope of subsequently buying licences on the cheap from their owners, and hence a decision needs to be made as to whether trading in licences is to be permitted. Normally, failure to make specified down-payments or to meet specified targets for network roll-out, or any transfer of licence ownership to an existing licensee, results in the return of a licence to the government. Given that up to several billion dollars may have been laid out to buy a licence, attempts to reclaim licences are likely to end up in the courts (as in Italy and the USA). For this reason, considerable thought is now being given to the legalizing of spectrum trading – see below and Faulhaber (2006), Forge and Blackman (2006), Valletti (2001), Xavier and Ypsilanti (2006) and Weiss (2006).

A final issue in a SABA is whether to introduce 'pay-your-bid' pricing whereby each licence is sold at the exact price offered by the successful bidder. The alternative is to use a uniform pricing method whereby every licence is sold at the same price, possibly that of the lowest successful bidder or the highest unsuccessful bidder. However, whereas these methods would clearly produce significantly different outcomes under a sealed-bid auction, they do not tend to make much difference in a SABA since bidders have little incentive to raise their offers by more than the minimum increment specified by the rules.

A more general issue is that there is a trade-off to be considered between openness and privacy. Although the ascending-bid auction is much more open than the sealed-bid variant, it nevertheless retains an important element of privacy – especially if a series of auctions is taking place – in that the winning bidder is not forced to reveal how much it was ultimately prepared to pay. There is also an obvious trade-off between increments and speed. If increments are set relatively low, an auction with lots of eligible bidders can potentially drag on for hundreds of rounds. On the other hand, the imposition of large increments designed to speed up the process may cause bidders to drop out too quickly, with a potential loss of revenue to the auctioneer. This is fostered by the fact that a lengthy auction leaves time for bidders to return to their shareholders and/or the financial markets to seek permission to commit additional monies to the auction. Given the existence of this and other trade-offs, it is understandable that there is no definitively optimum way to design an auction and, hence, in practice, no two are ever exactly alike.

A detailed discussion of the economics of auctions is to be found in the work of Paul Klemperer (listed at http://www.nuff.ox.ac.uk/economics/people/klemperer) and specifically in Klemperer (1998, 2004). In a series of letters published in the *Financial Times* of November 2002, he set out to rebut the three main criticisms levied at auctions, namely that carriers paid more for licences than they thought they were worth; that this expenditure reduced the overall level of investment in 3G; and that it also destroyed the market value of licensees. He argued, first, that all parties – including carriers and financial markets – were happy with the prices paid at the time and that Hutchison Whampoa even sold on part of its UK licence at a profit; secondly, that although 2G incumbents had a strong incentive to bid 'whatever it took', the ascending-bid auction format meant that it was the losers who set the final prices and these were hardly likely to value the licences more than the 2G incumbents – in any event, incumbents always had the possibility of withdrawing and subsequently taking over successful new entrants (Whalley and Curwen 2006); thirdly, that once the licence fees have been paid they become history and cannot be recovered by cutting back on future investment – indeed, there is an incentive to invest as quickly as possible in order to recover the licence costs; and, finally, that many carriers in the USA, which did not pay out allegedly excessive licence fees for 3G, lost even more market value than European licensees. Overall, all that happened was a massive transfer from shareholders to governments which was both equitable (since it reflected licensees' valuation of the assets) and efficient (in that it constituted a form of lump-sum taxation). Support for the view that successful bidders did not overpay for licences even in the UK is to be found in Basili and Fontini (2003). An analogy is sometimes drawn with the housing market where rents are not determined by the prices paid for the houses being rented. Nevertheless, it remains the case that if an operator can never raise prices to a level that, sooner or later, recovers all of its costs including the initial outlay on the licence, then in principle it becomes insolvent. In such a case, the licence will probably get sold on at a knock-down price which allows the new licensee to make a profit at the existing level of prices.

A Brief History of Spectrum Assignment

The early history of spectrum licensing is largely concerned with its experience in the USA (Plott 1997; Rosston 2003; Weber 1997). Prior to the late 1980s, the regulator, the Federal Communications Commission (FCC), issued wireless licences via beauty contests in the course of which prospective licensees sought to persuade the FCC that they would put the spectrum to better uses than other applicants. The process proved to be too long-winded and lacked transparency, so in 1982 Congress ordered the FCC to switch to a lottery.[5] This was a speedier process, but one that was even less efficient as it induced

5 See, for example, McMillan (1994), Milgrom (1998), Guala (2001) and Scanlan (2001). The Federal Communications Council's authority to use auctions was contained in the Omnibus Budget Reconciliation Act of 1993 (as revised in 1997 and again in March 2000). Auctions were permitted for mutually exclusive, initial (but not renewal) assignments. The

applications from parties who intended to sell on any licences they gained at a profit. As a result, in 1993, Congress asked the FCC to devise a better method, and in September of that year the FCC published a Notice of Proposed Rule Making which sought, in addition to the above-mentioned objectives, to protect the interests of smaller, non-metropolitan operators. The resultant rule book, which relied heavily upon so-called 'game theory' (French 2009; Milgrom 1998), was published in July 1994 and underpinned a series of 2G spectrum auctions through to the end of 1997. While these were generally considered to have been successful,[6] support was increasingly forthcoming for the SABA (Cramton 1997 and 1998; McAfee and McMillan 1996).

For its part, the European Commission initially requested Member States to award universal mobile telecommunications systems (UMTS) licences free of charge (although annual fees would, as in the case of 2G, customarily be levied) – a policy effectively followed by Finland when it became the first Member State to assign 3G spectrum in 1999. At the time, there was concern that the costs of rolling out 3G networks would be so high as to make it dangerous for the ongoing economic health of licensees if they had also to pay for the licences themselves. However, the potential pot of gold represented by licence fees was sufficient to prevent other Member States from rushing to follow the Finnish example.

Initial Examination of the Data

It goes without saying that the major lessons, when attempting innovatory approaches to a matter such as the assignment of spectrum for high-speed licences, are to be learned from the early attempts. These mainly took place in Western Europe, so what follows largely concentrates on that region. When 3G licences were first made available in Western Europe, three basic methods were employed: auctions; beauty contests (invariably associated with some kind of fee – lump-sum and/or annual – of highly variable value) and, less commonly, beauty contests followed by an auction. However, irrespective of the method used, two particular issues need to be addressed: first, why were there so few bidders compared to available licences on most occasions? Secondly, why did the amount raised per head of population (pop) vary so widely? To answer these questions we may begin by noting the sequence of licence awards since, in practice, this played a fairly crucial role (van Damme 2002a). It is helpful in this respect to divide up the overall period into two main parts. During the first, in the spring of 2000, a beauty contest was held in Spain and a long-winded auction took place in the UK. During the summer there were auctions in the Netherlands and Germany and in October there was a beauty contest-cum-auction in Italy. The second phase began with the beauty contest-cum-auction in Austria and the auction in Switzerland, and culminated in the failed beauty contest in France. There was subsequently a modest third phase incorporating auctions in Greece and Denmark and beauty contests in Luxembourg and Ireland, as well as the second

FCC was directed to ensure that small businesses, minorities, women and rural operators ('designated entities') had the opportunity to participate.

6 Scanlan (2001) argues that, although these auctions were flawed, they nevertheless raised large sums of money ($23 billion), involved few defaults and assigned the spectrum to those wishing to put it to the most valuable uses.

attempt at a beauty contest in France, before the licensing process effectively petered out at the end of 2002.

PHASE 1

The individual cases are dealt with in detail elsewhere – see, for example, Binmore and Klemperer (2002), Börgers and Dustmann (2002) and Klemperer (2002b). For our purposes we may note initially that four licences were offered in Spain, and hence one was guaranteed to a new entrant of whom three entered the beauty contest. The government insisted on a speedy roll-out and, partly as a result, charged little more than \$100 million per licence. When it became clear that much larger sums were being raised in the UK there was a public outcry in Spain to which the Spanish government responded by unilaterally raising the annual fee of \$5 million by a factor of 30 and, when this in turn caused protests by the licensees, Spain became the first EU country officially to delay the formal date for launching services (by six months) and lowered the revised annual fee considerably. The new entrant (Xfera) nevertheless responded by effectively putting itself into hibernation.

The UK government took a different approach altogether, actually asking economists to design an optimum structure for an auction based on the principles of game theory. This evolved into a SABA once it became government policy to offer five licences and hence to reserve a licence (carrying the largest amount of spectrum) for a new entrant. Because a participant could only bid for a single licence in any one round of bidding, one new entrant would always be left in the bidding. Further, since the new entrant would also get roaming rights on existing 2G networks, the process was inherently attractive to new entrants even though nobody expected them to win more than one licence, and nine duly entered the auction. The auction was set to end when only five bidders remained, with each paying their standing high bid. In practice, it took 94 rounds of bidding before the first bidder dropped out, and 150 rounds over seven weeks to complete the auction. Not surprisingly, the eventual sum raised, at \$34.2 billion including the unpaired spectrum, hugely exceeded initial expectations. The winner of the new entrant's licence, Telesystem International Wireless, was overstretched financially and fairly promptly sold the licence on to Hutchison Whampoa (which in turn sold a 35 per cent stake to partners KPN and DoCoMo, both of whom subsequently sold their stakes back at a loss).

In the light of the UK experience, the Netherlands auction, based on similar principles, was expected to raise up to \$10 billion. Unusually, however, there were five 2G incumbents and the government chose to issue only five 3G licences, some providing more spectrum than others. Although eight bidders qualified to enter the auction, the prospects for new entrants, already fairly dim, were rapidly diminished when the weakest 2G incumbent, Ben Nederland, teamed up with Deutsche Telekom. NTL then withdrew because its part-owner, France Télécom, was already involved in an incumbent, and was followed by Hutchison Whampoa which had formed a partnership with incumbent KPN. Predictably, the surviving new entrant, VersaTel, was chased from the scene before the sums bid got out of hand and only \$2.5 billion was raised in total.[7]

7 For a detailed analysis of the Dutch auction see van Damme (2002b). The relatively small sum raised could partly be attributed to the fact that bidders chose not to bid in every round during the early stages – a particularly effective tactic because, believing that it could not fix legally binding minimum prices, the government inserted a clause into the auction rules stipulating that reserve prices would be reduced every time there was no bid. But, in any event, it was

The German government decided to be innovatory despite the use of a SABA (Scheurle 2002). Rather than predetermine the size of the licences, it offered 12 equal-sized blocks of spectrum of which up to three could be acquired by a single bidder. Hence, between four and six licences would be issued with the regulator assigning the spectrum at the end of the auction. The right to roam on existing 2G networks was not automatic, which discouraged new entrants, and there was a clear possibility, if not a probability, that the incumbents would simply mop up the spectrum by buying three blocks apiece. The fact that a bidder could end up with the high standing bid on a single block – insufficient to operate a network – was also discouraging to new entrants. Hence, although 11 bidders were short-listed, only seven actually participated with some of the remainder forming consortia with other applicants. Most bidders initially went for three blocks, but when the price per block rose above $3 billion all bar T-Mobile and Mannesmann (now Vodafone) switched to two blocks. At this price level, new entrants were expected to withdraw as they had done in the UK, but the determination of Group 3G (Orange, Telefónica and Sonera) to soldier on soon appeared to induce the incumbents to cut back their bids to two blocks apiece, and the bidding process petered out with all licences ending up costing the same at roughly $7.6 billion. This represented slightly less per pop than in the UK.

The Italian government decided on yet another licensing variant, initially imposing a much more stringent qualification process prior to a SABA for five licences, one more than the number of incumbents. The licences were of equal size with the remaining 10 MHz paired reserved for new entrants. However, the minimum bid was fixed at $1.95 billion compared to a mere $50 million in Germany. As in the immediately preceding cases, there was extensive reorganization among potential bidders, with, for example, Hutchison Whampoa beating off Deutsche Telekom to head up Andala and, with one potential entrant disqualified and another failing to pay its deposit, only six bidders turned up. Unfortunately, one of these, the weakest incumbent Blu, was riven by internal strife and it withdrew after only 11 rounds, leaving new entrants Andala and IPSE 2000 to claim the extra spectrum at $683 million apiece for 5 MHz paired. Less than one-third of the price per pop was achieved compared to Germany and the UK.

Lessons from Phase 1

At the time, it was fairly widely agreed that the UK auction had been a resounding success: large sums of money had been raised; there was little evidence of collusion; licensees had probably paid close to the maximum valuations they had put on the licences and, hence, the licences appeared to be in the hands of those who valued them most highly; and a new entrant had been licensed. The reservations, mostly relating to the high prices paid, came later – see, for example, Börgers and Dustmann (2001). Surprisingly, however, the auctions in Germany and the Netherlands failed to replicate all of the apparently desirable attributes of that in the UK. In particular, there was no spectrum set aside for a new entrant. As a result, incumbents had a strong incentive to shut out new entrants and new entrants had a strong incentive to bid in conjunction with an incumbent or to join other bidding consortia before the auctions began.

always clear that VersaTel would be forced to drop out because it, uniquely, would be faced with the cost of acquiring 2G customers, and it was perhaps surprising that it survived for 13 days and 305 rounds of bidding.

So why was so much money raised in Germany? It was argued that this had more to do with irrationality on the part of participants than with the auction design, given that the two biggest incumbents began by driving prices up while bidding for the maximum permitted amount of spectrum, only to allow the auction to end abruptly with all bidders obtaining an equal share – on the face of it the worst possible outcome from the incumbents' viewpoint. The main response was an argument to the effect that the German auction was in fact better designed than the one in the UK because bidders were left free to decide for themselves how much spectrum to acquire. Whereas a UK-style SABA would quickly teach potential new entrants that incumbents had too much to lose and would therefore bid whatever was necessary to win licences, this would not be the case, as proven by events, in a German-style SABA. Indeed, had this structure been adopted in the UK, then the outcome would probably have been more licensees, more competition and more revenue.[8] On the other hand, bidders appeared to have clear preferences as to which spectrum bands they wanted to win, even if it paid them to bid tactically in the initial stages, and they were unhappy at having to leave the spectrum assignment to the regulator at the end of the auction – in practice, it took over one year to sort out the assignments. The virtues of the German-style SABA would subsequently be put to the test in Austria (see below).

That the Netherlands should copy the UK-style SABA was much harder to defend. New entrants had learned to be cautious from the UK experience, but their caution was much amplified in this case because not only were there as many incumbents as licences, but no one seemed to mind when major foreign operators (Deutsche Telekom and DoCoMo) teamed up with incumbents. Once it became clear that only one new entrant was expected to turn up, there appeared to be clear advantages in a switch over to a sealed-bid Dutch auction in the hope of arousing interest in other potential entrants and getting incumbents to bid relatively high for fear of not winning a licence. However, this did not happen.

The Italian government appeared to learn some lessons from the Netherlands experience (Scandizzo and Ventura 2006). It utilized a much stricter qualification process – effectively a beauty contest – and indicated that if the number of ensuing qualifiers was no greater than the number of licences then the latter would be reduced in number. Unfortunately, even if the consequent reduction in competition was ignored, this still meant that there would be only one more bidder than the number of licences, so a Netherlands-style denouement could surely be expected to ensue. Given that independent new entrants proved, predictably, to be scarce on the ground, the government should probably have resorted to a Dutch auction and was possibly fortunate that Blu survived long enough to bid, and hence that the high minimum licence fee set would be exceeded – if only just.

PHASE 2

The Austrian auction was effectively copied from that in neighbouring Germany, but with a more stringent initial beauty contest to determine qualification. With four incumbents

8 For various papers containing a detailed analysis of the German auction which contest the view that the result was in any way 'bizarre', contact wolf@wiwi.humberlin.de. See also Grimm, Riedel and Wolfsetter (2002) and the ensuing commentaries by Klemperer and Ewerhart and Moldovanu in the same volume.

and two powerful potential new entrants controlled by Hutchison Whampoa and Telefónica constituting the qualifiers, each quickly realized that it would be very costly to outbid the others for a three-block licence, especially given the relatively small size of the market. It accordingly came as little surprise that the bidding was perfunctory, lasting only 14 rounds and barely exceeding the minimum reserve set per block before the market was carved up into six equal-sized licences. There was even less interest expressed in the unpaired spectrum sold separately. It was rumoured that the bidding, cursory as it was, was prolonged to give the illusion of competition in order to avoid the auction being annulled.

A month later the Swiss government reverted to a UK-style SABA. With more licences than incumbents and a meagre $29 million reserve price, there was a rush to qualify. However, five new entrants among the 10 qualifiers promptly decided that the market wasn't that important to them and dropped out, while TDC moved to amalgamate large stakes in two of the remaining bidders into TDC Schweiz. Left with only four bidders for four licences, the government tried to renegotiate the terms of the auction, but, faced with the threat of legal action and in the absence of evidence of collusive behaviour, the government was obliged to proceed and effectively hand over the licences at the reserve price. The incumbents then refused to grant the new entrant, Telefónica subsidiary UMTS Switzerland, the right to roam on their 2G networks.

The three beauty contests at the end of 2000, in Sweden (Andersson, Hultén and Valiente 2005; Bjorkdahl and Bohlin 2002), Norway and Portugal, predictably raised fairly small sums of money. The auction in Belgium in early 2001 might have been expected to do better, but a combination of a deteriorating economic environment, a small market and two dominant incumbents in Proximus and Mobistar discouraged any new entry. The three incumbents accordingly got their licences at the reserve price and the fourth licence was left unsold. However, the price per pop was still three times that in Switzerland.

The experience of this phase already seemed to be fairly depressing, but worse was to come in France in May 2001. During the initial negotiations after the German licensing concluded, the three incumbents expressed vociferous opposition to a UK-style SABA given the worsening prospects for 3G, and the government proved to be agreeable to a beauty contest. However, it still wanted to raise very substantial sums of money and hit on the device of fixing a $4.5 billion licence fee, with half payable during the first two years of the relatively short 15-year licences. The potential new entrants considered this to be excessive and dropped out. Furthermore, incumbent Bouygues Télécom was deserted by partner Telecom Italia and felt unable to risk the licence fee by itself, and hence it also dropped out at the last minute leaving the government with egg on its face and two unsold licences. It responded by extending the licences to 20 years and redefining the fee as $550 million payable in 2002 plus 1 per cent of 3G revenues (excluding handset sales) thereafter, and submissions for requalification were set back to May 2002.

Lessons from Phase 2

The main lesson from Phase 2 was arguably that insufficient attention had been paid to the lessons of Phase 1! Given the worsening prospects for 3G, there could be little prospect of further excessive rewards from a UK-style SABA. The Belgian government did well under the circumstances by setting a fairly high, but realistic, minimum fee and giving the licences to the incumbents. The Swiss government, in contrast, mistakenly took the view that a low minimum would sustain entry and competitive bidding, but should have realized by that time that individual prospective new entrants would either drop out or form consortia and hence should have opted for a Dutch auction. The truth of that lesson was demonstrated by the comparative success in Denmark. As for the Austrian government, it evidently assumed that the auction methodology with 12 blocks for sale had been demonstrated to be sound by the sums raised in Germany, whereas, as noted above, most analysts thought that the methodology was flawed and that, if it suggested that all bidders would anyway end up with two blocks, there was no point in pushing up the price in the first place by going for three.

PHASE 3

This began with an auction in Greece that mirrored that in Belgium, with only the three incumbents showing any real interest in bidding and an almost identical price per pop being achieved. The Danish government initially looked to be in trouble as a licence had just been handed back in Norway, suggesting that those in comparable Denmark were worth very little. However, it switched its method to a sealed-bid Dutch auction with licensees all paying the least amount offered by a successful bidder, subject to a minimum of $58 million. In the event, the appearance of a powerful new entrant, Hutchison Whampoa/Investor AB, induced incumbents to take no chances with minimum bids and the eventual cost per licence was a respectable $118 million – with Telenor, obliged to bid without 2G partner BellSouth, failing to outbid the new entrant.

There was subsequently something of a hiatus prior to the three beauty contests in mid-2002. The first, consisting of an attempt by the French government to sell the two licences left unsold at the first attempt, was never going to attract applications at the original price, and as a result the government offered the licences at a much reduced upfront fee of $550 million, hoping to recover some of the lost income by also imposing an annual fee set at 1 per cent of annual turnover. This was necessarily offered retrospectively to the existing licensees, but in the event only one new applicant, the third incumbent Bouygues Télécom, came forward, leaving one of the original four licences still unsold. Not surprisingly, the governments in both Luxembourg and Ireland also failed to attract four applicants, but were doubtless happy enough to attract three apiece.

Lessons from Phase 3

By the time of Phase 3 it had become clear that a Dutch auction was the sensible approach. The Greek government chose to ignore this, but protected itself with a reasonably high minimum bid. The Danes took heed and realized twice as much per pop. The subsequent beauty contests were all unsuccessful in some respects, with the French government having to accept that even an apparently sharp reduction in the fee was insufficient to attract new entrants. However, given that neither Luxembourg nor Ireland had four incumbents, the fact that both attracted a new entrant could be deemed success of a kind even if the licences were fairly cheap per pop compared to Denmark.

Lessons from Western Europe

As we have discovered, the amounts raised per pop were hugely variable. It is hardly surprising that the sums raised in the UK and German auctions could not be replicated elsewhere in view of the deteriorating economic environment, but that does not fully explain the sharp drop from the $550+ per pop in those cases to the next highest at $180 per pop in Italy, let alone the $10 per pop in Spain, and there were certainly a lot of disappointed politicians staring at uncovered budget deficits. Clearly, the bursting of the dotcom bubble was also a factor forcing down the value of licences, as was the increasing realization that the technology was going to be late onto the market, particularly in respect of handsets. Further, it has to be borne in mind that regulators were proving somewhat unwilling to alter the terms of the licences in relation to network roll-out, although some concessions in respect of network sharing were forthcoming, so the best bet in terms of cost reduction for operators was to concentrate upon keeping down the cost of licences. This obviously became easier over time as potential bidders learned to play the game better. It was also the case that those operators that had failed to get licences early on (Bijwaard, Janssen and Massland 2008) would not have access to the potential economies of scale arising from a pan-European footprint.

But even after taking all the above into account, some auctions clearly did not meet reasonable expectations. Our analysis of the various phases gives us clues as to why this was. First, it is evident that there was something of a first-mover advantage. Although there had been some prior experience of auctions in the USA, European operators had to learn as they went along – which they did. They learned most obviously that incumbents could be expected to win licences because they had more to lose than new entrants, and that for new entrants to bid up to anything like the maximum that incumbents were prepared to pay would be self-defeating since they would have to add on the costs of establishing a foothold in the 2G market. Hence, if they were really serious about entry, they had to reduce the number of potential bidders and share the costs of bidding by forming joint ventures. Subsequent upon the UK auction, the largest number of bidders to participate was seven – that is, fewer than two per licence on offer.

Licence design also had a part to play. Whereas there was general agreement on the virtues of a SABA for auctions, the desirability of having at least one more bidder than the number of licences should have seemed fairly self-evident – but it was disregarded on several occasions with generally predictable results. As for the innovatory approach in Germany, involving multiple small blocks, this was arguably a failure despite the high licence fees since they appeared to reflect misjudgements by participants. Certainly, despite the views of those who argued that, in reality, a German-style licence would even have increased revenues in the UK, the Austrian experience (and that in Greece) offered little support for deviating from the UK line because the participants had learned that two blocks were better than three. In any event, it could be argued that the Germans and Austrians were applying a procedure designed for auctioning heterogeneous lots to circumstances where the lots were homogeneous, and hence it has been argued that it would have been far simpler just to require bidders to express their demand at the ruling price and to keep raising that price until demand became equal to supply – a method that broadly accorded to the approach used in Italy. Otherwise, the most obvious lesson was to switch to a Dutch auction when potential bidders disappeared, but only the Danes learned that lesson and profited from it despite a deteriorating environment.

There were also lessons to be learned about the fixing of minimum prices. On the face of it, a low minimum should attract entry, but if entry is anyway expected to be limited it stands to reason that there is no point in giving away licences to incumbents at rock bottom prices. The Spanish government certainly had no need to be so generous since there were more applicants than licensees, but the real irony was the complete role reversal in the case of France when the minimum fee was set so high in relation to recent experience that it was a disaster in the making almost from the word go.

There were also lessons to be learned about the role of governments. Most were willing to play straight and fair – but not all. For example, the Spanish government's attempt to raise the licence fee retrospectively (and massively) may technically have been legal, but it was certainly morally dubious. It is of interest that, in December 2002, the Spanish government announced some concessions, demonstrating that, whatever the issues may have been about the original price of the licences, the price was no longer regarded as cheap. The Italian government did at least have the apparent good grace to accept that it had set the rules and hence should abide by them. However, the fact that the antitrust authorities, as well as magistrates in Rome, promptly launched investigations into 'irregular' behaviour, and that the government tried to force Blu to forfeit its deposit for 'not trying', indicated a desire for revenge on the part of the authorities for being outwitted.

Finally, it may be asked whether there was a 'winner's curse' associated with the auctions (Cartelier 2003; Mackley 2008). It was widely believed that this was the case in the UK and Germany because the bids eventually rose to a level that suggested the winning bidders would lose money if they used the licences to build new networks. Perhaps so, came the reply, but it had to be borne in mind that participants had plenty of time to work out in advance exactly what they were willing to pay, and knew throughout the auction exactly how much each unit of spectrum would cost and hence exactly what it was worth to other well-informed bidders. If, indeed, they paid too much, they therefore had nobody to curse but themselves.[9]

Part of the difficulty lay in determining whether high licence fees would necessarily result in higher prices for services compared to where licences were cheap. It was argued, on the one hand, that this was not logical since, in principle, licence fees, which are sunk costs – once the money has been spent it cannot be recovered – could not affect pricing decisions which are based upon a comparison between marginal revenue and marginal cost – that is, what the market will bear at the time. Indeed, high fees would provide a spur to a rapid roll out of networks in order to recoup the investment as quickly as possible. It was pointed out, on the other hand, that in practice all licensees would have an interest in passing on any costs they could, and insofar as they were successful in passing on licence fees in higher prices, there would be a consequent downward pressure on demand where auctions resulted in high fees. This would mean, in turn, that the objectives discussed above are inherently contradictory, and that an auction cannot bring about an optimum industrial structure from the point of view of public welfare.

9 It must also be noted, however, that continuing to bid when the original upper price limit thought to be prudent has been breached is fostered by the tendency of managers to believe that, since they are better than the management of other bidders, they must be able to make as much profit even if paying a slightly higher price.

Experience Beyond Western Europe

Although there were almost as many 3G licensing events outside Western Europe as within it during the early 2000s, there are far fewer lessons to learn from them (although the 2G experiences in the likes of Nigeria (Doyle and McShane 2003; Lee 2003) and India (Jain 2001) are of interest). This is primarily because a very significant proportion of the licences were awarded without any kind of competitive process taking place. The most advanced country, Japan, gave the licences without payment to the 2G incumbents in the predictable absence of outside bidders (Vodafone having taken a stake in J-Phone) (Srivastava 2001). It is possible to relate this to the fact that Japan was the first country to roll out 3G networks, albeit initially very modest ones. However, DoCoMo already had enormous experience with i-mode, its version of 2.5G, so it was almost certain to be the first to roll out, irrespective of the licensing process. The United Arab Emirates (UAE) also gave away its licence, as did Thailand. In addition, Australia and Canada, with their highly clustered populations, chose to introduce a complex licensing process as much regionally as nationally based, whereas New Zealand simply lacked the population to justify sizeable bids although the auction attracted 20 applicants and was very active.

The highest amount per pop paid as an upfront fee – $63 – was obtained in Taiwan where a fifth licence (on a different wavelength) was conjured up by closing down a 2G network. This was the first and – disregarding the situation in Malaysia where the licensing process was heavily 'stage managed' by the government and not truly competitive despite the surplus of applicants – up to that point in time the only occasion in Asia that there were more applicants than national licences (seven compared to five) so a SABA became possible and bidding was fairly fierce even if, predictably, the four incumbents ended up with the favoured licences with IMT-2000 spectrum. In the Czech Republic, Hong Kong, Israel, Singapore and Slovenia the number of applicants either equalled, or fell short of, the number of licences, so the auctions were of no real significance and the minimum prices paid tended to fall in the $30–40 per pop range.

The failures in Eastern Europe are of interest because they appear to have been a response to the increasingly low amounts raised in Western Europe. Poland, ostensibly using a beauty contest in November 2000 after several false starts, managed to raise a reasonable $45 per pop by being a first-mover and setting the minimum fee at a relatively high level when one considers that there were only three applicants for five licences. It took the Czech Republic three attempts to get its auction under way (in November 2001), and it still only attracted two bids for three licences and $21 per pop. Slovenia also failed initially before obtaining a single bid for three licences in November 2001. Subsequently, Estonia managed only to award the three available licences to the 2G incumbents at the minimum price of $4.7 million apiece. Evidently, Western European operators without an equity stake in incumbents intended to stay well clear given the relatively unsophisticated nature of the markets.

South Korea is of some interest because, first, two of the licences were for the preferred W-CDMA (wideband code division multiple access) technology and one for cdma2000, and, being limited to three, all carried a relatively large amount of spectrum (as in Japan). Unlike those in Japan, however, the licences came with a high fee attached and the cdma2000 licence initially failed to attract a bidder, although it was later assigned to the only interested party, thereby raising the overall price per pop to roughly $60 (as in Taiwan). In Hong Kong, only four of the six incumbents (all bar one, unprofitable) participated in the auction[10] – partly because of onerous conditions concerning the provision of access for MVNOs (mobile virtual network operators) – and the fee was set as a percentage of revenues (subject to a minimum during the first five years), so the price per pop is uncertain. It is nevertheless of interest that it was only in Hong Kong, where the government was not in urgent need of revenue from the licences, that an innovatory approach was used with the express object of keeping down costs to the licensees while promoting the entry of MVNOs.[11]

Analysis

Given the highly variable circumstances among countries in the rest of the world compared to the relative homogeneity among the countries in Western Europe, it is not surprising that it is easier to trace the development of thinking about 'best practice' in Western European countries. Much of this thinking has been concerned with the use of auctions as a method for assigning licences, and, in that regard, it is evident that the handling of auctions has at times been rather clumsy and hence that the intended objectives have only partially been met. Naturally, it is possible to fall back upon the fairly sharp change in the overall economic environment commencing in 2000 to justify why mistakes were made. However, it is also clear that the conceptual approach was at times somewhat flawed and that governments were either unwilling or unable to show much flexibility in the face of adversity, even if not everyone is persuaded by the argument that the unheralded consequences of auctions can be attributed to poor design (Noam 2003).

By 2005, the pros and cons of different approaches for assigning spectrum had been thoroughly analysed and there seemed to be little that could be added to the debate,[12] so one must presume that governments subsequently made what appeared to them at the time to be rational decisions. In practice, as is shown in Chapter 3, they frequently chose to eschew both auctions and beauty contests in preference for outright assignments, often without any element of competition whatsoever. It goes without saying that this is unlikely to prove either efficient or equitable (and hence it has been subject to legal challenges on a number of occasions), but outside Western Europe and North America transparency is not necessarily a hallmark of the spectrum assignment process.

The outstanding issue is how governments will deal with the forthcoming availability of swathes of spectrum (see Chapter 1) that, in total, represent roughly twice as much bandwidth as was sold off during 2000–01. Much of this is being earmarked for LTE (long-term evolution) and hence there is a general expectation that auctions will be favoured. However, there is likely to be one significant difference next time around because from an operator's viewpoint the need for additional spectrum largely arises from a need to keep up with the growth in data transfer. In other words, the cost of acquiring additional spectrum is essentially a network maintenance cost which is unlikely to add much to an

10 One incumbent had financial problems and was perceived as a weak player, while another did not participate at the request of a major shareholder (Yan 2004).

11 See Ure (2003). Although royalties had previously been used in Europe as a component of beauty contests, only Hong Kong used them as part of an auction. Klemperer argues (see letter in the *Financial Times* of 9 December 2002) that royalties based on revenues are equivalent to a sales tax which would raise prices, depress profits and discourage investment.

12 But see, for example, Bazelon (2009) and Freyens (2009).

operator's market value. By implication, bids in any auction are likely to be on the low side. However, 3G incumbents have a considerable incentive to secure the spectrum and are mostly well stocked with cash reserves, so new entry is unlikely unless regulators specifically reserve licences to that end.

Spectrum Trading

As noted, the up-rating of existing networks has shown only too clearly that there is a need for additional spectrum. However, it is well known that in almost every country there is to be found some unused and/or underused spectrum that could, albeit with some effort, be used more efficiently if it were to be (re)assigned to new high-speed service provision. But reassignment implies the need for secondary trading, and that is by no means a straightforward matter.

Spectrum trading is an issue that has become increasingly topical at both national and supranational levels, driven in part by the launch of the likes of HSPA (high-speed packet access). The debate, both theoretical and practical, dates back to roughly 2002. At a practical level, the issue was examined in depth at this early stage in the UK[13] and, as noted below, has now been authorized there in relation to certain classes of wireless licences. The general view expressed there was that trading must take place within a regulatory framework designed to promote competition, prevent harmful interference and ensure compliance with international obligations. Therefore, spectrum trading can be characterized by mode, duration and extent. By mode is meant 'how, and whether, trade is organized to permit reconfiguration and change of use, as well as change of ownership; by duration is meant the length of time of a transaction; and by extent is meant the 'degree to which rights and obligations are transferred' (Radiocommunications Agency 2002: 3).

The EU Framework for Electronic Communication Services Directive (European Commission 2002) which came into force in July 2003 (although it did not come into law in every Member State by the due date), lays down a regulatory framework for the EU which states in Article 9 that 'Member states shall ensure that an undertaking's intention to transfer rights to use radio frequencies is notified to the national regulatory authority for spectrum management'. It goes on to say that Member States must also ensure that 'any transfer takes place in accordance with procedures laid down by the national regulatory authority and is made public'; that 'National regulatory authorities shall ensure that competition is not distorted as a result of any such transaction'; and that '[w]here radio frequency use has been harmonised through the application of Decision No 676/2002/EC (Radio Spectrum Decision) or other Community measures, any transfer shall not result in change of use of that radio frequency.'

Spectrum trading formed the major theme of an EU workshop in September 2002 where it was indicated that the newly established Radio Spectrum Policy Group would take trading into account, but it was also observed that it was up to individual regulatory authorities to interpret the new regulatory framework rather than the European

13 For the initial analysis see *Review of Radio Spectrum Management*, published by the DTI and HM Treasury on 6 March 2002, and Radiocommunications Agency (2002).

Commission and that there were wide differences of view as to how best to approach the issue.

The first country in Europe to permit spectrum trading was expected to be Spain which hoped to have the relevant legislation in place to permit periodic sales of spectrum by August 2003. France also intended to introduce similar legislation at roughly the same time, but, in practice, it was Italy that was first to achieve this landmark – with Telefónica immediately offering the IPSE 2000 licence to interested parties. However, it was the UK where the regulator, the Office of Communications (Ofcom), was first to spell out a programme for trading spectrum allocated for different purposes, commencing with that for paging services and fixed wireless at the end of 2004, for emergency services in 2006 and for mobile telephony in 2007 (Ofcom 2004a). Opinion among the pre-enlargement Member States was divided. According to a report by Analysys et al. (Analysys et al. 2004), half wanted to introduce spectrum trading, 32 per cent had no intention of doing so and the rest were undecided. Part of the reason is that there was some disagreement about what the term meant and how it would ultimately be interpreted by the European Commission. The report started out by posing three main questions:

- What types of uses should be allowed? – *Allocation* of spectrum.
- Who should be allowed to operate the frequencies? – *Assignment* of spectrum.
- To what extent should decisions on allocation and assignment be made by the state (*centralized* decision-making) or devolved to users (*decentralized* decision-making)?

The traditional European approach favoured centralized decision-making, but the increasing complexity of issues meant that this was no longer efficient and hence the spectrum management authorities (SMAs) began to experiment in a variety of ways: Some spectrum was allocated for unlicensed use, allowing multiple low-powered users to share the same spectrum – the *commons* model (Benkler 2002; Reed 2002; Forge and Blackman 2006; Freyens 2009) – while some countries used auctions to introduce a *market-driven* model (Hazlett 2001). Although the commons model is essentially an open-access model, it is nevertheless necessary for participants to honour some agreed technical standards or etiquettes that serve to restrict interference to acceptable levels.

The really critical question is whether there will be a 'tragedy of the commons', that is, the available spectrum will become overused as each new participant seeks to gain some limited benefit with the result that there is a corresponding, and much larger overall, loss of benefit to all existing participants (Freyens 2009: 132). If this is the case, many participants will be forced to acquire spectrum in the market in order to guarantee quality of service. However, a market-driven model may itself suffer from a 'tragedy of the anti-commons' whereby available spectrum is underused, in part because of high transaction costs (Xavier and Ypsilanti 2006: 38).

Although the theoretical debate continued to wage in Europe, SMAs remained in ultimate control, and hence imbalances in the demand for, and supply of, spectrum remained in existence. A particular problem was that SMAs were focused upon the issue of interference avoidance rather than on maximizing the economic benefits of spectrum use, leading in many cases to valuable spectrum lying unused.

One possible interpretation is that spectrum trading could be made obligatory but in conjunction with liberalization – that is, the ability to use the spectrum for purposes other than that for which it was originally issued. Alternatively, trading could be made obligatory but without the possibility of alternative use. Finally, Member States could be allowed discretion over whether to allow trading, but again only for the original use. Given its privatization background, the UK was likely to opt for authorization plus liberalization, which was the solution favoured by the report, and there can be no doubt that this could trigger innovative usage patterns for vacant spectrum. Dissenters tended to emphasize the possibility of interference with the signals of licensees, mixing up technologies within a specific spectrum band and differences in licence conditions between spectrum users.

Given the anticipated liberal inclinations of the UK, it is worth considering the subsequent opinions expressed by Ofcom. The consultation statement (Ofcom 2004b) started out by noting that the administrative allocation of spectrum had been appropriate at times when there had been relatively few uses and users and the spectrum manager had as good an understanding of the market as other participants. This 'command and control' approach could still be used, but had to be considered in relation to two other alternatives, namely 'market mechanisms' whereby the market manages the spectrum within the boundaries of the regulator's licences and 'licence-exempt use' or the so-called 'spectrum commons' whereby no one is in control but the possibility of interference ultimately dictates spectrum use. This encompassed roughly 5 per cent of spectrum in the UK at the time.

According to Ofcom calculations, making 800 MHz of spectrum available for licence-exempt use would provide every person with 100 MB of short-range services. Since 600 MHz had already been made available, the extra provision was, accordingly, not excessive. However, care would be needed to avoid difficulties where signals crossed international boundaries, where international mobility was critical and where there were legally binding EU harmonization measures in force. With those qualifications in mind, the mechanisms to achieve a more market-driven spectrum use policy would be spectrum trading and a facility for users to change technology as necessary.

A further issue that Ofcom examined was whether the introduction of new technologies, such as ultra-wideband (UWB), made it desirable to divide up the available spectrum other than by frequency. However, this appeared to offer few advantages. The final 100-page document (Ofcom 2005) set out to introduce mechanisms that would reduce 'command and control' to cover as close as possible to 22 per cent of spectrum use in 2010, with 'market mechanisms' to cover 71 per cent and 'licence exempt' to cover 7 per cent. The key mechanisms would be spectrum trading and liberalization of spectrum use. Ofcom intended to implement liberalization by first considering all requests to vary licence conditions in the light of its statutory duties and other factors (in particular, interference). Secondly, Ofcom intended to vary existing licences to make them less usage- and technology-specific.

However, this raised a number of challenging technological questions which the document set out to address. For example, the onset of UWB, authorized in the USA, appeared to suggest that spectrum should be divided by power rather than frequency, but in the event Ofcom decided to stick with frequency division. Meanwhile, regulators in a limited number of other countries such as Australia (Freyens 2009), Japan, New Zealand and the USA were grappling with similar issues (Xavier and Ypsilanti 2006: 36, 44–45).

Although both theoretical and practical issues associated with spectrum trading have now been extensively thrashed out (Wellenius and Neto 2006; Xavier and Ypsilanti 2006), the main conclusion appears to be that the matter cannot be decided on the basis of theory

– as Freyens argues (2009: 135), 'the notion of a continuum of spectrum management options rests uneasily with the literature, partly due to the ideological divide' – and that it would be foolish for regulators to introduce an inflexible system in case it proved to be difficult to correct if unsuccessful. Hence, the cautious Ofcom-style approach is likely to prevail, if only because no one truly trusts regulators to get it right first time in a world of imperfect information and flawed decision-makers.

References

Analysys Consulting, DotEcon and Hogan & Hartston. 2004. *Study on Conditions and Options in Introducing Secondary Trading of Radio Spectrum in the European Community*. A Report for the Radio Spectrum Policy Unit of the Information Society Directorate-General of the European Commission. London: Analysys, May.

Andersson, P., Hultén, S. and Valiente, P. 2005. Beauty contest licensing lessons from the 3G process in Sweden. *Telecommunications Policy*, 29(8), 577–93.

Bajari, P. and Yeo, J. 2009. Auction design and tacit collusion in FCC spectrum auctions. *Information Economics & Policy*, 21(2), 90–100.

Basili, M. and Fontini, F. 2003. The option value of the UK 3G telecom licences. *info*, 5(3), 48–52.

Bazelon, C. 2009. Too many goals: Problems with the 700 MHz auction. *Information, Economics & Policy*, 21(2), 115–27.

Benkler, Y. 2002. Some economics of wireless communications. *Harvard Journal of Law & Technology*, 16(1), 25–82

Bijwaard, G., Janssen, M. and Massland, E. 2008. Early mover advantages: An empirical analysis of European mobile phone markets. *Telecommunications Policy*, 32(3–4), 246–61.

Binmore, K. and Klemperer, P. 2002. The biggest auction ever: The sale of the British 3G Telecom Licences. *The Economic Journal*, 112(March), C74–C96.

Bjorkdahl, J. and Bohlin, E. 2002. Financial analysis of the Swedish 3G licensees – where are the profits? *info*, 4(4), 10–16.

Börgers, T. and Dustmann, C. 2001. *Strange Bids: Bidding Behaviour in the United Kingdom's Third Generation Spectrum Auction*, CEPR Discussion Paper No. 3072. Available at http://www.papers.ssrn.com (accessed 27 August 2009).

----- 2002. Rationalizing the UMTS spectrum bids: The case of the UK auction. *Ifo Studien*, 48(1), 77–109.

Brusco, S., Lopomo, G. and Marx, L. 2009. The 'Google effect' in the FCC's 700 MHz auction. *Information, Economics & Policy*, 21(2), 101–14.

Bulow, J. and Klemperer, P. 1996. Auctions versus negotiations. *The American Economic Review*, 86(1), 180–94.

Cartelier, L. 2003. Auctions versus beauty contests: The allocation of UMTS licences in Europe. *Annals of Public and Cooperative Economics*, 74(1), 63–85.

Cave, M. and Valetti, T. 2000. Are spectrum auctions ruining our grandchildren's future? *info*, 2(4), 347–50.

Cramton, P. 1997. The FCC spectrum auctions: An early assessment. *Journal of Economics & Management Strategy*, 6(3), 431–95.

----- 1998. Ascending auctions. *European Economic Review*, 42(3–5), 745–56.

Cramton, P., Ingraham, A. and Singer, H. 2008. The effect of incumbent bidding in set-aside auctions: An analysis of prices in the closed and open segments of the FCC Auction 35. *Telecommunications Policy*, 32(3–4), 273–90.

Doyle, C. and McShane, P. 2003. On the design and implementation of the GSM auction in Nigeria – the world's first ascending clock spectrum auction. *Telecommunications Policy*, 27(5–6), 383–405.

European Commission. 2002. EC Framework Directive 2002/21/EC of the European Parliament and the Council of 7 March 2003 on a common regulatory framework for electronic communications networks and services (Framework Directive), *Official Journal of the European Communities*, L108/33, 24 April.

Faulhaber, G. 2006. The future of wireless telecommunications: Spectrum as a critical resource. *Information Economics & Policy*, 18(3), 256–71.

Forge, S. and Blackman, C. 2006. Spectrum for the next radio revolution: The economic and technical case for collective use. *info*, 8(2), 6–16.

French, R. 2009. Governance and game theory: When do franchise auctions induce firms to overbid? *Telecommunications Policy*, 33(3–4), 164–75.

Freyens, B. 2009. A policy spectrum for spectrum economics. *Information, Economics & Policy*, 21(2), 128–44.

Fuentelsaz, L., Maícas, J. and Polo, Y. 2008. The evolution of mobile communications in Europe: The transition from the second to the third generation. *Telecommunications Policy*, 32(6), 436–49.

Grimm, V., Riedel, F. and Wolfsetter, E. 2002. The third generation (UMTS) spectrum auction in Germany. *Ifo Studien*, 48(1), 123–43.

----- 2007. 3G mobile telecommunications licenses in Europe: A critical review. *info*, 9(1), 35–44.

Guala, F. 2001. Building economic machines: The FCC auctions. *Studies in History and Philosophy of Science*, 32(3), 453–77.

Hazlett, T. 2001. The wireless craze, the unlimited bandwidth myth, the spectrum auction *faux pas*, and the punchline to Ronald Coase's 'big joke': An essay on airwave allocation policy. *Harvard Journal of Law & Technology*, 14(2), 335–545.

Jain, R. 2001. Spectrum auctions in India: Lessons from experience. *Telecommunications Policy*, 25(10–11), 671–88.

Klemperer, P. 1998. Auction theory: A guide to the literature. *Journal of Economic Surveys*, 13(3), 227–86.

----- 2002a. How (not) to run auctions: The European 3G telecom auctions. *European Economic Review*, 46(4–5), 829–45.

----- 2002b. Some observations on the British 3G telecom auction: Comments on Börgers and Dustmann. *Ifo Studien*, 48(1), 115–20

Klemperer, P. 2002c. Letter to the Editor. *Financial Times*, 9 December.

----- 2004. *Auctions: Theory and Practice*. Princeton, NJ: Princeton University Press.

----- listed at http://www.nuff.ox.ac.uk/economics/people/klemperer (accessed 27 August 2009).

Lee, D. 2003. Lessons from the Nigerian GSM auction. *Telecommunications Policy*, 27(5–6), 407–16.

Mackley, J. 2008. European 3G auctions: Using a comparative event study to search for a winner's curse. *Utilities Policy*, 16, 275–83.

McAfee, R. and McMillan, J. 1996. Analyzing the airwaves auction. *Journal of Economic Perspectives*, 10(1), 159–75.

McMillan, J. 1994. Selling spectrum rights. *Journal of Economic Perspectives*, 8(3), 145–62

----- 1995. Why auction the spectrum? *Telecommunications Policy*, 19(3), 191–99.

Melody, W. (2001) Spectrum auctions and efficient resource allocation: Learning from the 3G experience in Europe. *info*, 3(1), 5–10.

Milgrom, P. 1998. Game theory and the spectrum auctions. *European Economic Review*, 42(3–5), 771–78.

Noam, E. 1997. Beyond spectrum auctions. *Telecommunications Policy*, 21(5), 461–75.

----- 1998. Spectrum auctions: Yesterday's heresy, today's orthodoxy, tomorrow's anachronism. *Journal of Law & Economics*, 41(S2), 765–90.

----- 2003. The third way for spectrum. 13 March. Available at http://news.ft.com (accessed 14 March 2003).

OECD. 2001. *Spectrum Allocation: Auctions and Comparative Election Procedures*. DSTI/ICCP/TISP(2000)12/FINAL, 13 December. Paris: OECD.

Ofcom. 2004a. *A Statement on Spectrum Trading: Implementation in 2004 and Beyond*. London: Ofcom, 6 August.

Ofcom 2004b. *Spectrum Framework Review: A Consultation on Ofcom's Views as to how Radio Spectrum should be Managed*. London: Ofcom, 23 November.

Ofcom. 2005. *Spectrum Framework Review*. London: Ofcom, 28 June.

Plott, C. 1997. Laboratory experimental testbeds: Applications to the PCS auction. *Journal of Economics & Management Strategy*, 6(3), 605–38.

Pritchard, W. 1997. The communications spectrum: Frequency allocations and auctions. *Acta Astronautica*, 40(208), 447–53.

Radiocommunications Agency. 2002. *Implementing Spectrum Trading*. London: Radiocommunications Agency, July.

Reed, D. 2002. Comments for FCC Spectrum Task Force on Spectrum Policy. 10 July. Available at http://gullfoss2.fcc.gov (accessed 10 November 2009).

Rosston, G. 2003. The long and winding road: The FCC paces the path with good intentions. *Telecommunications Policy*, 27(7), 501–15.

Scandizzo, P. and Ventura, M. 2006. Bids for the UMTS system: An empirical evaluation of the Italian case. *Telecommunications Policy*, 30(10–11), 533–51.

Scanlan, M. 2001. Hiccups in US spectrum auctions. *Telecommunications Policy*, 25(10–11), 689–701.

Scheurle, K. 2002. Spectrum auctions and regulation in telecommunications. *Ifo Studien*, 48(1), 73–75.

Srivastava, L. 2001. 3G mobile policy: The case of Japan. *info*, 3(6), 461–75.

The Economist. 2000. Giddy bidding. 15 April, 36.

Ure, J. 2003. Deconstructing 3G and reconstructing telecoms. *Telecommunications Policy*, 27(3–4), 187–206.

Valletti, T. 2001. Spectrum trading. *Telecommunications Policy*, 25(10–11), 655–70.

van Damme, E. 2002a. The European UMTS-auctions. *European Economic Review*, 45(4–5), 846–56.

----- 2002b. The Dutch UMTS-auction. *Ifo Studien*, 48(1), 175–200.

Webb, W. 2009. An optimal way to licence the radio spectrum. *Telecommunications Policy*, 33(3–4), 230–37.

Weber, R. 1997. Making more from less: Strategic demand reduction in the FCC spectrum auction. *Journal of Economics & Management Strategy*, 6(3), 529–48.

Weiss, M. 2006. Secondary use of spectrum: A survey of the issues. *info*, 8(2), 74–82.

Wellenius, B. and Neto I. 2006. The radio spectrum: Opportunities and challenges for the developing world. *info*, 8(2), 18–33.

Whalley, J. and Curwen, P. 2006. Third generation new entrants in the European mobile telecommunications industry. *Telecommunications Policy*, 30(10–11), 622–32.

Yan, X. 2004. 3G licensing in Hong Kong: The debate. *Telecommunications Policy*, 28(2), 213–26.

Xavier, P. and Ypsilanti, D. 2006. Policy issues in spectrum trading. *info*, 8(2), 34–61.

CHAPTER 3

Licensing of High-Speed Data Networks

Introduction

Since its inception at the start of the 1980s, the mobile telecommunications industry has changed beyond all recognition. The number of mobile users has grown to such an extent that there have for some years been more mobile than fixed subscribers (ITU 2003) – the current count of mobile subscribers is said to be roughly four billion – and competing mobile services are available in most markets (ITU 2008). As the number of mobile users has grown, and markets have been liberalized, international operators such as Vodafone and France Télécom have emerged with multinational footprints (Curwen and Whalley 2008).

Alongside the structural changes that the industry has experienced, there have been technological changes as discussed in Chapter 1. However, for any services to be offered operators need to obtain a licence. Although the licensing of high-speed data networks has been investigated in specific countries and regions,[1] it has not been examined globally. One consequence of this is that there is no definitive overview of the variety of licensing methods that have been adopted, while another is that it is not even known for certain how many networks have been licensed.[2] This chapter seeks to remedy these shortcomings with respect to high-speed mobile telecommunication licences. Using data from the licensing of high-speed mobile networks in more than 130 countries, this chapter proposes a taxonomy of licensing methods as well as highlights the similarities and differences that exist across the globe.

Methodology

Because spectrum is finite, and certain spectrum bands are subject to excess demand, it has long been the custom to assign spectrum via licences although defining property rights for spectrum is a complicated matter, especially when considered in an international/supranational context (Gruber 2004; Faulhaber 2006). Among the issues to be addressed are:

---

1 Examples of country-specific investigations into 3G licensing include Hong Kong (Ure 2003), Lithuania (Korsakaite and Lazauskaite 2006), Sweden (Andersson, Hultén and Valiente 2005) and the UK (Basili and Fontini 2003). Although it is possible to identify country-specific studies, it is harder to identify those that have examined licensing across an entire region. Those regional studies that have been undertaken either focus on Europe – for example, Gruber (2007) and Whalley and Curwen (2006) – or combine Europe together such other countries as have issued licenses – for example, Bauer (2002) and Curwen (2002).

2 Neither of these difficulties has stopped the likes of Cheng et al. (2003) and Kam (2006) from claiming that auctions, especially those in Europe, contributed to the telecoms crash that occurred between 2000 and 2002.

- the size of the spectrum band to be assigned to each licence
- the area to be covered by each licence
- the life span of each licence
- the uses to which each licence may be put
- whether each licence is to be transferable or tradable.

However, before one can study such particulars it is necessary to compile a database of licences on a worldwide basis.

We begin with a simple premise: for every launch there must be an explicit or implicit licence. By implicit licence we mean that the operator has overlaid a new technology onto an existing network without the need to amend the terms of the existing licence – standard practice where cdma2000 1x networks are involved, but by no means unknown in relation to W-CDMA (wideband code division multiple access). Given that 2.5G/2.75G networks do not need to be licensed, they have no further role to play in what follows. This means that the analysis that follows covers only W-CDMA/HSDPA (high-speed downlink packet access) and cdma2000 1xEV-DO Rel. '0' (or occasionally Rev. 'A' because increasingly operators are skipping Rel. '0') even though in the latter case it has not been the custom to issue explicit licenses in the majority of cases.

The database for high-speed mobile network licences was compiled as follows. Commencing in 1998, relevant data were collected on an ongoing basis, with information relating to both licences and launches being downloaded from a wide range of websites. Commencing in August 2008, the public domain databases were searched to compile a complete list of launches, followed by a rigorous Internet search where there were conflicts between the data on licences and the data on launches. At this point, all so-called 'wireless in the local loop' (WLL) networks were excluded on the grounds that they are not fully mobile. A second detailed Internet search was then instigated, mainly involving the websites of operators and regulators, to check for evidence of licences where there had as yet been no matching launches, with a view to resolving any remaining discrepancies that had arisen solely for definitional reasons and to add missing details on licences. Throughout 2009, ongoing searches were executed to update the database such that it now relates to the position as at 31 December 2009.

Certain issues needed to be addressed in relation to the identification of networks. As the list of licences and associated launches was being compiled, it became clear that discrepancies existed between the original names recorded as the licensees and those currently in use. This complicated the data collection process, and reflected the extensive merger and acquisition activity that had occurred throughout the mobile telecommunications industry,[3] as well as operators changing their names for a variety of reasons.[4] Operators themselves can be the source of confusion by applying for licences using the names of existing or newly created subsidiaries or via joint ventures/consortia. Furthermore, for a variety of reasons, the preferred name of a network may evolve over time without due recognition being taken of this by the media. Thus, for example, Celtel

3 See, for example, Curwen and Whalley (2008) for a discussion of merger and acquisition (M&A) activity within the mobile telecommunications industry.

4 It is by no means uncommon to find three or four names being used in the public domain in connection with a single network. The names may refer to the majority owner of the network, to a minority owner whose name is nevertheless used because it is better known than that of the other owners, to the traditional name of the network itself or to the brand name of the main service(s).

evolved into MTC Kuwait which evolved into Zain. Finally, although a name may appear to have been used consistently, this may disguise a somewhat different reality – for example, the current AT&T Wireless is a quite different organization from that which existed under the same name five years ago.[5]

Licensing data were collected covering more than 140 countries. Network launch data were also collected, enabling those networks where a launch has yet to occur to be identified. In some cases no specific date could be attributed for the award of a licence because, for example, operators required neither a licence nor additional spectrum at the time of market entry in order to provide high-speed data services, while in other occasional cases it was not possible to confirm the claimed date of the licence award. Such occasional cases have either been treated as unconfirmed or the most probable date has been used. A further complication is that the date on which a licence is officially awarded is often not the date when it is handed over (in return for a payment of some kind) – for example, Vodafone Qatar was awarded its licence in December 2007, but did not receive it until July 2008. Wherever possible, the date of the award is used below.

It is customary for licences to be national in scope, but there are significant exceptions to this practice in countries such as Australia, Canada, New Zealand, Russia and the USA. Where there are an excessive number of regional/local licences and no realistic prospect of the operators becoming national, the number of licensees has been trimmed to ensure consistency with the other entries. In practice, however, such action only has a significant effect in two cases, namely Russia, where more than 20 regional 1xEV-DO networks have launched, and the USA, where multiple regional W-CDMA and 1xEV-DO networks have launched.

It was also of some importance to consider whether there are distinct differences between the patterns of network licences in different regions of the world. The difficulty here is that there is no established method for the division of the world into regions. In the most basic sense, it is possible to use continents, but that only really works if there is a clear physical dividing line such as applies to Africa, and it does not work at all well for either Europe or Asia. Furthermore, the dividing lines between North America, Central America and South America are unhelpful for our purposes since the relevant dividing line clearly runs between the USA and Mexico – that is, between North America and Latin America. As for Europe, it is a moving feast, with the eastern border of Europe marching steadily ever further to the east.[6]

But neither Russia nor Turkey is geographically European, both consisting of a relatively small Europeanized western section with a much larger Asian section to the east. Indeed, Turkey in many ways looks to sit most comfortably with neither Europe (although it has applied for EU membership) nor Asia, but rather the Middle East, and the same could in principle be said about Egypt, which is often considered as part of the Middle East. Further, North Cyprus is treated as part of Turkey as it is under Turkish control and the operators are Turkish; and Kosovo is left out due to a widespread refusal to accept it as a sovereign state.

5 See, for example, Whalley and Curwen (2007) for a description of telecommunications restructuring within the United States.

6 Indeed, if the entry list to the Eurovision Song Contest is anything to go by, Belarus, Moldova and the Ukraine, not to mention Russia and Turkey, are all happy to gather under the European umbrella when it suits them.

A further complicating issue is how islands/island groups should be treated. There are almost 100 nations and other entities that are islands. Many of these are rather obscure, for example, the Ashmore and Cartier Islands, Svalbard and Wallis and Futuna, but a small number have a role to play in our analysis. Their treatment is addressed in the second footnote of Table 3.1. What therefore appears as Table 3.2 is not intended to be treated as definitive, but rather as a sensible distribution to facilitate the analysis of high-speed mobile licences. It is important to emphasize that the above decisions are not driven by any political agenda. There has to be some element of arbitrariness because there is no single agreed regional distribution of countries. In any event, the distribution is not a critical feature of the research.

Table 3.1 Regional grouping of countries that have licensed high-speed networks[1, 2]

Region	**Countries**
Western Europe	Andorra, Austria, Belgium, Bulgaria, Cyprus (South), Czech Republic, Denmark, Estonia, Finland, France, Germany, Greece, Guernsey, Hungary, Iceland, Ireland, Isle of Man, Italy, Jersey, Latvia, Liechtenstein, Lithuania, Luxembourg, Malta, Monaco, Netherlands, Norway, Poland, Portugal, Romania, Slovakia, Slovenia, Spain, Sweden, Switzerland, UK
Eastern Europe & CIS	Armenia, Belarus, Bosnia-Herzegovina, Croatia, Georgia, Macedonia, Moldova, Montenegro, Russia, Serbia, Tajikistan, Ukraine, Uzbekistan
Middle East	Azerbaijan, Bahrain, Iran, Israel, Jordan, Kuwait, Oman, Palestine, Qatar, Saudi Arabia, *Syria*, Turkey, UAE
Asia Pacific	Afghanistan, Australia, Bhutan, Brunei, Cambodia, China, *Fiji*, French Polynesia, *Guam*, Hong Kong, India, Indonesia, Japan, *Laos*, Macau, Malaysia, Maldives, Mongolia, Nepal, New Zealand, North Korea, Philippines, Singapore, South Korea, Sri Lanka, Taiwan, Thailand, Vietnam
North America	Bermuda, *British Virgin Isles*, Canada, Puerto Rico, USA, *US Virgin Isles*
Latin America	Argentina, Aruba, Barbados, Belize, Bolivia, Brazil, Cayman Isles, Chile, Colombia, Dominican Republic, Ecuador, El Salvador, French Guiana, French West Indies, Guatemala, Honduras, Jamaica, *Mexico*, Netherlands Antilles, Nicaragua, Panama, *Paraguay*, Peru, Trinidad & Tobago, Uruguay, Venezuela
Africa	*Angola*, Botswana, *Cameroon*, *DR Congo*, Egypt, Equatorial Guinea, *Ethiopia*, Gabon, Gambia (The), Ghana, Kenya, *Lesotho*, Libya, Malawi, Mauritania, Mauritius, Morocco, Mozambique, Namibia, Nigeria, Réunion (La), Rwanda, *Senegal*, Seychelles, South Africa, Sudan, Tanzania, Uganda, Zimbabwe

Source: Compiled by authors from a wide variety of sources.

Notes: [1] Entries in italics relate to countries where, for a variety of reasons, it is not possible to establish a reliable date for the licences. [2] Treatment of islands: The status of individual islands or island groups is rather problematic. In essence there are four considerations: the location of the government, the extent of local autonomy, the independence of the operator and the degree of physical separation. However, all of these factors rarely point in the same direction, so high-speed licences are deemed to have been awarded or not in the following cases.

Isle of Man: A Crown Dependency of Great Britain and self-governing (counted).

Réunion (La): Officially an overseas *département* of France and one of the 26 French *régions*, but geographically part of Africa (counted).

North Cyprus: Not recognized internationally so counted as part of Turkey.

Altogether, 151 countries are listed of which 137 are associated with known licence dates – this is referred to henceforth as regional allocation 1 (RAI). The largest share of the 137 countries are in Western Europe (36, or 26 per cent) and a case could arguably be made out for lumping together Eastern Europe & CIS with the Middle East, and North America with Latin America, in order to create five regions with a more equal spread of numbers of licensed countries – referred to as regional allocation 2 (RA2) below.

Analysis

It is possible to analyse the licensing of high-speed mobile networks along seven dimensions, namely the regional grouping of licences, the issuing of licences over time, the method of grouping the licences, the number of licences issued by each country, the length of licences, the spectrum assigned in each case, and, finally, the cost of licences.

REGIONAL GROUPING OF LICENCES

Of the 137 countries issuing licences on known dates, 85 among them licensed operators on only one occasion, 39 on two occasions, 11 on three occasions, one on four occasions (France) and one (the somewhat complicated case of Indonesia) on five occasions – making 205 occasions in all. Multiple licensing has been rare in both Latin America and Africa – in the former case because many licences were initially implicit and were not made explicit until much later and, in the latter case, because the entire process is relatively recent. Not surprisingly, multiple licensing has been far more common in Western Europe where 17 of the 36 countries have issued licences on two or more occasions. However, it is Eastern Europe & CIS that has shown the most variability, with the 13 countries comprising one offer in four cases, two offers in five cases and three offers in four cases. This gives us Table 3.2 as follows.

Table 3.2 Number of occasions on which licences have been issued

Western Europe	Eastern Europe & CIS	Middle East	Asia Pacific	North America	Latin America	Africa
57	26	16	42	5	29	30

Reverting to RA2 does little to downplay the relative significance of Western Europe, although it does almost even out the three intermediate regions (all fall in the 34 to 42 range) while leaving Africa as the laggard (at least for now).

One obvious hypothesis is that the countries that have issued licences on more than one occasion have been attempting to sell off licences that were offered but not awarded first (or second) time around. Alternatively, it can be hypothesized that the subsequent licensing occasions were to dispose of returned or revoked licences.

Altogether, of the 52 occasions on which a country issued licences for a second time, roughly one half of these represented circumstances in which all of the original licences on offer had been sold, and hence the second set of licences was completely independent, while the other half represented attempts to sell previously unsold or returned/revoked licences. Of the latter occasions, some 13 took place in Western European countries and a further three in Eastern Europe & CIS. While it is easy to demonstrate that the number of returned/revoked licences in Western Europe is closely related to the fact that this region witnessed a high proportion of the early occasions on which licences were explicitly awarded, the tendency for licences to be left in the hands of the regulators is worthy of further analysis.

Table 3.3 accordingly examines a sample of the occasions noted above in order to shed some light on the matter. It excludes those relatively straightforward occasions on

Table 3.3 Illustrative secondary licence sales, Western Europe and Eastern Europe & CIS

Country	Number Sold	Date	Comment
Czech Republic	2/3	12/01	Arguably the third rather than first attempt as no applications had been made in either 09/01 or 10/01.
	1/1	02/05	Eventual success.
Denmark	4/4	09/01	TeliaSonera bought Orange Denmark and returned the licence
	1/1	12/05	in 02/05 which was duly resold.
Estonia	3/3	07/03	All goes well.
	1/1	11/06	ProHolding (Bravocom) bought this licence in 01/07 after it had been offered to three other applicants, none of which managed to satisfy the terms of the licence.
France	2/4	07/01	Beauty contest but massive fee puts off all bar two incumbents.
	1/2	09/02	Eventually persuading the third incumbent to apply.
	0/1	10/07	But struggling to attract an acceptable new entrant.
	1/1	12/09	New entrant appears for smaller, cheaper licence.
Ireland	3/4	06/02	No interest from new entrants.
	1/1	11/05	The regulator withheld the licence. This was eventually awarded to the runner-up in 03/07.
Latvia	2/3	09/02	One licence was set aside for a new entrant, but none applied.
	1/1	05/05	Eventual success.
Macedonia	1/4	01/08	Incumbent T-Mobile refuses to bid.
	1/3	12/08	But later changes its mind although no interest from new entrants.
	0/3	08/09	Absolutely no interest from anyone.
Norway	4/4	12/00	All goes well.
	1/2	09/03	The attempted resale of licences returned in 08/01 and 11/02.
	1/1	12/07	Another attempt at the resale of a returned licence.
Romania	2/4	11/04	One incumbent declines to bid.
	2/2	01/07	But later changes its mind and is joined by a new entrant.
Slovenia	1/3	11/01	Previous attempt failed in May because two applicants were needed.
	2/3	09/06	Cut-price offer attracts one new entrant.
	1/1	04/08	2G new entrant awarded free licence.

Source: Compiled by authors from a variety of sources

which all licences bar one were awarded at the initial time of asking and the one left not awarded was duly awarded at the second time of asking. However, it should be noted that the shortage of applicants first time around could be ascribed either to a reluctance to bid by either an incumbent or a new entrant. Table 3.4 is accordingly restricted to the 10 occasions on which matters were less straightforward, all bar one involving Western European operators. The circumstances are quite varied in practice, although one fairly common theme is the difficulty in attracting new entrants. Given the problems faced by even the third-largest established operator in most European markets (see Chapter 10), it is, perhaps, surprising that any new entrants ever come forward, although these do tend to be operating in another part of the telecommunications sector with an established brand name.

LICENCE ISSUE OVER TIME

Let us now turn to the 137 countries with known licence dates and consider the year in which they first licensed an operator.[7] This is shown on a regional basis in Table 3.4. It is notable that half the Western European countries had issued licences before the end of 2001, at which point only Asia among the other regions had even begun to show any real interest in licensing. Predictably, activity in Western Europe then tailed off fairly sharply while Latin America got into its stride. Interestingly, the Asia-Pacific region never built on its comparatively early start, in part because intra-country competition was generally not that fierce and hence 3G was not seen as a desirable method of gaining competitive advantage until the technology had been proven to work properly.[8] Africa, predictably, has been the last region to enter the fray, but, in part because of the large number of countries on the continent, has caught up rapidly since 2006.

Table 3.4 Pattern of regional licence issue over time: Number of countries

Period	Western Europe	Eastern Europe & CIS	Middle East	Asia Pacific	North America	Latin America	Africa
1999	2	0	0	0	0	0	0
2000	12	0	0	3	0	0	0
2001	7	0	1	4	1	0	0
2002	4	1	0	1	0	1	0
2003	3	0	1	3	0	4	0
2004	2	1	2	0	0	3	4
2005	4	3	1	2	2	2	0
2006	1	2	3	8	1	4	7
2007	1	4	1	1	0	7	6
2008	0	1	0	2	0	2	4
2009	0	1	3	1	0	1	2
Total	36	13	12	25	4	24	23

Source: Compiled by authors from a variety of sources

7 Bearing in mind that there may be a gap of several months between the date on which a licence is awarded and that on which it is officially handed over.

8 Although Japan and South Korea are notable for their early innovative activity.

METHOD OF GROUPING

Numerous studies have examined the 3G licensing process. Some of these, particularly earlier studies such as Cave and Valetti (2000), Börgers and Dustmann (2003) and Cartelier (2003), adopted a simple dichotomy to identify the method of assigning licences.[9] This dichotomy contrasted some variant of an auction, involving competitive bidding until the number of bidders is reduced to equal the number of licences, with some variant of a beauty contest, whereby an administrative body such as a national regulator specifies a set of attributes and scores the applications according to how well they satisfy them. Normally, but by no means universally, the highest-scoring applicants are awarded the licences.

Gradually researchers began to develop a more nuanced understanding of licensing methods. Although Goodrick & Srivastava (2002) added 'hybrid' to the aforementioned dichotomy, details of what this entailed in practice were not provided, while Bauer (2002) added 'bidding' to describe the Swedish 3G case. Curwen (2002), in contrast, highlighted the variation that can occur even within the simple dichotomy of auctions and beauty contests, with the latter, for example, sometimes involving an up-front fee or being used in conjunction with an auction. Others allowed for the possibility of tenders, whereby bidders make usually a singled, sealed bid with the highest bids winning the licences. However, in some cases, as indeed is true of auctions, a small number of factors other than price might be taken into account when choosing the licensee(s) so the highest bid is not necessarily successful.

Nevertheless, while these clear-cut alternatives simplified the analysis, they disguised the reality which has become increasingly complex over time. To reflect this, this chapter introduces an original methodology for licensing methods. In addition to auction (AU), beauty contest (BC) and tender (TE), it first, adds the possibility that a form of beauty contest is instigated prior to either an auction (BC/AU) or a tender (BC/TE) to whittle down the initial list of suitable bidders. Secondly, it adds an assignment (AS) which comes in three variants: (1) the licences are awarded in combination with 2G licences; (2) the licences are awarded to specified operators without any form of competition taking place; and (3) the licences are awarded to all of the applicants because there are at least as many licences available as there are applicants. In the last case, it follows that the assignment is made after an auction (AU/AS), tender (TE/AS) or beauty contest (BC/AS) has been instigated. This does not preclude tenders being submitted at above the minimum price. Equally, a shortfall of applicants in a beauty contests does not mean that they will all receive licences by default.

Somewhat similar to an assignment is the use of licensing *per se* (LI), the difference being that licensing covers either a general permission for any existing operator to provide high-speed data services over its existing spectrum, or the introduction of amendments to existing regulations governing technologies already in use with a view to the provision of high-speed data services by incumbents or new entrants. Finally, there is first-come first-served (FCFS), which means what it says.

---

9 The dichotomy is perhaps best illustrated by Gruber (2004) who states that one half of Europe opted for auctions and the other half for beauty contests. The dichotomy has also been used in recent studies such as Gruber (2007), as well as Fuentelsaz, Maicas and Polo (2008).

It should also be borne in mind that all of the above can be associated with three different fee structures: (1) an upfront fee paid in return for the handing over of the licence; (2) an annual fee thereafter, usually dating from the launch of the network; and (3) both of the above. It is unusual for the fee structure to play a major role in its own right, but it is worth observing that beauty contests can sometimes involve no upfront fees at all (as in Japan), whereas at other times the fees can be so high as to influence the outcome. A classic example is France in July 2001 when the upfront fee was set at a staggering $4,520 million per licence – a figure far higher than had been achieved in Italy and not all that far short of the sums bid in the infamous German and UK auctions.[10] Not surprisingly, only the 'big two' incumbents could be induced to apply, and even a reduction to $550 million upfront plus 1 per cent of turnover (applied retrospectively to existing licensees) induced only the third incumbent to apply the second time around.[11] The third offer attracted no bidders at all.

Despite the much improved sophistication of the above methodology, it should nevertheless be borne in mind that some cases are extremely difficult to pin down. For example, in Malta, the incumbent, Vodafone and new entrant, Mobisle Communications, claimed to have been awarded licences in September 2000 under the terms of the Communications Act 2000. The regulator rejected this, and it was not until August 2004 that it agreed to licence both operators. A third operator, PrimaCell, also expressed an interest in a licence, and the regulator was forced by the European Commission to rescind the incumbents' first right of refusal on licences on the grounds that the fixed price offered to them was the minimum set for a third licence. Accordingly, the fixed price was now attached to three licences which would be assigned on the basis of a beauty contest. Three applications were duly made (the third by 3G Telecommunications), so the beauty contest was cancelled in April 2005. However, whereas the incumbents were awarded their licences in August 2005, the third was not handed over until August 2007.

Table 3.5 Primary method for assigning licences

Region	Total	AU	BC	TE	FCFS	AS	LI
Western Europe	36	16	13	2	0	5	0
Eastern Europe & CIS	13	1	1	3	0	7	1
Middle East	12	3	1	4	0	4	0
Asia Pacific	25	5	4	3	0	12	1
North America	4	3	0	0	0	0	1
Latin America	24	3	1	2	2	9	7
Africa	23	2	2	2	1	15	1
Total	137	33	22	16	3	52	11

Source: Compiled by authors from a variety of sources

10 The UK auction raised $34.2 billion, while the German auction raised $45.8 billion (Curwen 2002: 17). Four French licences, if all sold, would have raised $18.1 billion.

11 See, for example, Whalley and Curwen (2006: 625), for further details.

That said, Table 3.5 initially concentrates upon the primary method irrespective of whether it was used in isolation or combined with other methods, and only the first occasion on which a country issued licences is included. The first point is that whereas anyone reading exclusively the early literature on licensing would almost certainly be of the impression that the beauty contest is the commonest form of licence award, in reality it is currently an assignment with an auction in the runner-up spot. This is because assignments have been a favoured method in every region of significance – although failure to attract potential licensees has also been an issue – bar Western Europe, where both auctions and beauty contests prevail. The interesting question this generates is whether the relative paucity of auctions and beauty contests outside Western Europe is a reaction to their perceived failures there or a simple expressed preference for the choice of licensees to be made by governments and/or regulators.

Licensing is popular only in Latin America, reflecting the tendency for existing 2G licensees to be given the right to provide high-speed data services over existing spectrum even if they have shown no interest in so doing. For its part, first-come first-served remains something of a curiosity.

In Table 3.6, the potential combinations of different methods are taken into account. As shown, 110 licence awards use a single method while 27 use combinations of methods. The majority of the latter, predictably, are in circumstances where an auction or beauty contest failed to attract as many applicants as there were licences and hence all applicants were licensed.

Table 3.6 All methods for assigning licences

Region	AU	BC	TE	FCFS	AS	LI	AU/ AS	BC/ AS	TE/ AS	BC/ AU	BC/ TE
Western Europe	7	8	1	0	5	0	6	5	1	2	1
Eastern Europe & CIS	1	1	2	0	7	1	0	0	1	0	0
Middle East	0	0	4	0	4	0	3	0	0	0	1
Asia Pacific	3	3	3	0	12	1	2	1	0	0	0
North America	3	0	0	0	0	1	0	0	0	0	0
Latin America	3	0	1	2	9	7	0	0	1	0	1
Africa	1	1	2	1	15	1	1	1	0	0	0
Total	18	13	13	3	52	11	12	7	3	2	3

Source: Compiled by authors from a variety of sources

In Table 3.7, all subsequent attempts to award licences beyond the initial attempt in any individual country are added in, bringing the total up to 204. Predictably, most second and third rounds of licensing used the same primary method as was adopted on the first occasion. What Table 3.8 shows clearly is that only in Western Europe was any interest shown in relatively complex methods of licensing, and even there it was very modest at best. But it is also notable that there were many occasions in the region on which a shortage of qualified applicants was manifested. It may be argued, naturally, that the same phenomenon is not present in other regions because so many licences were awarded without any form of competitive element.

Table 3.7 All methods for assigning licences on all occasions

Region	AU	BC	TE	FCFS	AS	LI	AU/AS	BC/AS	TE/AS	BC/AU	BC/TE
Western Europe	10	9	8	0	8	0	9	8	1	2	2
Eastern Europe & CIS	3	1	3	0	16	1	0	0	2	0	0
Middle East	1	0	4	0	6	0	4	0	0	0	2
Asia Pacific	3	5	7	0	21	1	3	2	0	0	0
North America	4	0	0	0	0	1	0	0	0	0	0
Latin America	3	0	2	2	13	7	0	0	1	0	1
Africa	1	3	2	1	18	1	2	1	0	0	0
Total	25	18	26	3	82	11	18	11	4	2	5

Source: Compiled by authors from a variety of sources

NUMBER OF LICENCES ISSUED/NUMBER OF COUNTRIES

Table 3.8 examines the number of licences issued, restricting itself to the first occasion on which licences were offered in each country. Where the licences were a mix of national and regional licences, the countries have been omitted to ensure comparability. The 'multi' designation usually refers to a provision for any qualified applicant to be awarded a licence.

Table 3.8 Number of licences issued/number of countries

Number of licences	Western Europe	Eastern Europe & CIS	Middle East[1]	Asia Pacific	North America	Latin America	Africa
1	3	5	7	8	0	8	11
2	2	2	3	4	0	4	5
3	8	5	2	6	1	2	3
4	18	1	0	3	0	2	1
5	3	0	0	2	0	0	0
6	2	0	0	0	0	1	0
Multi	0	0	0	1	0	6	3
Total countries	36	13	12	24	1	23	23
Total licences	130	28	19	>56	3	>36	>34

Source: Compiled by authors from a variety of sources

Note: [1] Turkey, where the first assignment of one licence was subsequently cancelled and three issued on a later date, is counted as three. Iran, which issued one licence but subsequently cancelled it, is counted as one.

As can be seen, the majority of licences were issued in Western Europe, with a very clear preference for either four licences (50 per cent of countries) or three licences (22 per cent of countries). No other region exhibits anything comparable. In Eastern Europe & CIS there was a clear preference for no more than three licences and only one was on offer in almost 40 per cent of cases. This was closely replicated in the Middle East and there was a broadly similar pattern in the Asia Pacific region where 75 per cent of countries offered no more than three licences. Indeed, both Latin America and Africa would have been highly comparable were it not for a certain tendency to offer multiple licences, largely in relation to cdma2000 1xEV-DO overlays.

Logically, therefore, Western Europe must be viewed as the exception with other regions declining to replicate its many early experiences of licensing. This is explained in part by the absence of monopoly and duopoly provision in most of Western Europe, as well as by a marked preference for awarding licences via a competitive process of some kind. Elsewhere in the world, there was a distinct tendency to favour the original – and often state-owned – incumbent and a reluctance to introduce significant numbers of competitors in one fell swoop.

It is possible to approach this issue from a different viewpoint, namely the desire by countries to increase competition through the issue of more 3G licences than there were 2G licences at the time. As shown in Table 3.9, every Western European country issued at least as many 3G licences as there were operational 2G licensees. That this was broadly true of the Middle East is a bit misleading because, unlike in Western Europe, there was only one 2G operator in roughly half of the cases. Although there were almost always multiple 2G operators in Eastern Europe & CIS, there was clearly little interest in introducing further competition via 3G licensing, and this is replicated by the experience of the Asia-Pacific region. The picture presented in Latin America and Africa is also somewhat misleading because whereas multiple licences may suggest a willingness to introduce new entrants via 3G licensing, the reality is more in tune with a desire to allow all 2G operators to provide high-speed data services should they so wish. On that basis, the pattern in Latin America and Africa no longer appears all that unusual.

Table 3.9 Licences issued by country: 3G versus 2G

Number of licences	Western Europe	Eastern Europe & CIS	Middle East	Asia Pacific	North America	Latin America	Africa
Fewer 3G than 2G	0	6	3	10	0	8	11
Same 3G as 2G	14	4	7	11	1	8	5
More 3G than 2G	22	3	2	3	0	7	7

Source: Compiled by authors from a variety of sources

LENGTH OF LICENCES

Information on the length of licences is often difficult to pin down, and the sample set of 103 countries accordingly uses the RA2 format. As can be seen in Table 3.10, Western Europe has almost uniformly adopted a 15-year or 20-year format, the odd man out being Norway at 12 years. The average is relatively high at 17.1 years. This can be compared with the Asia-Pacific region which exhibits a broadly similar pattern but with more anomalies in the form of, for example, Macau (8 years) and both Koreas (25 years). Cambodia, at 30 years, has (with Russia) issued the longest licences in the world.

Table 3.10 Length of licences

Region	Country Number	10-year Number	15-year Number	20-year Number	Shortest (years)	Longest (years)	Average (years)
Western Europe	36	1	17	17	10	20	17.1
E Europe, CIS, Middle East	17	3	7	4	7	30	15.4
Asia Pacific	23	2	9	7	8	30	17.3
Americas	20	11	6	1	5	20	12.1
Africa	12	1	9	1	10	25	15.9
Total	108	18	48	30	5	30	15.7

Source: Compiled by authors from a variety of sources

Africa, where information is particularly elusive, is apparently a region of 15-year licences – the only exceptions being Uganda (10 years) and Morocco (25 years). Eastern Europe together with the CIS and Middle East also exhibit a preference for 15-year licences, but with more variability. Finally, the Americas manifest a definite preference for relatively short licences.

MEGAHERTZ AWARDED

Each high-speed data licence issued carries a specified amount of spectrum measured in megahertz (MHz). For the delivery of services using the technologies discussed in this chapter, the spectrum is always paired – that is, it provides both a downlink and an uplink. As noted previously, the original assignments of spectrum for 3G envisaged 75–80 MHz paired (including guard bands) for W-CDMA – enough for four 20 MHz licences, five 15 MHz licences or six 10/12 MHz licences. Because cdma2000 1xEV-DO is normally overlaid on existing 2G networks with highly variable amounts of spectrum, the amount of spectrum officially allocated to high-speed data services is available in only a small number of cases.

The situation is complicated by the fact that a significant number of countries chose not to pre-assign a uniform amount of spectrum to each licence on offer because, for example, they did not wish to pre-specify the number of licences to be awarded (as in Germany), they wanted to favour a new entrant (as in the UK) or they required certain licensees to be accessible to MVNOs.[12] In certain other cases, notably Canada and the USA, the spectrum was assigned in regional as well as national parcels, with the possibility of the regional parcels subsequently being conjoined to form national licences. As a result, the sample set for Table 3.11 is relatively small at 79 countries and the RA2 format is accordingly used.

12 For example, in the Irish beauty contest one of the conditions associated with the Class A licence was that it had to be opened up to MVNOs, and to this end it was assigned more spectrum than the other three licences.

Table 3.11 Megahertz awarded

Region	Country Number	5 MHz	10 MHz	15 MHz	20 MHz	Smallest MHz	Largest MHz	Average MHz
Western Europe	31	1	3	22	5	5	20	15.0
Eastern Europe, CIS, M/East	10	0	0	8	2	15	20	16.0
Asia Pacific	14	0	3	6	5	10	20	15.7
Americas	15	2	3	9	1	5	20	11.9
Africa	9	2	2	5	0	5	15	11.7
Total	79	5	11	50	13	n/a	n/a	14.3

Source: Compiled by authors from a variety of sources

In the light of this small sample, it is probably unwise to conclude other than that 15 MHz paired is the preferred assignment in every region, and especially in Western Europe, with, understandably, little interest shown in less than 10 MHz paired. However, it is worth noting that exploiting 2G spectrum for high-speed data services is not a zero-sum game – high-speed data occupy much more spectrum space than simple telephony/SMS (short messaging service). Hence, spectrum outside the 2G bands will sooner or later (and in most cases very shortly) need to be allocated specifically for high-speed data services in regions such as Latin America and Africa. Conversely, because the original IMT-2000 bands are filling up very rapidly in Western Europe, the main issue in that region is, currently, how to re-farm the 2G spectrum for the provision of high-speed data services.[13]

COST OF LICENCES

The aspect of licensing about which least is known for certain concerns the financial costs involved. For whatever reason, outside Western Europe it is by no means unusual to keep such matters private. In practice, this probably reflects the fact that in a good many cases no fees are charged so the matter is not considered worthy of comment, but it would be unwise to equate an absence of information with a zero fee.

The manner of presenting financial information is also problematic. Clearly, the simplest approach is to report the cost per licence at the time of its award. However, this is not very meaningful *per se* because it takes no account of population numbers: a licence in Monaco is hardly going to fetch a comparable amount to one in Germany. Reporting the fee on a per capita (per pop.) basis largely overcomes this difficulty, but there are two further issues. In the first place, the value of money declines over time due to inflation: $1 million in 2000 is worth more than $1 million in 2008. This is not of much import on an annual basis since inflation has been relatively modest since 1999, but is potentially significant if comparing fees during the entire period 1999–2008. Secondly, fees are reported in a wide range of currencies, although the dollar and euro predictably dominate. Over time, the value of currencies in relation to one another can vary quite widely, as has, indeed, been the case for the dollar–euro exchange rate.

Ideally, therefore, all fees should be reported on an inflation-adjusted and exchange-rate adjusted basis per head of population. Technically, this is relatively straightforward,

13 For an overview of spectrum re-farming see, for example, El-Mohgazi, Whalley and Curwen (2008).

but it still leaves three further difficulties. In the first place, the number of licences on offer has been highly variable and, in theory at least, the fewer the number of licences on offer the higher will be the cost to the winner(s). Secondly, each licence has not conferred an equivalent amount of spectrum, and hence the fee should be adjusted to take account of this. Thirdly, the length of licences has not been uniform, and a 20-year licence can reasonably be expected to cost more than a 10-year licence.

This tells us that the only truly meaningful comparisons must include adjustments for inflation, exchange rates, population size, number of licences, spectrum width and licence length. This is technically feasible, but, ultimately, rather pointless (and hence it has not been done here) because there is little evidence that individual countries conducted this exercise before determining what to charge, The reason for this is that the history of 3G/3.5G licensing has not been conducted on a level playing field. For example, the huge fees realized in the German and UK auctions had an immense impact upon the fee structure of licence offers during the ensuing five years. More recently, the desire by mobile operators from oil-rich countries to build international empires has led to licence offers bearing no resemblance to prior experience in comparable countries. Furthermore, a complicating factor is that a number of these offers have involved a combination of 2G and 3G licences, and it is not a straightforward process to separate them out.

The above can be formalized as follows. The fee charged can be associated with a number of factors including those related to market forces, the uniqueness of the offer and the methodology employed. Into the first category fall issues pertaining to competition and to whether or not the technology has been pre-specified. The second category primarily concerns the potential threat of new entrants from an incumbent's perspective and, from a new entrant's perspective, the probability that an equivalent opportunity will arise in the future. The final category has been discussed in detail above.

The UK, German and Italian auctions aside – the high price in the first French auction was later much reduced retrospectively – one may ask whether any other auctions raised more than $100 per pop. The Netherlands auction of July 2000 produced $154 per pop, but it was also very early in the sequence. Otherwise the answer is negative, with the least shortfall visible for the operators involved in the complicated Hong Kong offer of October 2001, which raised $94 per pop, and those involved in the third offer in Saudi Arabia – the only one with competing bidders – which raised $91 per pop and per pop per licence.

Lowering the yardstick by half again, to $50 per pop and $25 per pop per licence, makes little difference. Other than Hong Kong and Saudi Arabia, the only qualifiers are Austria ($85 per pop in November 2000), Denmark ($88 per pop in September 2001), Ireland ($67 per pop in June 2002), Malta ($52 per pop in August 2005), Montenegro ($53 per pop in May 2007), Panama ($51 per pop and $25 per pop per licence in May 2008), Slovenia ($45 per pop per licence in November 2001) and Taiwan (($63 per pop in February 2002). Only in Ireland and Malta were these relatively (but not excessively) high fees the result of beauty contests.

As noted, the licence fees are not the entire story as roll-out costs also need to be met, although the revenues from the first stages of roll-outs can be used to part-finance the later stages. In summary, therefore, it is fair to argue that the German and UK auctions did indeed have a dramatic impact upon behaviour in subsequent offers, and that irrespective of how many adjustments are made, operators (and regulators) currently take a conservative stance on the value of 3G licences. In any event, it may sometimes be cheaper simply to buy a 3G licensee that has written off its licence in its accounts to (near-)zero.

Conclusion

Information on the licensing of high-speed data networks is available in the public domain on a widely dispersed basis and contains significant inaccuracies. Furthermore, no one has previously attempted to gather it together into a single database. This chapter sets out to remedy this deficiency in relation to the two main high-speed technologies in current widespread usage, namely W-CDMA and cdma2000 1xEV-DO Rel. 0, and also in respect of the more recent HSDPA and 1xEV-DO Rev. A.

Given that the award of a licence is a physical event – at a minimum, a licence must be handed over on an identifiable date – it is truly surprising how little information about licences is to be found on the websites of regulators and operators, at least outside Western Europe. And even then only the bare bones of the licence may be announced. Furthermore, licences may be implicit rather than explicit where overlays are involved. For these reasons it is arguably too complex a task to produce an absolutely accurate list of licences with all of the details of individual licences fully pinned down. Nevertheless, this chapter hopefully represents a major step forward in our knowledge about high-speed data network licensing.

Given that 3G licensing largely originated in Western Europe and subsequently spread out around the world, the analysis has been conducted on a regional basis in order to provide a proper historical (but up-to-date) perspective. Because, for example, there were significant discrepancies between existing methodologies for dividing up the world into regions, and no method was designed specifically to take account of mobile telephony, this chapter has had to develop its own approach. This has proved to be somewhat problematic in relation to North America because, although it contains two huge countries, their respective approaches to mobile licensing are both highly atypical. That said, it is to be hoped that this new regional map will be helpful to future researchers.

It has also been necessary to develop a new methodology for licence awards to take account of the inadequacies of an approach originally based upon a simple auction/beauty contest dichotomy. In reality, even the much expanded methodology used in this chapter cannot fully represent the subtle differences between countries, but there is always the risk that further subdivision may add little to the analysis.

On a broader level, it is possible to argue that the early start of licensing in Western Europe, and in particular the experiences of licensing in Germany and the UK, heavily influenced subsequent events, although it may equally be argued that this influence was felt more strongly in Western Europe in later years rather than in other regions where regulators/governments often appear to have paid little attention to practices elsewhere. Other events, such as the bursting of the TMT (technology, media and telecommunications) bubble in 2000–02, cannot be wholly disregarded, but it is not the commonality that is striking but rather the diversity.

It may finally be noted that the analysis indicates that it is misleading to argue that 3G licensing has been used to enhance competition in mobile telecommunication markets based solely on the number of licenses that have been awarded. Even within the supposedly pro-competitive region of Western Europe, more countries have awarded the same number of 3G as 2G licences as have awarded more 3G than 2G licences. Moreover, events within Western Europe have demonstrated that it is not easy for 3G-based new entrants to enter the market and prosper. The picture is further confused by the fact that some countries have allowed 2G operators to deliver high-speed data services. As a result, markets in some African and Latin America countries are likely to be more competitive than is implied by a simple count of the number of high-speed licences.

References

Andersson, P., Hultén, S. and Valiente, P. 2005. Beauty contest licensing lessons from the 3G process in Sweden. *Telecommunications Policy*, 29(8), 577–93.

Basili, M. and Fontini, F. 2003. The option value of the UK 3G telecom licences. *info*, 5(3), 4852.

Bauer, J. 2002. Impact of license fees on the prices of mobile voice service. *Telecommunications Policy*, 26(7–8), 417–34.

Börgers, T. and Dustmann, C. 2003. European telecom licences. *Economic Policy*, April, 215–68.

Cartelier, L. 2003. Auctions versus beauty contests: The allocation of UMTS licences in Europe. *Annals of Public and Cooperative Economics*, 74(1), 63–85.

Cave, M. and Valetti, T. 2000. Are spectrum auctions ruining our grandchildren's future? *info*, 2(4), 347–50.

Cheng, J., Tsyu, J. and Yu, H-C. 2003. Boom and gloom in the global telecommunications industry. *Technology in Society*, 25(1), 65–81.

Curwen, P. 2002. An analysis of the licensing process for third-generation mobile communications. *info*, 4(6), 16–26.

Curwen, P. and Whalley, J. 2008. *The Internationalisation of Mobile Telecommunications: Strategic Challenges in a Global Market*. Cheltenham, UK: Edward Elgar.

El-Moghazi, M., Whalley, J. and Curwen, P. 2008. Is re-farming the answer to the spectrum shortage conundrum? Department of Management Science Working Paper 08(03). Available at http://www.strath.ac.uk/mansci (accessed 15 August 2009).

Faulhaber, G. 2006. The future of wireless telecommunications: Spectrum as a critical resource. *Information Economics and Policy*, 18(3), 256–71.

Fuentelsaz, L., Maicas, J. and Polo, Y. 2008. The evolution of mobile communications in Europe: The transition from the second to the third generation. *Telecommunications Policy*, 32(6), 436–49.

Goodrick, J. and Srivastava, L. 2002. Licensing of third generation (3G) mobile. *info*, 4(1), 48–56.

Gruber, H. 2004. Is spectrum still a constraint for mobile telecommunications? *info*, 6(5), 289–97.

---- 2007. 3G mobile telecommunications licenses in Europe: A critical review. *info*, 9(6), 35–44.

International Telecommunication Union. 2003. *World Telecommunications Development Report – Access Indicators for the Information Society*. Geneva: ITU.

International Telecommunications Union. 2008 *Trends in Telecommunication Reform 2008: Six Degrees of Sharing*. Geneva: ITU.

Kam, J. 2006. No pain, no gain: Rethinking the telecoms crash. *Technology Analysis and Strategic Management*, 18(5), 497–514.

Korsakaite, D. and Lazauskaite, V. 2006. Licensing of 3G mobile communications: Efforts to balance interests of the budget and the market. *Organizacijø Vadyba*, 39, 85–102.

Ure, J. 2003. Deconstructing 3G and reconstructing telecoms. *Telecommunications Policy*, 27(3–4), 187–206.

Whalley, J. and Curwen, P. 2006. Third-generation new entrants in the European mobile telecommunications industry. *Telecommunications Policy*, 30(10–11), 622–32.

---- 2007. Whatever happened to the Baby Bells? Internationalisation and de-internationalisation in the telecommunications industry. *Minnesota Journal of Law, Science & Technology*, 8(1), 149–73.

CHAPTER

4 High-Speed Data Network Launches

Introduction

Since the licensing of third-generation (3G) mobile telecommunications commenced around the turn of the millennium, a wide-ranging and increasingly extensive body of research has emerged. A significant proportion of this research has focused on the evolution of the structure of the mobile telecommunications industry,[1] or on the licensing processes used and their consequences.[2] Other researchers, in contrast, have concentrated on the technologies that have been adopted by mobile operators, the services that have been developed and how these have been introduced into the marketplace.[3]

However, researchers have largely overlooked the launch of mobile telecommunications networks. Through focusing on the use of 3G to enter new markets within Europe, Whalley and Curwen (2006) highlighted that, for a variety of reasons, some licences have not resulted in the actual launch of the mobile telecommunication services that they were intended to deliver, while Gruber (2007: 37) noted that there had been a general tendency in Europe to delay both the roll-out of networks and the subsequent launches of services. But even when a network operator has definitely launched its services, it is often unclear when this launch actually occurred. Although there are several public domain websites containing data on network launches, which are maintained by industry-funded organizations, a cursory glance reveals a whole host of discrepancies between their respective lists.

Although data on network launches of all types are far from clear-cut, this chapter is directed exclusively towards the analysis of high-speed data network launches, a phenomenon largely confined to the past decade – that is, our analysis focuses only on those technologies frequently labelled as 2.75G, 3G or 3.5G. Although the primary focus is on 3G, there are still a large number of networks that have yet to reach that stage in their evolution, which, *inter alia*, is why 2.75G needs to be covered in order to provide a full picture. Equally, 3.5G has recently become commonplace, with the consequence that it, too, needs to be included in the analysis.

---

1 See, for example, Bijwaard, Janssen and Maasland (2007), Curwen and Whalley (2008) or Fuentelsaz, Maicas and Polo (2008).

2 Börgers and Dustmann (2003), among others, have examined the 3G licensing processes used by European Union countries, while Björkdahl and Bohlin (2002), Dunnewijk and Hultén (2007), Gruber (2007), Macklay (2008) and Ure (2003) have all examined different aspects of the aftermath of 3G licensing.

3 The general standardization process has been explored by, for example, Funk and Methe (2001) and Gandal et al. (2003). Steinbock (2003) placed standards within the broader context of changes in the wireless value chain to highlight the changes that were occurring and Saugstrup and Henten (2006) investigated competition between the two main 3G standards. Mobile innovation and business models have also been examined by researchers such as, for example, Bouwman et al. (2008).

Methodology

PROCESS

The database for high-speed mobile network launches was compiled as follows. Commencing in 1999, relevant information was collected on an ongoing basis from a wide range of websites and industry sources. This was periodically cross-checked against the main public domain databases, but with only those discrepancies that could be speedily resolved being dealt with at the time. Commencing in August 2008, a more extensive attempt to resolve discrepancies was instigated. Public domain databases were used to compile a single master list of alleged launches, in relation to which a rigorous Internet search was undertaken to resolve discrepancies between the relevant databases.

The two lists of launches, one compiled over a lengthy period based on announcements in the media (including the Internet) and the other compiled exclusively at the end of 2008 (then updated to cover 2009), were then compared and roughly 200 discrepancies noted. At this point, all so-called 'wireless in the local loop' (WLL) networks were identified and excluded on the grounds that they lacked full mobility (Ballon 2007). A second detailed Internet search was then instigated, mainly involving the websites of operators, equipment vendors and regulators, to try to resolve those discrepancies which could not be ascribed solely to definitional matters. In those cases where a definitive answer was still not forthcoming, certain launch dates were chosen on a 'best judgement' basis that reflected the authors' experience of the sector on the one hand and the weight of available data on the other. As a consequence, while the research has not been able to generate an absolutely accurate list of launches, it is undoubtedly the most reliable that has been compiled to date.

SOURCES

The most comprehensive ongoing public domain sources for cdma2000 launches are www.cdg.org and www.3gtoday.com which lists 'wireless service providers now offering commercially available 3G services to end users through commercially available 3G devices'. For EDGE (enhanced data (rates) for GSM evolution) launches, the most comprehensive source is www.gsacom.com. Ongoing summaries of the situation with respect to W-CDMA (wideband code division multiple access) can be found at www.3gamericas.org which uses the following dichotomy: 'in service' means that the operator has commercially launched its network to both consumer and enterprise markets, with handsets available in retail outlets; 'pre-commercial' means that the operator has launched limited non-commercial trials, including those with 'friendly' users (this category includes data cards targeted at the corporate market); 'planned/in deployment' means that the network is either being built or is at the network planning stage. For its part, www.gsacom.com publishes a useful series of W-CDMA Fact Sheets that incorporate 'commercial' launches defined as meeting a set of criteria as follows: (1) anyone can subscribe to the service (as against a trial); (2) W-CDMA handsets or terminals are available to users; (3) W-CDMA handsets are sold rather than rented to users; (4) the operator has made a public announcement; (5) the service is charged for. Finally, all live GSM-based networks are listed at www.gsmworld.com.

It is important to emphasize that, partly because of the use of different definitions for the term 'launch', and partly due to a reliance upon different sources for information on launches, there are serious discrepancies between the various sources noted above. Hence, one source should never be used in isolation. One of the main reasons why great care must be taken in the use of these databases is that operators tend to notify their compilers of intended launch dates which may, or may not, be adhered to. However, because the intended launch date is often then simply left in the database, it is (not altogether unreasonably) assumed by those consulting the databases that launches did indeed occur as notified even though, in practice, launch delays have been, and remain, fairly commonplace.

This is a particular problem in relation to EDGE launches. The main source of data, gsacom.com, includes a number of dates that are unquestionably incorrect, and hence it cannot be relied upon. Furthermore, most of the recent alleged launches in the database are given as '2008' or '2009' without a month being specified. Because EDGE is an overlay, it attracts much less media attention than W-CDMA/HSDPA (high-speed downlink packet access), and although recent extensions of GSM (global system for mobile) networks have tended to come EDGE-enabled, that does not prove that EDGE is available. In counting EDGE launches, therefore, this chapter prefers to err on the side of caution by excluding alleged launches that remain unverifiable despite an independent search by the authors.

ISSUES

A significant problem arises, as noted above, over what is meant by a launch. In general, launches are said to be 'commercial' when services are made available to subscribers via handsets. However, that was not necessarily, and indeed generally was not, the first date at which the services could be accessed. In the first place, it is customary for operators to run pre-commercial trials with a limited number of customers, possibly their own employees, who get the services for free. More significantly, an initial launch may be directed at the enterprise (business) market with access via data cards plugged into computers. The cards are available to non-enterprise customers – who will, however, mostly be put off by the high costs involved – and hence it is reasonable to list the launch date for data cards, assuming that it precedes that for handsets, as the official launch date, and that is the practice chosen in compiling the database of launches used in this chapter. This means that there will be some discrepancies compared to tables prepared solely on the basis of launches involving handsets. It may finally be noted that initial launches are beginning to appear that involve so-called 'dongles', but for our purposes these can be treated as equivalent to data cards.

There is a second significant problem that arises when counting the number of launches. In a country such as the USA, a wide range of regional and local operators have launched high-speed networks, but there are only a small number of truly national operators. The issuing of licences to non-national operators has also occurred in countries such as Brazil, Canada and Russia. It would accordingly appear sensible to restrict the analysis to national operators, but this is by no means a straightforward matter. Several countries such as India and Nigeria have introduced unified licensing schemes with the result that regional operators are qualified to apply for high-speed data licences irrespective of whether they intend to expand their coverage across the entire country, and since this in turn depends, in part, upon whether they ever get to receive the requisite spectrum, which is by no means guaranteed, it is unclear whether to include them.

However, it would be unsatisfactory to include these while ignoring similar networks in Russia and the USA. The sensible approach accordingly appears to be to limit countries with large numbers of *potential* national licensees to those operators that are *already* more or less established as national players with relatively large subscriber bases compared to their competitors. Thus we get:

- Brazil (4): América Móvil, Telefónica, Telecom Italia, Telemar (Oi)
- Canada (3): Bell Mobility, Rogers Wireless, Telus Mobility
- India (6): Bharti, BSNL, Idea Cellular, Reliance, Tata, Vodafone Essar
- Indonesia (5): Excelcomindo, Hutchison CP, Mobile-8, Indosat, Telkomsel
- Nigeria (3): Globacom, MTN, Zain
- Puerto Rico (5): América Móvil, AT&T, Centennial, Sprint Nextel, T-Mobile
- Russia (3): MegaFon, MTS, VimpelCom
- Ukraine (3): Astelit, Kyivstar, MTS
- USA (4): AT&T, Sprint Nextel, T-Mobile, Verizon Wireless.

India is now left as the only country in the world with more than five qualifying networks. This is not unreasonable given that it has the second-largest number of subscribers of any country in the world. The fact that no mention has been made of the country with the largest number, China, simply reflects the effective historical duopoly of the Hong Kong-listed operators China Mobile and China Unicom.

Finally, as noted in the previous chapter, the identification of networks presents a significant problem. It is by no means uncommon in public domain material to find three or four different names used in connection with a single network. These may relate either to the majority owner of the network, to a minority owner whose name is nevertheless used because it is better known than that of the other owners, to the traditional name of the network itself or to the brand name of the main service(s). Operators themselves can be the source of added confusion by applying for licences using the names of existing or newly created subsidiaries or via joint ventures. Furthermore, for a variety of reasons, the preferred name of a network may evolve over time without due recognition being taken of this by the media. Thus, for example, Celtel has evolved into MTC Kuwait, which, in turn, has evolved into Zain. Finally, although a name may appear to have been used consistently, this may disguise a somewhat different reality: for example, the current AT&T Wireless is a quite different organization from that which existed under the same name five years ago (Curwen and Whalley, 2008: Chapter 8). As a consequence of the above, matching a licensee to a network that has launched is not always a straightforward process.

REGIONAL ALLOCATION OF COUNTRIES

It is of interest to consider whether they are distinct differences between the patterns of network launches in different regions of the world. However, the lack of an agreed method of dividing the world into regions is problematic.[4] In the most basic sense it is possible to use continents, but this only really works if there is a clear physical dividing

4 Compiling the list of countries applicable to individual regions is also problematic. CIA (2007), for example, omits the French West Indies and Palestine.

line such as applies to Africa or runs between North and South America. In contrast, it does not work at all well for either Europe or Asia.

In any event, the dividing line between North and Central America is meaningless for our purposes since the relevant dividing line clearly runs between the USA and Mexico – that is, between North and Latin America. As for Europe, it is a moving feast, with the eastern border of Europe steadily marching ever further to the east. But neither Russia nor Turkey is geographically European, with both consisting of a relatively small Europeanized western section with a much larger Asian section to the east. Indeed, Turkey in many ways looks to sit most comfortably with neither Europe nor Asia, but rather the Middle East.

The same could in principle be said about Egypt, which is often included under the Middle East heading. Indeed, in some lists, it is unambiguously so included because the Middle East's southern boundary is drawn as an almost straight line below the southern borders of the countries on the northern shores of Africa (with, slightly oddly, the Sudan occasionally thrown in for good measure) – see, for example, GSA (2008), http://en.wikipedia.org/wiki/Middle_East or www.britannica.com. As for Asia, its sheer diversity leads to it being divided up in too many ways to mention.

A further issue relates to the treatment of islands/island groups. There are roughly 100 nations and other entities that are islands. Many of these are rather obscure – for example, the Ashmore and Cartier Islands, Svalbard and Wallis and Futuna – and others have fewer than one thousand subscribers, but others have a role to play in our analysis.[5]

What, therefore, appears as Table 4.1 is certainly not intended to be treated as definitive, but it represents a sensible distribution bearing in mind that the focus is upon mobile telephony launches and not, for instance, geopolitics. It should also be borne in mind that many of the listed countries have yet to issue high-speed network licences, and hence will play no further part in the analysis.

That some countries will play no role in our analysis is demonstrated by the fact that whereas Table 4.1 contains 220 entries, the number of countries with operational high-speed networks is somewhat lower – for example, Asia Pacific contains a number of islands without launches.

Table 4.1 Regional allocation of countries[3]

Region	Countries
Western Europe	Andorra, Austria, Belgium, Bulgaria, Cyprus (South), Czech Republic, Denmark, Estonia, Faroe Isles, Finland, France, Germany, Gibraltar, Greece, Guernsey, Hungary, Iceland, Ireland, Isle of Man, Italy, Jersey, Latvia, Liechtenstein, Lithuania, Luxembourg, Malta, Monaco, Netherlands, Norway, Poland, Portugal, Romania, San Marino, Slovakia, Slovenia, Spain, Sweden, Switzerland, UK
Eastern Europe & CIS[1]	Albania, Armenia, Azerbaijan, Belarus, Bosnia & Herzegovina., Croatia, Georgia, Kazakhstan, Kyrgyzstan, Macedonia, Moldova, Montenegro, Russia, Serbia, Tajikistan, Turkmenistan, Ukraine, Uzbekistan
Middle East	Bahrain, Iran, Iraq, Israel, Jordan, Kuwait, Lebanon, Oman, Palestine, Qatar, Saudi Arabia, Syria, Turkey, UAE, Yemen

5 This matter is addressed in note 3 to Table 4.1.

Table 4.1 *Concluded*

Asia Pacific	Afghanistan, American Samoa, Australia, Bangladesh, Bhutan, Brunei, Cambodia, China, Cook Islands, Fiji, French Polynesia, Guam, Hong Kong, India, Indonesia, Japan, Kiribati, Laos, Macau, Malaysia, Maldives, Marshall Isles, Micronesia, Mongolia, Myanmar, Nepal, New Caledonia, New Zealand, Norfolk Island, North Korea, North Marianas, Pakistan, Palau, Papua New Guinea, Philippines, Samoa, Singapore, Solomon Isles, South Korea, Sri Lanka, Taiwan, Thailand, Tonga, Vanuatu, Vietnam
North America	Bahamas, Bermuda, British Virgin Isles, Canada, Greenland, Puerto Rico, USA, US Virgin Isles
Latin America	Anguilla, Antigua & Barbuda, Argentina, Aruba, Barbados, Belize, Bolivia, Brazil, Cayman Isles, Chile, Colombia, Costa Rica, Cuba, Dominica, Dominican Republic, Ecuador, El Salvador, Falkland Isles, French Guiana, French West Indies, Grenada, Guatemala, Guyana, Haiti, Honduras, Jamaica, Mexico, Montserrat, Netherlands Antilles, Nicaragua, Panama, Paraguay, Peru, St Kitts & Nevis, St Lucia, St Vincent & Grenadine, Suriname, Trinidad & Tobago, Turks & Caicos, Uruguay, Venezuela
Africa[2]	Algeria, Angola, Benin, Botswana, Burkina Faso, Burundi, Cameroon, Cape Verde Isles, Central African Republic, Chad, Comoros Isles, Congo-Brazzaville, Côte d'Ivoire, Djibouti, DR Congo, Egypt, Equatorial Guinea, Eritrea, Ethiopia, Gabon, Gambia, Ghana, Guinea, Guinea-Bissau, Kenya, La Réunion, Lesotho, Liberia, Libya, Madagascar, Malawi, Mali, Mauritania, Mauritius, Mayotte, Morocco, Mozambique, Namibia, Niger, Nigeria, Rwanda, São Tomé, Sénégal, Seychelles, Sierra Leone, Somalia, South Africa, Sudan, Swaziland, Tanzania, Togo, Tunisia, Uganda, Zambia, Zimbabwe

Source: Compiled by authors from a wide variety of sources

Notes:

[1] Kosovo excluded as not internationally recognized.

[2] Western Sahara is disputed by Morocco and Algeria-backed Polisario Front and not counted separately.

[3] Treatment of islands: the status of individual islands or island groups is rather problematic. In essence there are four considerations: the location of the government, the extent of local autonomy, the independence of the operator and the degree of physical separation. It should be borne in mind that high-speed data licences are scarce (so far) in relation to the following:

Corsica: A French *région* but also, in law, a *collectivité territoriale*; it is not practical to distinguish it from the French mainland so not counted.

Faroe Islands: Autonomous province of Denmark so counted.

Greenland: Autonomous province of Denmark so counted, but geographically in North America.

Guam: Governed independently so counted.

Isle of Man: A Crown Dependency of Great Britain and self-governing so counted.

La Réunion: Officially an overseas *département* of France and one of 26 French *régions*, but clearly a part of Africa geographically so counted.

Mayotte: Same status as for Corsica, but geographically part of Africa with different set of operators to mainland France so counted.

North Cyprus: Not recognized internationally so counted as part of Turkey.

Saipan: Counted as part of North Marianas.

Sardinia: Part of Italy with some regional autonomy, but not warranting a separate inclusion.

Solomon Islands: Governed independently so counted.

Analysis

The database that results from the methodology described above can be analysed in a number of ways. In the subsections below, the analysis initially focuses on launches over time before moving on to examine launches by region.

PATTERNS OF LAUNCHES OVER TIME

It is now possible to examine the pattern of network launches during the period 2000 to the end of 2009. The annual distribution of network launches by region and by technology is shown in Table 4.2 while Table 4.3 charts the cumulative number of launches by technology over the period in question. As only a handful of launches occurred during the two-year period of 2000–01, they are treated as a single unit in Table 4.2. It is, however, worth enumerating these launches as follows: in 2000, LG Telecom and SK Telecom in South Korea launched cdma2000 1xRTT, and in 2001, KTF in South Korea and Vivo in Brazil launched cdma2000 1xRTT while NTT DoCoMo launched W-CDMA in Japan.

From Table 4.2 it can be seen that the pattern of network launches differs both within and between regions. Within each of the regions that we have identified, differences occur with respect to both the technology adopted as well as the pattern of network launches over time. Within Western Europe, networks using EDGE and W-CDMA were typically launched earlier than those using HSDPA, but, overall, there were almost as many HSDPA as W-CDMA network launches (97 compared to 105).

Of the three most common technological choices in Western Europe, EDGE is by far the least popular overall although it was briefly in favour as a result of initial delays in the launch of 3G networks. In contrast, EDGE has been the most popular single technological choice for networks launched in four regions: Eastern Europe & CIS, Middle East, Latin America and Africa. Looking across all seven regions, EDGE network launches are heavily concentrated during the five-year period 2004 to 2008 inclusive, with HSDPA launches dominating the latter part of the period. Interestingly, within two regions, Asia Pacific and Western Europe, the peak year for HSDPA network launches was 2006 and in both cases the second-busiest year for launches was 2007. This can most readily be ascribed to the fact that it became possible to overlay a W-CDMA network with HSDPA at modest expense during 2005, and the ability to offer a faster service rapidly became a means of distinguishing one network from those of its competitors which were quick to respond in the majority of cases. In this context it is worth reminding ourselves that once a network provider has put together a turnkey package, roll-out need take only a matter of weeks.

Table 4.2 Distribution of launches by technology and by year to the end of 2009

	2000–2001	2002	2003	2004	2005	2006	2007	2008	2009
Western Europe									
EDGE	0	0	4	16	25	14	9	3	1
W-CDMA	0	0	9	39	17	22	13	4	1
HSDPA	0	0	0	0	1	54	36	5	1
cdma2000 1xRTT	0	1	0	1	1	0	0	2	2
cdma2000 EV-DO '0'	0	0	0	1	2	1	2	2	0
cdma2000 EV-DO 'A'	0	0	0	0	0	0	3	2	0
Eastern Europe & CIS									
EDGE	0	0	1	5	12	4	5	6	0
W-CDMA	0	0	0	0	2	5	6	6	5
HSDPA	0	0	0	0	0	3	6	6	6
cdma2000 1xRTT	0	0	5	4	2	0	1	0	0
cdma2000 EV-DO '0'	0	0	0	0	0	1	3	0	1
cdma2000 EV-DO 'A'	0	0	0	0	0	0	1	0	0
Middle East									
EDGE	0	0	1	3	8	3	6	4	0
W-CDMA	0	0	2	2	0	5	5	3	7
HSDPA	0	0	0	0	1	6	7	3	7
cdma2000 1xRTT	0	1	1	0	0	1	0	0	0
cdma2000 EV-DO '0'	0	0	0	1	0	0	0	0	0
cdma2000 EV-DO 'A'	0	0	0	0	0	0	0	0	0
Asia Pacific									
EDGE	0	0	2	12	4	11	4	9	2
W-CDMA	1	1	3	3	15	9	7	9	12
HSDPA	0	0	0	0	0	22	14	12	11
cdma2000 1xRTT	3	2	6	6	4	1	3	1	1
cdma2000 EV-DO '0'	0	2	1	1	0	4	5	2	5
cdma2000 EV-DO 'A'	0	0	0	0	0	2	1	2	3
North America									
EDGE	0	0	6	0	2	0	0	1	0
W-CDMA	0	0	0	1	1	2	0	2	2
HSDPA	0	0	0	0	1	2	1	2	2
cdma2000 1xRTT	0	8	2	0	0	2	1	0	0
cdma2000 EV-DO '0'	0	0	1	1	5	3	0	0	0
cdma2000 EV-DO 'A'	0	0	0	0	0	2	5	0	0
Latin America									
EDGE	0	0	3	15	16	5	8	9	1
W-CDMA	0	0	0	0	0	1	10	26	5
HSDPA	0	0	0	0	0	2	10	25	5
cdma2000 1xRTT	1	6	8	1	1	10	2	2	0
cdma2000 EV-DO '0'	0	0	1	2	4	6	1	0	0
cdma2000 EV-DO 'A'	0	0	0	0	0	1	1	2	0
Africa									
EDGE	0	0	0	6	6	21	24	22	5
W-CDMA	0	0	0	2	1	5	4	14	9
HSDPA	0	0	0	1	0	5	5	12	9
cdma2000 1xRTT	0	0	0	1	2	2	6	0	0
cdma2000 EV-DO '0'	0	0	0	0	0	1	3	1	1
cdma2000 EV-DO 'A'	0	0	0	0	0	0	0	2	1
Total									
EDGE	0	0	17	57	73	58	56	54	9
W-CDMA	1	1	14	47	36	49	45	64	41
HSDPA	0	0	0	1	3	94	79	65	41
cdma2000 1xRTT	4	18	22	13	10	16	13	5	3
cdma2000 EV-DO '0'	0	2	3	6	11	16	14	5	7
cdma2000 EV-DO 'A'	0	0	0	0	0	5	11	8	4
Total for year	5	21	56	124	133	238	218	201	105

Source: Compiled by authors from a wide variety of sources

Table 4.3 shows the cumulative number of network launches by technology type since 2000. It is evident from this that while there have been more EDGE-based high-speed network launches than those based on any other technology, verifiable EDGE launches have been few and far between during 2009. Given the availability of relatively cheap turnkey packages, it is not surprising that operators are typically now launching W-CDMA/HSDPA rather than EDGE in the likes of Latin America. It is also of note that, despite the early lead established by cdma2000 1xRTT networks, there have been noticeably fewer cdma2000 network launches of any type overall compared to EDGE, W-CDMA or HSDPA although cdma2000 remains the technology of choice for WLL networks and those utilizing the 450 MHz band.

Table 4.3 Annual accumulator 2000–09

Technology	2000	2001	2002	2003	2004	2005	2006	2007	2008	2009
EDGE	0	0	0	17	74	147	205	261	315	324
W-CDMA	0	1	2	16	63	99	148	193	257	298
HSDPA	0	0	0	0	1	4	98	177	242	283
cdma2000 1xRTT	2	4	22	44	57	67	83	96	101	104
cdma2000 1xEV-DO '0'	0	0	2	5	11	22	38	52	57	64
cdma2000 1xEV-DO 'A'	0	0	0	0	0	0	5	16	24	28
2.75G	2	4	22	61	131	214	288	357	416	428
3G	0	1	4	21	74	121	186	245	314	362
3.5G	0	0	0	0	1	4	103	193	266	311
Total	2	5	26	82	206	339	577	795	996	1,101

Source: Compiled by authors from data in Table 4.2

The cumulative number of network launches by technology type since 2000 is shown in Table 4.3. Given the sequential nature of the technologies, it is perhaps no surprise that there are more 2.75G launches than either 3G or 3.5G launches. Having said this, the recent sharp increase for 3.5G highlights its rapid roll-out, particularly when compared to that of 3G.

EFFECT OF ALLOWING FOR REGIONAL NETWORKS

It is necessary to remind ourselves briefly at this point that an adjustment was made above to exclude network numbers in individual countries where those numbers were excessive for the purpose of making meaningful inter-country comparisons. In practice, the adjustment needed to take account of regional networks is modest for Brazil, Canada, India, Indonesia, Nigeria, Puerto Rico and the Ukraine. However, in Russia (where there are over 30) and the USA (where there are roughly 60) the number of launched regional operators is extremely large. It should also be borne in mind that certain databases represent China Mobile, China Telecom and China Unicom as a large number of regional networks rather than as one national network apiece, thereby considerably inflating recorded network numbers. From the point of view of a technology forum, the objective in so doing is clearly to maximize the number of networks claimed to be using their technology.

The effect of including all of these networks would be to add roughly 200 networks to the total, primarily cdma2000 1xRTT, but also 1xEV-DO and EDGE. This is because the great majority of the excluded networks in Russia and the USA use cdma2000 technology, and while it may pay advocates of this technology, and especially the CDMA Development Group (CDG), to use the regional data, it would be misleading for our purposes. The simple fact is that, so far as high-speed network launches with full mobility and (near-) national coverage are concerned, cdma2000 is very much a minority technology in most parts of the world. Indeed, especially in Latin America, GSM overlays of CDMA have become commonplace.

PATTERNS OF LAUNCHES BY REGION

The number of networks that have launched one or more high-speed technologies varies across the seven regions that we have identified above, from 119 networks in Western Europe to 23 networks in North America. In between these two extremes are to be found Asia-Pacific with 111 networks, Africa with 101 networks, Latin America with 78 networks, Eastern Europe & CIS with 54 networks and the Middle East with 31 networks. It is, however, worth reiterating that the low figure for North America is somewhat artificial because the numerous regional networks are not being counted. Moreover, these figures are network totals that do not shed any light on which generations of which technologies are being used, and it must be borne in mind that a very large number of operators have launched high-speed networks using more than one technology. Hence, for example, almost every W-CDMA network in Western Europe is overlaid with HSDPA and there are only a handful of networks exclusively using EDGE. The total number of launches by region, taking into account the use of multiple technologies by operators, is shown in Table 4.4.

Table 4.4 Total network launches by region: 2000–2009

	Western Europe	Eastern Europe & CIS	Middle East	Asia Pacific	North America	Latin America	Africa
EDGE	72	33	25	44	9	57	84
W-CDMA	105	24	24	60	8	42	35
HSDPA	97	21	24	59	8	42	32
cdma2000 1xRTT	7	12	3	27	13	31	11
cdma2000 1xEV-DO Rel. '0'	8	5	1	20	10	14	6
cdma2000 1xEV-DO Rev. 'A'	5	1	0	8	7	4	3
2.75G	79	45	28	71	22	88	95
3G	113	29	25	80	18	56	41
3.5G	102	22	24	67	15	46	35
Total	294	96	77	218	55	190	171

Source: Derived by authors from data in Table 4.3

What Table 4.4 confirms is that, as previously noted, the W-CDMA/HSDPA route has become increasingly dominant with EDGE increasingly used in recent years as a means of filling in the outer reaches of networks where it is not economic to build 3G base stations. In contrast, cdma2000 barely gets a look-in for fully mobile 3G/3.5G, and insofar that it does is partly through the use of the 450 MHz band in the former Eastern Europe. Although it remains reasonably popular in the modern version of Eastern Europe as well as the CIS, the concern in such countries is predominantly to obtain roaming rights across Western Europe, which necessitates the use of W-CDMA/HSDPA technology.

It is worth reiterating that the use of technology is historically biased by the choices made by the original networks since interconnectivity/roaming is a paramount consideration. Africa has, as noted in Table 4.4, shown a marked preference for the route based upon GSM, but with a greater emphasis than elsewhere upon EDGE since it is an upgrade requiring no new licences and is fast enough for most purposes, especially bearing in mind that relatively few Africans can afford the expense of buying and using 3G handsets. In contrast, Latin America was traditionally dominated by CDMA-derived technology, partly due to its historical association with the USA, but a significant number of networks have converted their networks to GSM, despite the huge cost, in order to enjoy the benefits of better interconnectivity and a better, and cheaper, range of 3G handsets.

The only region where cdma2000 remains deeply embedded is North America. Currently, as a result of a lengthy process of M&A activity (Curwen and Whalley, 2008: Chapter 8), there are four national operators of which two – Sprint Nextel and Verizon Wireless (despite being 45 per cent owned by GSM operator Vodafone) – use cdma2000, whereas AT&T and T-Mobile use EDGE/W-CDMA/HSDPA. The regional operators largely use cdma2000. This is not going to change, and it is broadly the same in Canada with Rogers (GSM) versus Bell and Telus (cdma2000) although the last two both launched HSDPA in November 2009

LAGS BETWEEN LICENCE AND LAUNCH OF 3G NETWORKS

A further issue to consider is the gap in months between the issuing of a licence and the launch of the corresponding network. Since EDGE does not need a licence, what follows for the GSM route is restricted to W-CDMA launches or, if HSDPA has been introduced before W-CDMA, the launch date for HSDPA. The situation is less clear-cut for cdma2000 1xEV-DO since it has frequently not been licensed as such, and hence a variety of methods has had to be used to calculate the notional licence date which is, in many cases, either that of the original 2G network or the date on which the government changed the rules to permit the use of cdma2000 technologies. However, in 21 cases, no specific date can be ascribed to the issue of the licences.

All regional licences have been excluded, as have all licences in the USA due to the ambiguities inherent in matching licence and launch dates in respect of 3G. On that basis, the sample set contains 367 W-CDMA licences and 88 cdma2000 1xEV-DO licences. The 367 W-CDMA licences have led to 275 launches as of the end of 2009, while the 88 cdma2000 1xEV-DO licences have led to 61 launches.

In principle, it is to be expected that the lags between licence and launch dates would be quite large where the licences were issued at the beginning of the process – commencing in Finland in March 1999 in respect of W-CDMA – and quite small in very recent times. This is because there is a learning curve involved in rolling out a 3G network of base stations, as a result of which the equipment vendors will eventually be able to put together a package of all the requisite technology that can be delivered to a licensee on a ready-to-use (turnkey) basis. Other factors are more ambiguous. When 3G licences were first issued they were very highly valued as they were expected to lead to a massive upsurge in high-priced data transfer.[6] Hence, the sooner a network could be rolled out, the sooner revenue generation would commence. However, the reality was that customers were slow to make use of the services on offer, often because their quality was unacceptable, and as a result licensees became much more cautious about the speed of roll-outs, especially bearing in mind that roll-outs would incur huge expenditures over and above any licence costs.[7]

If five years (60 months) is taken as a yardstick for a seriously delayed launch – it should be borne in mind that licences almost always included a series of geographic or population coverage requirements by specified dates which included much shorter delays than this in respect of the initial launch date – then 21 W-CDMA launches, but only three cdma2000 1xEV-DO launch, were heavily delayed. These were as follows for W-CDMA (country; licence date; licensee; (delay in months)):

- Belgium; February 2001; Orange (67), KPN (88)
- Canada; February 2001; Rogers Wireless (69)
- Czech Republic; December 2001; T-Mobile (60)
- Denmark; September 2001; TeliaSonera (75)
- Finland; March 1999; TeliaSonera (67), Elisa (68), DNA Finland (81)
- Guatemala; March 2003; América Móvil (61), Millicom (66)
- Isle of Man; April 1999; Telefónica (formerly mmO_2) (79)
- Israel: December 2001; Pelephone (85)
- Liechtenstein; March 2001; Orange (71), Telekom Austria (71)
- Monaco; June 2000; Monaco Telecom (72)
- Netherlands; July 2000; T-Mobile (66)
- New Zealand; January 2001; TNZ (94)
- Poland; December 2000; Orange (61)
- Puerto Rico; October 2001; AT&T (61)
- Spain; March 2000; TeliaSonera (81)[8]
- Switzerland; December 2000, TDC (60)

The three launches for 1xEV-DO were Telekom Austria in Belarus, which was licensed in August 2002, but launched 81 months later in May 2009; Tatem Telecom in the DR Congo, which was licensed in September 2001, but launched 66 months later in March 2007; and LG Telecom in South Korea, which was licensed in August 2001, but launched 80 months later in April 2008.

---

6 See, for example, UMTS Forum (1999) for an overview of the diverse array of new, more lucrative, services that it was expected would be introduced.

7 *The Economist* (2001: 71) estimated that, within the EU, roughly the same amount would need to be spent on constructing 3G networks by mobile operators as they spent on licences.

8 See Whalley and Curwen (2008) for further details.

The preponderance of early dates (2000–01) clearly indicates second thoughts on the part of some network operators. In contrast, many operators have since early 2006 taken to rolling out and testing networks in anticipation of a predictable licensing date (which nevertheless often ends up being postponed), and hence have been able to switch on their networks (almost) immediately upon the receipt of a licence. Although some of the sample sizes shown in Table 4.5 are too small to be reliable, it is interesting to observe that the two technologies show a similar pattern overall and that the typical delay appears to be between two and two and a half years. The relatively long delay in the case of W-CDMA in Western Europe is a straightforward function of the early licensing regime fostered by the European Union (Curwen 2002), while the relatively short delay in Africa reflects the much more recent introduction of high-speed networks.

Table 4.5 3G launch delays by region[1]

Region	No. of launches W-CDMA	Average delay – months	Shortest and longest delays	No. of launches 1xEV-DO	Average delay – months	Shortest and longest delays
Western Europe	102	36	0 & 88	7	19	0 & 36
Eastern Europe & CIS	26	12	1 & 48	4	35	7 & 81
Middle East	20	21	0 & 85	1	33	33 & 33
Asia Pacific	53	27	1 & 94	12	26	0 & 80
North America	3	57	41 & 69	4	39	6 & 57
Latin America	30	19	0 & 66	8	20	1 & 46
Africa	32	14	0 & 23	4	27	8 & 66
Total	266	27	0 & 94	40	26	0 & 80

Source: Compiled by authors

Note: [1] Not all launches have known corresponding licence dates

There are also some cases where licences have been awarded, but as yet there have been no corresponding launches. In Table 4.6, each year is split into two halves (H1 and H2) in order to show the increase in W-CDMA licensing that occurred after 1 July 2007. Taking the four-year period 1 January 2006 to December 2009, there were 63 licensed W-CDMA networks that had yet to launch together with 17 cdma2000 1xEV-DO networks. Prior to 2006, there were only 29 such networks in total. However, given the typical condition that a launch must occur within three years from the licence date, this number may still appear to be surprisingly large.

In practice, however, regulators have never revoked licences immediately in respect of networks that have not been launched on time (although revocations have eventually taken place on numerous occasions), and there have been several occasions when appeals against the original licence conditions have allowed the launch process to be delayed. Hence, unlaunched licences dating back to 2004 are no great surprise. As for the cases prior to 2004, of which there are 10 (excluding the Netherlands Antilles), it is probable that no launch will ever occur, although the example of Xfera (the brand of TeliaSonera) noted above might suggest to the contrary.

Table 4.6 Number of licensed networks not yet launched

	2000	2001	2002	2003	2004	2005	2006	2007	2008	2009
H1 W-CDMA	1	2	7[1]	2	2	3	6	5	9	7
H2 W-CDMA	0	1	0	0	3	3	10	10	7	9
H1 1xEV-DO	0	0	1	0	0	1	4	3	0	0
H2 1xEV-DO	0	0	0	1	0	2	3	3	4	0

Source: Compiled by authors

Note: [1] Including five technology-neutral licences issued in the Netherlands Antilles.

RECENT DEVELOPMENTS

At this point it is worthwhile to examine the launch of high-speed networks beyond 3.5G even though these are as yet somewhat thin on the ground. In the first place, it is possible to take note of the improvement in the performance of the uplink in a HSPA (high-speed packet access) network – in other words, the specific introduction of uprated HSUPA (high-speed uplink packet access). Because this is an overlay it is not necessarily reported in the media and, therefore, records are somewhat erratic. So far as can be traced, there have been 72 upgrades to basic HSUPA, of which 46 occurred in Western Europe and 15 in the Asia-Pacific region. The records indicate that 30 occurred in 2007, 26 in 2008 and 15 in 2009 which, superficially, does not appear to be likely as one might have expected the numbers to build up year on year. However, www.gsacom.com provides no dates and www.3gamericas.com only provides 2009 dates that include March, June and September – in other words, projected dates not actual launch dates.

Aside from database imprecision, another possible explanation is that operators' thoughts are turning increasingly to HSPA+ which involves upgrades to both the downlink and the uplink. According to GSA (2009), three operators had launched HSPA+ by May 2009 – Telekom Austria, Telstra and StarHub (Singapore) – with a further 25 operators testing, trialling or deploying the technology. Subsequently, there have been launches by Telekom Austria (Bulgaria); Bell, Rogers and Telus (Canada); OTE, Telefónica and Vodafone (Greece); SmarTone and Telstra (Hong Kong); Indosat (Indonesia); Telecom Italia (Italy); eMobile (Japan); Zain (Kuwait); PTC and Vodafone (Poland); Portugal Telecom, Sonae.com and Vodafone (Portugal); OTE (formerly Zapp) (Romania); Telefónica and Vodafone (Spain); and Swisscom (Switzerland).

Conclusion

This chapter has sought to examine high-speed network launches. To do so, a database of high-speed network launches has been developed that overcomes the significant discrepancies in the existing public domain databases. Although a rigorous methodology has been adopted, it should nevertheless be noted that the numerous definitional issues that need to be addressed mean that the term 'definitive' must be treated with caution in relation to the database. Furthermore, the approach has been conservative, preferring to omit rather than include launches for which unverifiable claims are made in public domain databases or the media. Having said this, in the light of the extraordinary and unpredictable problems encountered in ascertaining launch dates, this database is much the most reliable of those currently available.

From the above analysis, it is possible to draw a series of conclusions. The first conclusion that can be drawn is that licensing a high-speed network does not necessarily result in its subsequent launch. The analysis has demonstrated that some networks have not been launched, at least not yet, and that those subject to considerable delays of more than 60 months were licensed at the start of the 3G licensing process. As delays of any kind reduce competition within mobile telecommunications markets, it is likely that consumers have not benefited – through the enhanced competition and innovation that result from more operators being present in the market – to the extent that was anticipated when the licensing process occurred.

A second conclusion that can be drawn is that across the seven regions that were identified, differences regarding when the networks were launched can be distinguished. Over the 10-year period under examination, two-thirds of the networks that were launched did so within the four-year period 2004 to 2007 inclusive. The number of launches peaked in 2006 at 238, then fell back modestly until the end of 2008. In comparison, 2009 showed little activity, suggesting that the era of 3G/3.5G is largely over. While this is true at the aggregate level, variations nevertheless occur at the level of the individual regions with, for example, an upward trend being observable in Latin America over the entire period until the end of 2008.

Differences also occur with respect to the technologies that have been chosen by mobile operators. Western Europe, for example, has overwhelmingly favoured the EDGE/W-CDMA/HSDPA route while the cdma2000 route is largely associated with North America when the numerous regional networks are taken into account. Having said this, cdma2000 has been adopted, albeit less extensively, in several other regions such as Latin America and Asia Pacific, although a perceptible trend has merged in Latin America over the past few years to overlay CDMA networks with GSM and its higher-speed developments.

References

Ballon, P. 2007. Changing business models for Europe's mobile telecommunications industry: The impact of alternative wireless technologies. *Telematics and Informatics*, 24(3), 192–205.

Bijwaard, G., Janssen, M. and Maasland, E. 2008. Early mover advantage: An empirical analysis of European mobile phone markets. *Telecommunications Policy*, 31(3–4), 246–61.

Björkdahl, J. and Bohlin, E. 2002. Financial analysis of the Swedish 3G licensees. *info*, 4(4), 10–16.

Bouwman, H., de Vos, H. and Haaker, T. 2008. *Mobile Service Innovation and Business Models*. Berlin: Springer-Verlag.

Börgers, T. and Dustmann, C. 2003. Awarding telecom licences: The recent European experience. *Economic Policy*, April, 215–68.

CIA. 2007. *The World Factbook*. Available at http://www.cia.gov//library/publications (accessed 15 August 2009).

Commission of the European Communities. 1997. Communication from the Commission to the Council, the European Parliament, the Economic and Social Committee and the Committee of the Regions: Strategy and policy orientation with regard to the further development of mobile and wireless communication (UTS). COM(97) 513 final. Brussels: CEC.

Curwen, P. 2002. *The Future of Mobile Communications: Awaiting the Third Generation*. London: Palgrave.

Curwen, P. and Whalley, J. 2008. *The Internationalisation of Mobile Telecommunications*. Cheltenham, England: Edward Elgar.

Dahlman, E., Parkvall, S., Sköld, J. and Berning, P. 2007. *3G Evolution: HSPA and LTE for Mobile Broadband*. London, Academic Press.

Dunnewijk, T. and Hultén, S. 2007. A brief history of mobile communication in Europe. *Telematics and Informatics*, 24, 164–79.

Fuentelsaz, L., Maícas, J. and Polo, Y. 2008. The evolution of mobile communications in Europe: The transition from the second to the third generation. *Telecommunications Policy*, 32(6), 436–49.

Funk, J. and Methe, D. 2001. Market- and committee-based mechanisms in the creation and diffusion of global industry standards: The case of mobile communication. *Research Policy*, 30(4), 589–610.

Gandal, N., Salant, D. and Waverman, L. 2003. Standards in wireless telephone networks. *Telecommunications Policy*, 27(5–6), 325–32.

Gruber, H. 2007. 3G mobile telecommunications licenses in Europe: A critical review. *info*, 9(6), 35–44.

GSA. 2008. *EDGE Fact Sheet*. Available at http://www.gsacom.com (accessed 15 August 2009).

GSA. 2009. *GSM/3G Market/Technology Update: Evolution of Network Speeds*. May. Available at http://www.gsacom.com (accessed 15 August 2009).

Macklay, J. 2008. European 3G auctions: Using a comparative event study to search for a winner's curse. *Utilities Policy*, X(6), 275–83.

Saugstrup, D. and Henten, A. 2006. 3G standards: The battle between WCDMA and CDMA2000. *info*, 8(4), 10–20.

Steinbock, D. 2003. Globalisation of wireless value system: From geographic to strategic advantage. *Telecommunications Policy*, 27(3–4), 207–35.

The Economist. 2001. Pass the painkillers. 5 May. 71–2.

UMTS Forum. 1999. *The Future Mobile Market*, Report Number 8. Available at http://www.umts-forum.org (accessed 12 May 2000).

Ure, J. 2003. Deconstructing 3G and reconstructing telecoms. *Telecommunications Policy*, 27(3–4), 187–206.

Whalley, J. and Curwen, P. 2006. Third generation new entrants in the European mobile telecommunications industry. *Telecommunications Policy*, 30(10–11), 622–32.

CHAPTER 5 Characterizing Data Services

Introduction

The following three chapters will highlight the increased complexity that has emerged in the mobile telecommunications industry. With the emergence of high-speed networks and increased competition, mobile operators have attempted to move away from the provision of voice and basic data services like SMS (short message service) with the offer of an ever more diverse, and, hopefully, more lucrative, array of data services. In so doing, operators have sought to enhance their competitiveness through closer links with handset vendors on the one hand and content providers on the other. Handsets are clearly important as they are the primary means through which consumers communicate with one another and, perhaps more importantly, access the data-based services that mobile operators have enthusiastically promoted as high-speed networks are deployed.

The handset market is fiercely competitive. New companies have entered the market, while existing ones have struggled or even exited altogether. One aspect of this competition is the range of features offered by each handset,[1] while another is whether or not the handset is exclusive to a specific operator. Handsets are increasingly seen as a means of providing access to other services, some of which sit on the handset while others are accessed via the Internet, and thus have received considerable attention from companies like Google that see the handset as the means through which their services can be accessed. In other words, the handset has become a platform that supports the delivery of other services.

As a consequence of these and other developments, the mobile industry is substantially more complex than it was just a few years ago when it was characterized by a limited range of services and the presence of a handful of companies.[2] One consequence of this increased complexity is that operators are sometimes caught unprepared by the success of a specific service or handset. Alternatively, they seek to replicate the success of rivals through copying their strategies. This is not to suggest that analysts have neglected to predict how the mobile market may develop. It is clear from Fildes and Kumar (2002) that the efforts of analysts to understand how the demand for existing and future services may develop is both long-standing and wide-ranging in terms of the methodological

---

1 Interestingly Funk (2008) argues that technological developments in other parts of a broadly defined information technology (IT) sector, such as the semi-conductor industry, have resulted in changes within the mobile industry. Due to such linkages it is likely that companies in the mobile industry will be unaware of the pace and direction of technological changes, with the consequence that they may be caught unprepared by changes initiated by developments elsewhere. In other words, such linkages increase the uncertainty experienced by companies in the mobile industry.

2 Such complexity can be seen geographically, with the industry increasingly organized at a regional or global level, or in the way in which the fortunes of companies, and their host countries, can quickly change (Steinbock 2002, 2003).

approaches that have been adopted. This diversity is also evident from Anderson, Daim and Kim (2008) who identify a range of statistical techniques that have been used to forecast wireless technologies, while Lindgren, Jedbratt and Svensson (2002) include qualitative methods in their discussion of mobile marketing.

Thus, before progressing in the following three chapters to a discussion of change within the mobile market, it is useful to draw on this wide-ranging literature to outline several concepts that will aid our understanding of how the mobile market has changed.

Diffusion

The diffusion of new products is a widely researched issue with many different models evident in the literature. One of the best-known and earliest models is that of Bass (1969), which places the adoption of a new product at the intersection of actual and potential users. At the start of the process the number of innovative adopters is greatest, but their number falls over time. In contrast, the number of later adopters influenced by early adopters will increase over time. The Bass model has subsequently been extended and modified, in part because of the limitations of the original model that result from its market-based assumptions (Fildes and Kumar 2002).

How fast a particular product is adopted will be reflected in its diffusion curve as represented in Figure 5.1. The diffusion curve of a product charts its uptake over time, with this uptake often represented as a proportion of some theoretical maximum. A faster pace of uptake in the initial stages of a product's availability will result in a steeper diffusion curve. Geroski (2000) suggests that simpler technologies will diffuse into the marketplace faster than those that are more complex, though simplicity here is rather narrowly defined in terms of software knowledge. Gruber and Verboven (2001) investigate the diffusion of mobile telecommunications services in the EU, drawing three conclusions from their analysis. They conclude, first, that the switch from analogue to digital systems at the start of the 1990s, with its associated increase in spectrum capacity, played a significant role in encouraging the diffusion of mobile telecommunications. Secondly, they found that penetration levels in those EU Member States that were slow to issue 2G licences took a considerable time to catch up with those that licensed early. Thirdly, the introduction of competition into mobile markets also played a role, though interestingly this impact was found to be of less significance in encouraging mobile diffusion than the switch to digital technologies.

In contrast, the focus of Rouvinen (2006) is broader as the objective is to determine if differences exist between developed and developing countries. The analysis finds that differences do exist between these two groups, with, for example, the market-size effect and availability of prepaid services being more important for developing than developed countries. Boretos (2007) compares actual mobile phone diffusion patterns in key markets against the classic 'S' shaped diffusion curve. The analysis finds that the S-shaped models proposed compare favourably with what happened in practice, thereby enabling growth forecasts to be made. It is worth noting, however, that Geroski (2000) states that symmetric S-curves are rarely encountered in practice, with the later stages of diffusion often occurring more slowly than predicted by such a curve.

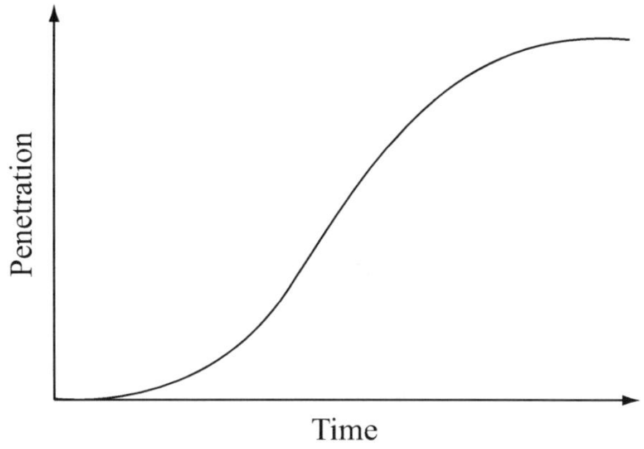

Figure 5.1 Diffusion curve

A similar comparative approach can be observed in Singh (2008), although this time the analysis focuses on a single country, India. The research concludes that high mobile growth rates there will continue until 2012–13, after which the pace of growth will slow until the market reaches 900 million subscribers by 2015–16. Interestingly, while acknowledging that such growth will have an impact on many aspects of the Indian mobile telecommunications market, Singh suggests that, after an initial decline, average revenue per user (ARPU) will stabilize between 2010–11 and 2015–16. In another example, Gamboa and Otero (2009) use a S-curve model in the context of the diffusion of mobile phones in Colombia. They conclude that although the mobile growth rate has slowed in recent years, the market will not become saturated until the second half of 2013.

The diffusion of specific services has also been examined in the literature. Chu and Pan (2008) extend Bass (1969) to include substitution and competition between products in order to forecast the diffusion of the mobile Internet in Taiwan. Their analysis found that GPRS (general packet radio service) and 3G,[3] the two competing mobile technologies that they examined, displayed quite different growth patterns during the period 2001 to 2013. They suggested that the number of GPRS Internet subscribers would rapidly rise before falling back, whereas the growth rate for 3G would be slower and less dramatic. As a result, they predict that during almost all of the period considered, GPRS would be the main technology through which subscribers would enjoy the mobile Internet although it would eventually be overtaken by 3G.

A different methodological stance is evident in García and Rendón (2009). After commenting on the socio-economic, educational, cultural and technological diversity that can be observed across Latin America, they develop a simulation model for Mexico to determine the diffusion patterns of a range of what they term cellular-based technologies and Wi-Fi over a 36-month period. Their simulation shows that whereas 3G and Wi-Fi will co-exist within the Mexican market, there will be considerably more Wi-Fi than 3G subscribers throughout the period covered by the simulation model.

3 Somewhat surprisingly, Chu and Pan (2008) group together PHS (personal handy-phone system) and 3G and collectively compare these two technologies against GPRS.

Many factors influence why a consumer adopts one particular service over another. It is sometimes argued that the decision to adopt one product over another is the result of a comparison being made between the products. Rogers (2003) suggests that the relative advantage that a new product may have over an existing one is a significant determinant of whether the product will be adopted or not. The most obvious relative advantage that a mobile phone has over a fixed connection is that it is portable. However, this advantage does not exist when the choice is between an existing mobile phone running over a 2G network and one using a 3G or higher-speed network. In both circumstances, consumers are geographically mobile, though this mobility may be circumscribed by the geographical scope of the network being used. Having said this, as 3G networks are more recent than 2G networks, a reasonable assumption to make is that their geographical coverage is narrower, but that this gap is likely to diminish and disappear altogether as their coverage is expanded.

As both 2G and 3G networks provide mobility, the next issue concerns other factors that may encourage adoption. For example, handsets today are equipped with a wide range of features, any one of which may encourage adoption. Economides and Grousopoulou (2009) use a sample of students to investigate the relative importance of handset features. While it is probably no surprise that they found that some features were more important than others, what may be surprising is the comparatively long list of features that were considered to be important, which comprise: battery life, Bluetooth, calendar, camera (video and photo), design and elegance, mp3 player, organizer and reminder, and storage memory. A shorter list of handset features – chat, common use of files, cryptography, encryption, teleconference, touch-screen, printing and voice commands – were identified as unimportant. Moreover, gender differences were identified in their analysis. For example, females placed greater importance on the design and appearance of the handset than males, with the converse being the case for battery life. However, although such differences exist, and Economides and Grousopoulou rightly draw attention to them, they note that the differences are not statistically significant.

Wei (2008) conducted a telephone survey to explore what factors encourage the use of mobile phones, finding that the desire to access news and the Internet while on the move encouraged use. The research also found that the more calls a subscriber makes and receives, the more likely is the use of other services. It is worth noting, however, that Wei suggests that the results of the research should be treated with caution given the sample size and composition. Services are also at the heart of the analysis undertaken by Bouwman et al. (2007) who examined the drivers of, and barriers to, service adoption in Finland. After examining a series of bundled service packages, they concluded that the inherent differences that exist within each of the six packages needs to be taken into account when any analysis of the drivers and barriers is made.

Given the above, it is possible to identify a series of key issues that can be taken into account when the following three chapters are being read. The first of these is that the pace of diffusion is important – a rapidly diffusing product soon produces a critical mass of users. In turn, this results in a second issue for consideration, namely network effects. Network effects are the benefits that an individual receives when others use the same product. Consequently, the more people that use a specific technology, the greater the benefits that can be enjoyed. Finally, as the number of users of a specific technology increases, this changes the economic characteristics of its production and consumption.

Standards

Tassey (2000: 588) states that an industry standard 'is a set of specifications to which all elements of products, processes, formats or procedures under its jurisdiction must conform'. Standards provide a way to specify quality as well as to encode and communicate information (Williams, Whalley and Li 2003) and specify those properties necessary to ensure that complementary products work with one another. Funk (2002) adds to this, highlighting the role of standards in defining interfaces between complementary products in industries such as telecommunications and broadcasting, whereas Windrum (2004) notes that standards have enabled complex technologies like the Internet to emerge without companies having to coordinate their products through a series of formal agreements.

Standards emerge in a variety of ways. David and Greenstein (1990) identify four different standardization processes. Market-based competition may lead to the emergence of unsponsored standards: that is, a standard where no single company or organization is large enough to impose its pricing or technological choices on other companies, organizations or users. In contrast, sponsored or proprietary standards emerge when a company or organization controls a set of assets. Through this control, the company or organization is able, for example, to determine whether the technology is or is not compatible with other technologies and shape the development of future standards. Standards may also emerge through agreement reached within voluntary organizations or from governmental edict. In contrast, Gandal, Salant and Waverman (2003) identify three different ways through which a standard may emerge. In the first place, a standard may be set primarily by the market and can be proprietary in nature. In other words, a *de facto* standard emerges. Standards may also emerge from voluntary industry agreements or be imposed. If the standard is imposed, this may be by a national body or as the consequence of agreement reached at a regional or global level.

The above hints at a distinction that is often made in the literature, namely between *de facto* and *de jure* standards. *De facto* standards are picked by the market whereas *de jure* standards are the result of the deliberations of a committee. Whereas Lea and Hall (2004) draw attention to the definitional ambiguity of *de facto* and *de jure*, Funk (2002) argues that such a distinction is of increasingly less value. He argues that as companies recognize the value of standards, they have thereby begun to collaborate in the development of *de jure* standards that subsequently compete against one another in the market. As a consequence, Funk proposes a hybrid model that combines elements of the market and committee approaches and explores their development and interaction through the example of mobile telecommunications. Depending on how the standard was initially derived, as well on as the geographical scope at which it is applied, the balance between market and committee will differ. For example, at a regional or large country level, the creation of a single standard through committees incorporating both operators and vendors has proved to be more effective in creating global standards than has competition between standards at either of these levels. Alternatively, within the context of a single standard, companies may also compete against one another in committees as well as in the market. Participation in committees not only ensures that a company understands the intricacies of the standard but also enables it to push for its own technologies to be included within the standard.[4]

According to Liebowitz and Margolis (2002), the utility of a standard rises with the number of people who use it. The user of a particular standard benefits from the presence of fellow users through, for example, having more fellow mobile subscribers to call (Rohlfs 2001). Another network effect is that prices may fall – as more people use a mobile phone, vendors will be able to achieve scale economies in the production of handsets with the consequence that their prices to operators will fall. Operators can then lower their own prices. In turn, this will encourage the further uptake of mobile handsets, increasing the size of the market and thus enabling yet more scale economies to be achieved by vendors. This illustrates what is known as positive feedback, which, as Shapiro and Varian (1999: 175) write, 'makes the strong to get stronger and the weak get weaker'. A virtuous cycle may occur, which is when, for example, prices fall through scale economies being achieved that in turn encourage the growth of the market and thus enable further scale economies to be obtained. However, Shapiro and Varian suggest that this may quickly turn into a vicious cycle. A vicious cycle could occur when, for example, a vendor's handset market share is falling, as this reduces its ability to achieve scale economies on the one hand and fund research and development to remain competitive on the other.

Irrespective of the benefits that may be derived from one standard or another, the possibility exists that a user could become locked into using one standard when alternatives exist. Although such 'lock in' can be advantageous for the sponsor of the technology, there is the possibility that users may be locked into using an inferior technology due to the impact that network effects have on user choice (Liebowitz and Margolis 2002). One frequently cited example of the widespread use of an inferior technology is that of the QWERTY keyboard (David 1985; Rogers 2003), although it has been suggested that the QWERTY keyboard is not, in fact, inferior compared to the Dvorak keyboard (Howells 2005). Assuming that there is some merit in these arguments, why has the QWERTY keyboard continued to be preferred? One explanation that has been suggested is that the costs to move between the QWERTY and Dvorak keyboards are too large to offset the improved efficiency that would result, while another is that untrained typists prefer to acquire skills in QWERTY as it is more marketable, and trained ones are proficient in QWERTY and not Dvorak (Shapiro and Varian 1999).

Another example of an inferior coming to dominate a particular market is that of VHS/Betamax. After a period where both technologies coexisted, VHS grew faster than Betamax to become the dominant standard due, in part, to the availability of pre-recorded content (Howells 2005). Interestingly, the alleged superiority of Betamax over VHS has been questioned in the literature, with Liebowitz and Margolis (2002) noting several areas where Betamax was deficient, and Rohlfs (2001) noting that neither was superior across the board.

Competition between Betamax and VHS is used by Rohlfs to illustrate the 'winner takes all' nature of some high-technology markets. Although technologies may coexist and compete against one another for a while, one of them is eventually able to grow more quickly than the others and thereby exert control over the entire market. In other words, 'winner takes all' is an extreme form of positive feedback that occurs where a single

4 For a discussion of market- and committee-based standard-setting mechanisms in the mobile industry, see Funk and Methe (2001).

technology out competes all others so that it is the only one left in the market (Shapiro and Varian 1999). Rohlfs (2001: 115) aptly notes that 'Betamax was not a bad product, but there was no second prize'.

A final point worth making is that the competition that occurs between technologies can be influenced by both serendipity and path dependency. Howells (2005), drawing on Cringely (1996), argues that the combination of IBM and Microsoft was an 'accidental event' that resulted in the combination of IBM PC and Microsoft DOS jointly controlling 40 per cent of the market. Subsequently, the dominant position of IBM PC/Microsoft DOS in the market was solidified as other companies began to produce software that was compatible with this standard. Microsoft's position was also aided by its strategy of ensuring that as successive generations of its operating system were launched, 'backward compatibility' with older versions of applications software was possible (Howells 2005). Backward compatibility would encourage users to upgrade to the most recent version of Microsoft's operating system safe in the knowledge that a wide range of software applications was available. In turn, the presence of a wider range of Windows-compatible software would encourage new users to opt for this operating system instead of those offered by rivals such as Apple. And as the number of Windows users grew, software developers would produce new applications for this operating system because their potential market would be maximized compared to developing software for other operating systems. While these network effects undoubtedly contributed to the success of Windows, Rohlfs (2001) identified another, albeit related, contributory factor, namely pricing. In the aftermath of the initial launch of MS DOS and then Windows, Microsoft set prices with a view to increasing the penetration of both of these products rather than to maximizing short-term profits.

It is often suggested that the success of the Windows operating systems over its Apple counterparts is another example of the market preferring an inferior technology. If we assume that Apple's operating system was indeed superior, then the preference for Windows illustrates what Liebowitz and Margolis (1995) call 'third-degree path dependence', which is where an initial decision results in an inefficient outcome being chosen. They also identify two other forms of path dependence, with one being where no inefficiencies occur (first-degree) and one where choices are inefficient with hindsight (second-degree).

From the above it is possible to draw several lessons that will help us understand the developments outlined in the following three chapters. The first of these is that once a critical mass of users has been achieved, further growth occurs as network effects start to come to the fore. Secondly, if two technologies are competing against one another, the pace at which they grow will play a key role in determining whether one is more successful than the other in the marketplace. The faster growth of one technology compared to another also results in economic changes, enabling scale economies to occur as well as encouraging producers of associated products to favour the faster-growing technology. A third lesson that can be drawn is that due to the presence of network effects, the technology favoured by the market may not necessarily be the best technology.

From Value Chains ...

The structure of industries and organizations can be understood with reference to the notion of a value chain. At its most basic, a value chain is a series of steps that transform inputs into products and services. Porter (1985) identifies two types of activities – primary activities and support activities – that occur within an organization as it transforms inputs into the products or services that it sells to its customers.

In all, five primary activities are identified: inbound logistics, operations, outbound logistics, marketing and sales, and service. Some of these are more or less self-explanatory, with inbound logistics being concerned with the receiving, storing and then distribution of those inputs required by an organization to produce its services or products. Conversely, outbound logistics deals with the storage and then distribution of the organization's final product or service to its customers. Within the organization, operations transform inputs into final products and services while marketing and sales promote the product so that consumers are aware of it. Finally, the service function installs the product or service and then supports its use by customers through the provision of after-sales care.

In addition to these five primary activities there are four supporting activities: procurement, technology development, human resources management and firm infrastructure. Procurement, which may occur either centrally or across the organization, acquires those resources that are used by the primary activities of the organization. The technology development support activity is widely defined, including all those activities in the organization that somehow involve technology with the consequence that research and development (R&D) is included as are those improvements that arise from process innovations. The firm infrastructure activity is also wide ranging, and broadly speaking organizes the other activities so that the company can achieve its objective. Thus, it includes the enterprise-wide information systems that may be used to manage inventory as well as the financial aspects of the organization. Finally, the human resource support activity recruits, trains and manages the employees of the organization.

Although our focus is on individual companies, it is useful to recount the aforementioned activities undertaken by a company for three reasons. First, these activities transform inputs into products and services and, in so doing, organizations add value to them. In other words, through adding value to the inputs they are able to sell the resulting products and services at a profit. Secondly, it is through these various activities that organizations compete against one another. The organization's competitiveness may be due to how the various activities are combined with one another or reflect their uniqueness. Thirdly, the use of the value chain model highlights the coordination that companies must undertake in order to deliver their products and services to their customers.

Given the above, an inevitable question to ask is what is the structure of the value chain for the telecommunications industry? A useful starting point when answering this question is Fransman (2002b) who divides the old pre-Internet telecommunications industry into three layers. In the first of these – the equipment layer – can be found switches, transmission systems and so forth, whereas the circuit switched network constitutes the second layer, network. The service layer is the third layer of the old industry and includes services such as voice and fax.

This basic division of the industry into equipment, infrastructure and services is also evident in Maitland, Bauer and Westerveld (2002). They divide the 2G mobile telecommunications value chain into four stages: network and handset equipment manufacture; network operation; network access provision; and, retail and handset service sales. Naturally, such a division of the mobile value chain reflects the presence and subsequent delivery of a narrow range of services such as SMS, and therefore does not take into account the wider range of services commonly associated with 3G. This limitation is acknowledged by Maitland, Bauer and Westerveld (2002), who draw on the Internet value chain, as well as the role played by third parties when implementing services, to suggest a more complex 3G value chain.

In their model there are now nine stages in the 3G value chain. The five additional stages are all located between handset and network equipment manufacturing and network operation stages, and highlight the broader range of roles that need to be undertaken in the 3G value chain. Yet another variation of the mobile value chain is suggested by Kuo and Yu (2006), who identify 11 different roles including equipment vendors, application platform vendors, and content aggregators and developers. The inclusion of additional roles within the mobile value chain reflects, to a greater or lesser extent, the broader array of services that can be delivered over 3G networks.

One consequence of this wider range of roles is that mobile operators are faced with the dilemma of deciding what their strategy should be. Mobile operators could focus their attention solely towards infrastructure, or alternatively they could seek to move into the provision of services themselves. Both of these are suggested by Maitland, Bauer and Westerveld (2002), as is the notion of the mobile operator generating revenues through entering into wholesale agreements with MVNOs (mobile virtual network operators). A somewhat different set of possible strategies can be found in Li and Whalley (2002). Drawing on Hagel and Singer (1999), they suggest that telecommunication operators could either focus on managing the customer relationship, developing new products and services or the provision of infrastructure. Although all three are presently undertaken within the same company, it is argued that a series of benefits – enhanced innovation, reduced prices and improved quality – would result if this was changed and provision was through several companies. An individual company would still require access to other parts of the value chain irrespective of which strategy is adopted. What results is a series of market-based relationships between numerous companies. In other words, the relative simplicity of the value chain, where activities are organized largely within a single company, would be replaced by myriad relationships. This increased complexity highlights another consequence of the emergence of a wider range of roles, namely the need to coordinate both within and between companies.

... to Value Networks and Ecosystems

The previous section demonstrates that the notion of a value chain can be applied either at the level of the industry or to a specific company. However, the increased complexity of the value chain that is emerging due to the switch to high-speed data networks, both across the telecommunications industry and within individual companies, necessitates a different approach being adopted

Li and Whalley (2002) argue that value chains are evolving into value networks as new business models emerge, due to the aforementioned deconstruction of telecommunication companies on the one hand and the widespread use of internet protocol (IP) on the other. While previously telecommunication companies occupied the ground between infrastructure companies and the end customer, the emergence of new business models have seen their replacement by a range of new companies such as software and financial intermediaries. In some cases these new companies have directly replaced the telecommunications operator, while in other cases the replacement has been by more than one company. For example, a reseller may replace the original telecommunications company and connect the infrastructure company with the end customer. Alternatively, an aggregator may perform this role, though given the nature of its business model this, by definition, necessitates the involvement of more companies between the infrastructure company and the end customer.

Interestingly, Li and Whalley (2002) argue that value chains and value networks are interwoven. They suggest that although the value chain has retained a key role in the telecommunications industry, value creation has migrated towards the value network with its numerous relationships and new business models. Moreover, as new commercial opportunities are more likely to be found closer to the final customer, the value network will develop unevenly in different parts of the telecommunications industry. This is clearly evident in their discussion of a mobile portal value network, which highlights not only the coordination that is required between a wide array of actors, but also the complexity of these relationships.

To understand the relationships which exist between companies, Feijóo et al. (2009), Fransman (2007a, 2007b) and Li (2009) use the notion of an ecosystem. Although the nature and scope of their analysis differs, they all highlight that an array of relationships exists within the ecosystem and thus demonstrate its complexity. According to Li there are three parts to a business ecosystem – symbiosis, platform and co-evolution. The multiple relationships which exist between companies are part of the first characteristic of an ecosystem that he identifies, as is the competition that may emerge between ecosystems. His second characteristic is that of a platform, which is defined as 'services, tools or technologies that other members of the ecosystem can use to enhance their own performance' (Li 2009: 380). Co-evolution can be interpreted as meaning that, acting together, the various members of the ecosystem develop new products and services, and thus individual companies gain from participating in the wider ecosystem. In other words, membership of the ecosystem is value-adding for an individual company. Interestingly, Li suggests that through a common vision the value of synergistic benefits for the ecosystem can be more than they would be for the sum of individual companies. His concept of an ecosystem is then applied to that which surrounds Cisco, identifying in the process the numerous relationships observable within this ecosystem across and within different sectors, but also how the desire for 'technical excellence' by Cisco also permeates into other companies (Li 2009: 385).

Feijóo et al. (2009) analyse the mobile-content market. Although they do not explicitly define what they mean by the term 'ecosystem', it is clear from their discussion that it includes the creation and distribution of mobile content through a variety of business models. Thus, relationships occur between content providers, third-party aggregators of content, mobile operators and the final consumer through a variety of business models. They argue that not all business models are appropriate for mobile content, and that different types of content are suitable to different business models. From their analysis they conclude that the mobile-content ecosystem is diverse and fragmented.

A broader definition of the constituent elements of an ecosystem can be found in Fransman (2007a, 2007b). His analysis of the ICT (information, computing and technology) sector identifies four constituent elements: network element providers; network operators; platform, content and applications providers; and, final consumers. While these represent four distinct layers in the ICT ecosystem, the boundaries between them are neither fixed nor solid. Moreover, companies may be present in more than one of these layers. Given this structure, Fransman (2007b: 94) then investigates how innovations emerge within the ICT sector and suggests that they are the product of six different symbiotic relationships between the aforementioned four constituent elements of the ecosystem. These six relationships are as follows:

- between network element providers and network operators;
- between network operators and content and applications providers;
- between content and applications providers and final consumers;
- between network element providers and final consumers;
- between network element providers and content and applications providers; and
- between network operators and final consumers.

Although Fransman explains and illustrates each of these six symbiotic relationships, it is worth noting that all of them are characterized by a series of flows. The buy/sell relationship that exists between companies is inevitably associated with a flow of financial resources from the buyer to the seller. This will differ depending on the symbiotic relationship in question, or it may be indirect. Significantly, there may be no actual financial transfers between the buyer and seller – for example, between content providers such as Google and final consumers. A second characteristic is that of the material flow – where inputs move from one company to another until they eventually reach the final consumer.

The third flow identified by Fransman is that of information. As transactions are consummated and materials flow between companies and to the final consumer, information is provided to the various involved parties. Given the complexity of the ecosystem with its numerous relationships, such information can be quite diverse in nature. It could, for example, be the sales-related information that a handset vendor receives from mobile operators. Alternatively, the information may arise from the relationship between the network operator and the final consumer and shed light on, for example, the level of mobile calls or the propensity of different age cohorts to use SMS. The final flow builds on the previous three and is innovation-related. Through combining the resources of the other three flows, companies are able to engage in innovative activity. Such innovation could, of course, occur at any point within the ecosystem and may be product- or process-related. If the resources are combined to engage in product innovation, a handset vendor could develop a new model or a mobile operator could launch a new service. In contrast, a process innovation would witness a mobile operator innovating to improve the efficiency of delivering its existing range of products or services.

Fransman places the ecosystem within a wider context. The symbiotic relationships that exist between the four layers of the ecosystem are influenced, to a greater or lesser extent, by environmental factors, of which four are identified: competition, regulation and competition law, financial institutions and other institutions. The presence of competition in the symbiotic relationship between, say, network operators and final consumers would influence not only the prices that can be charged, but also the extent to which they able and willing to bundle services or lock-in customers through the use of longer-term contracts. This relationship would also be influenced by regulation, which could, for example, limit the extent to which products may be bundled together by a mobile operator or long-term contracts imposed on subscribers.

Standardization is one of the other institutions mentioned by Fransman, and its presence will have an impact on the symbiotic relationship between network operators and content and application developers. Standardization would allow, for example, content and application developers to configure their products so that they run over more than a single network or draw on software developments elsewhere to speed up their own development process.

When we combine together the four layers of the ecosystem with the six symbiotic relationships and the external influences, it is clear that the ecosystem as envisaged by Fransman is complex. The ecosystem is also constantly changing, in part due to the interaction between these factors, but also due to innovation. While innovation is a broad concept and can refer to products or processes, Fransman makes the distinction between radical and incremental innovation. Radical innovations, such as the move from analogue to digital mobile technologies, are those that have a big impact on the ecosystem, while incremental innovations are more limited in their scope and thus have a smaller impact on the ecosystem. The process improvements that occur in the production of, say, handsets so that they are produced more efficiently provide an example of an incremental improvement.

Business Models

According to Johnson and Scholes (2002: 496), 'a business model describes the structure of product, service and information flows and the roles of the participating parties'. While a broadly similar definition is suggested by Timmers (1999: 32), his definition also draws attention to two additional aspects: the potential benefits for each of the companies in the business model and the sources of revenue. Elements of these two definitions are brought together in a third definition when Osterwalder and Pigneur (2004: 66f) write that 'a business model is a description of the value a company offers to one or several segments of customers and the architecture of the firm and its network of partners for creating, marketing and delivering this value and relationship capital, in order to generate profitable and sustainable revenue streams'.

Given this definition, it is no great surprise that Osterwalder and Pigneur (2004: 68) then go on to identify four components of a business model, namely product innovation, customer relationship, infrastructure management and financial management. These four components are, in turn, deconstructed as part of the development of an ontology of business models.[5] Not only does the ontology highlight the interaction that occurs both within and between the four components, but also the inherent complexity of business models. Such complexity is also evident in Ballon (2007), who identifies four different levels on which business models operate: value network, functional model, financial model and value proposition. The first two of these are termed 'control parameters' and the latter two 'value parameters', while three 'critical design parameters' are identified at each level.

5 See Ostewalder and Pignuer (2004) for details of this ontology and descriptions of the individual components.

Bouwman et al. (2008) identify a series of 'domains' – service, technology, finance and organization – which together form a business model. Not only do they demonstrate that each of these four domains is complex in their own right, but that this complexity is compounded through the interactions that occur between the four domains. It is also worth noting that they propose a dynamic business model, recognizing that business models have to be 'adapted over time in order to maintain a fit with the environment' (Bouwman et al. 2008: 64). In addition to changes in the environment that directly require the business model to be amended, changes in one of the four domains they identified may require changes in another.

The dynamic model that they propose also recognizes that products pass through different stages as they are developed and ultimately launched onto the market. In the first of the three stages they identify – technology R&D – they suggest that R&D will play a significant role in the development of the business model, while in the second stage – roll-out – regulators could influence how the business model develops. Regulators might be alarmed by the competitive implications of the new product, and thus impose limitations on the company. The regulator might, for example, prevent the company from bundling its new products or services with those that it already offers. This stage may also be influenced by the response of competitors. Once competitors become aware of the product's development, they could respond by reducing their own prices or developing a similar product themselves. Either response would necessitate a rethink of the business model. In the final stage – market – the focus is on retaining customers and tackling the response of competitors. As the product passes through each of these stages, changes will be required that take into account not only environmental developments, but also the interrelationships that exist between the four domains identified above.[6]

Timmers (1999) also highlights how business models are changing. After reviewing a wide range of electronic commerce business models, he categorizes them on the basis of how innovative they are, as well as the extent to which functions are integrated. With respect to innovation, a distinction is made between those business models that are effectively electronic ways of undertaking a task that already exists and those that use the Internet to replace an existing function or create a new one. In the case of integration, Timmers makes a distinction between those business models that integrate a single function and those that integrate several functions. The resulting matrix is shown in Figure 5.2.

This framework allows Timmers to identify two business model trends. The first of these trends is that integration of information flows has increased, while the second is the development of specialized and innovative services. The first trend is illustrated by the changes that many websites have experienced, from effectively being a site that advertised the existence of a company to then enabling would-be customers to search online catalogues and eventually allowing them to purchase the product online. The second trend is evident in the emergence of specialists who, for example, authenticate the privacy policies of websites or provide payment services. Taken together, these trends show that business models do not remain static, but rather change in response to technological advances. They also show that although function integration may be possible, some companies eschew this option in favour of being a specialist.

6 For a more detailed discussion of the dynamic nature of the model, see de Reuver and Bouwman (2008).

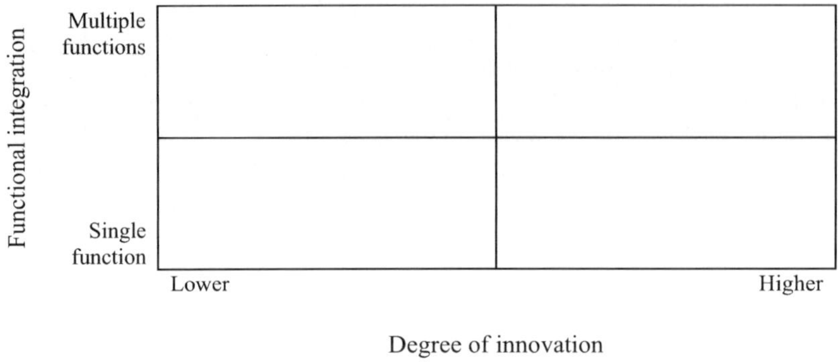

Figure 5.2 Categorizing business models

Source: Adapted from Timmers (1999: 42)

THE MOBILE BUSINESS MODEL

In many respects, the 'traditional' mobile business model was relatively straightforward: operators would provide their subscribers with a limited range of services – voice and basic data. However, this has changed as technological advances have been made. The emergence of new technologies has given rise to new services, ranging from music and wallpaper downloads on the one hand to mobile broadband, mobile-TV, mobile shopping and multi-media messaging on the other. One consequence of the wider range of services that are now available is that mobile operators are required to collaborate with other companies in their delivery, while another is that the failure of a 'killer application' to emerge forces them to offer a broad range of services to ensure that they are not placed at a competitive disadvantage if a 'killer application' does emerge.

Context-based services are one example of the new 3G-facilitated services that have emerged. Drawing on the framework of Bouwman et al. outlined above, de Reuver and Haaker (2009) explore the development of appropriate business models for such services. Their analysis highlights that in each of the four domains a series of issues need to be addressed. For example, in the technology domain the integration of different elements required to deliver the service needs to be addressed, while in the organizational domain the roles undertaken by the various companies participating in the business model need to be resolved. Interestingly, they state that some of the experts they interviewed suggested that 'powerful incumbents like telecom operators may try to slow down technological developments that may reduce their profits' (de Reuver and Haaker 2009: 245). This hints at one of the underlying tensions faced by mobile operators as they offer an ever-wider array of services, namely that the increased diversity of services available to customers does not necessarily mean that the total spend increases. If total spending does not increase, then as mobile operators launch new services they may in fact be cannibalizing their existing revenues and potentially undermining the commercial viability of the older services.

Braet and Ballon (2008) discuss the delivery of mobile-TV in Europe. Through elaborating a generic framework for mobile broadcasting, they show that cooperation between various companies is required if the service is to be brought to the market. It is also clear from their discussion that the delivery of mobile-TV is dependent on a complex value network that combines together a diverse range of companies with other stakeholders such as the state. Moreover, their conclusion that five critical issues can be identified in the European pilots that they analysed both suggests that no single mobile-TV business model is identifiable and that many issues need to be resolved before mobile-TV can become widely available.

Mobile-TV is one of the new services discussed by Funk (2007). Although his analysis initially focuses on the Japanese mobile commerce market, it then focuses on the fastest growing segment: the integration of mobile commerce with other media such as television and radio. According to Funk, the advent of 3G in Japan enabled one operator to expand its ringtone business to include CD-quality songs while improvements in handset technology have enabled radio tuners to be included in them. In other words, technological advances have created new commercial opportunities. It is, however, the area of mobile-TV where the most significant opportunities for integration are possibly to be found. Subscribers could vote for the contestants on a game show or for one particular ending of a drama over another. Irrespective of the actual voting that takes place, it will require cooperation between at the least the producers of the television programme, the television station and a mobile operator.

These examples, as well as the definitions noted above, help us to identify a series of salient lessons regarding business models. The first of these is that business models combine together a series of resources in order to deliver a particular product. The resources required will change over time, reflecting different stages in the development and delivery of products on the one hand and the technological and competitive landscape on the other. Secondly, the business model may combine together the resources of several companies. As a consequence, coordination issues are likely to come to the fore. The involvement of multiple companies in the delivery of a service also draws attention to the revenue model, not only in terms of the pricing strategy adopted, but also how any revenue generated is divided between the various participating companies. Related to this is a fourth point, namely that the business model needs to generate sufficient revenue to be both sustainable and profitable.

Conclusion

This chapter has outlined a series of concepts that will aid the understanding of how the mobile market has changed. One conclusion that can be drawn from the discussion of these concepts is that the mobile industry is characterized by its complexity. Related to this is a second conclusion, namely that the structure of the mobile industry has moved away from being a linear value chain to one that is better described by the notions of a value network or ecosystem. This partly reflects the emergence of a broad array of services that necessitates cooperation between companies to ensure their development and subsequent delivery, as well as the strategic decision of some to focus their efforts on just part of the mobile industry.

A third conclusion that can be drawn is that technological advances have also contributed to the mobile industry's complexity. It is due to technological advances that mobile operators are able to offer their subscribers a wide range of services. Not only does the adoption of standardized technologies enable companies to collaborate with one another in the delivery of these services, but it also allows subscribers to benefit through network effects. Having said this, standardization is not without its problems. Standardization allows subscribers to move between operators and undermines the efforts of companies to 'lock-in' their customers, which may be either the operator or subscriber, to a specific technology. We shall return to these themes in the following three chapters.

References

Anderson, T., Daim, T. and Kim, J. 2008. Technology forecasting for wireless communication. *Technovation*, 28(9), 602–14.

Ballon, P. 2007. Changing business models for Europe's mobile telecommunications industry: the impact of alternative wireless technologies. *Telematics and Informatics*, 24(3), 192–205.

Ballon, P. and Delaere, S. 2009. Flexible spectrum and future business models for the mobile industry. *Telematics and Informatics*, 26(3), 249–58.

Bass, F. 1969. A new product growth for model consumer durables. *Management Science*, 15(5), 215–27.

Blackman, C., Forge, S., Bohlin, E. and Clements, B. 2007. Forecasting user demand for wireless services: A socio-economic approach for Europe. *Telematics and Informatics*, 24(3), 206–16.

Boretos, G. 2007. The future of the mobile phone business. *Technological Forecasting and Social Change*, 74(3), 331–40.

Bouwman, H., Carlsson, C., Molina-Castillo, F. and Walden, P. 2007. Barriers and drivers in the adoption of current and future mobile services in Finland. *Telematics and Informatics*, 24(2), 145–60.

Bouwman, H., Faber, E., Haaker, T., Kijl, B. and de Reuver, M. 2008. Conceptualizing the STOF model. In Bouwman, H. et al. (eds). *Mobile Service Innovation and Business Models*. Heidelberg: Springer.

Braet, O. and Ballon, P. 2008. Cooperation models for mobile television in Europe. *Telematics and Informatics*, 25(3), 216–36.

Chu, C-P. and Pan, J-G. 2008. The forecasting of the mobile Internet in Taiwan by diffusion model. *Technological Forecasting and Social Change*, 75(7), 1054–67.

Cringely, R. 1996. *Accidental Empires: How the Boys of Silicon Valley Make Their Millions, Battle Foreign Competition and Still Can't Get a Date*. London: Penguin.

Curran, K., Woods, D. and Riordan, B. 2006. Investigating text input methods for mobile phones. *Telematics and Informatics*, 23(1), 1–21.

David, P. 1985. Clio and the economies of QWERTY. *American Economic Review*, 75(2), 332–37.

David, P. and Greenstein, S. 1990. The economics of compatibility standards: An introduction to recent research. *Economics of Innovation and New Technologies*, 1(1–2), 3–41.

Dunnewijk, T. and Hultén, S. 2007. A brief history of mobile communication in Europe. *Telematics and Informatics*, 24(3), 164–79.

Economides, A. and Grousopoulou, A. 2009. Students' thoughts about the importance and costs of their mobile devices' features and services. *Telematics and Informatics*, 26(1), 57–84.

Feijóo, C., Maghiros, I., Abadie, F. and Gómez-Barroso, J-L. 2009. Exploring a heterogeneous and fragmented digital ecosystem: Mobile content. *Telematics and Informatics*, 26(3), 282–92.

Fildes, R. and Kumar, V. 2002. Telecommunications demand forecasting – a review. *International Journal of Forecasting*, 18(4), 489–522.

Forge, S. and Bohlin, E. 2008. Managed innovation in Korea in telecommunications – moving toward 4G mobile at a national level. *Telematics and Informatics*, 25(4), 292–308.

Fransman, M. 2002a. Mapping the evolving telecoms industry: The uses and shortcomings of the layer model. *Telecommunications Policy*, 26(9/10), 473–83.

----- 2002b. *Telecoms in the Internet Age. From Boom to Bust to ...?* Oxford: Oxford University Press.

----- 2007a. *The New ICT Ecosystem. Implications for Europe*. Edinburgh: Kokoro.

----- 2007b. Innovation in the new ICT ecosystem. *Communications & Strategies*, 68(4), 89–110.

Fuentelsaz, L., Maícas, J. and Polo, Y. 2008. The evolution of mobile communications in Europe: The transition from the second to the third generation. *Telecommunications Policy*, 32(6), 436–49.

Funk, J. 2002. *Global Competition Between and Within Standards: The Case of Mobile Phones*. Basingstoke: Palgrave.

----- 2006. The future of mobile phone-based Intranet applications: A view from Japan, *Technovation*, 26(12), 1337–46.

----- 2007. The future of mobile shopping: The interaction between lead users and technological trajectories in the Japanese market. *Technological Forecasting and Social Change*, 74(3), 341–56.

----- 2008. Components, systems and technological discontinuities – lessons from the IT sector. *Long Range Planning*, 41(5), 555–73

Funk, J. and Methe, D. 2001. Market- and committee-based mechanisms in the creation and diffusion of global standards: The case of mobile communications. *Research Policy*, 30(4), 589–610.

Gamboa, L. and Otero, J. 2009. An estimation of the pattern of diffusion of mobile phones: The case of Colombia. *Telecommunications Policy*, 33(10/11), 611–20.

Gandal, N., Salant, D. and Waverman, L. 2003. Standards in wireless telephone networks. *Telecommunications Policy*, 27(5/6), 325–32.

García-Murillo, M. and Rendón, J. 2009. A model of wireless broadband diffusion in Latin America. *Telematics and Informatics*, 26(3), 259–69.

Geroski, P. 2000. Models of technology diffusion. *Research Policy*, 29(4/5), 603–25.

Gruber, H. and Verboven, F. 2001. The diffusion of mobile telecommunications services in the European Union. *European Economic Review*, 45(3), 577 88.

Hagel, J. and Singer, M. 1999. *Net Worth. Shaping Markets when Customers Take the Risks*. Boston, Mass: Harvard Business School Press.

Howells, J. 2005. *The Management of Innovation and Technology*. London: Sage Publications.

Johnson, G. and Scholes, K. 2002. *Exploring Corporate Strategy*. 6th edn. Harlow: Prentice Hall.

Kuo, Y-F. and Yu, C-W. 2006. 3G telecommunications operators' challenges and roles: A perspective of mobile commerce value chain. *Technovation*, 26(12), 1347–56.

Lea, G. and Hall, P. 2004. Standards and intellectual property rights: An economic and legal perspective. *Information Economics and Policy*, 16(1), 67–89.

Lemstra, W. and Hayes, V. 2009. License-exempt: WiFi complement to 3G. *Telematics and Informatics*, 26(3), 227–39.

Li, F. and Whalley, J. 2002. Deconstruction of the telecommunications industry: From value chains to value networks. *Telecommunications Policy*, 26(9/10), 451–72.

Li, Y-R. 2009. The technological roadmap of Cisco's business ecosystem. *Technovation*, 29(5), 379–86.

Liebowitz, S. and Margolis, S. 1995. Path dependence, lock-in and history. *The Journal of Law, Economics and Organization*, 11(1), 205–26

------ 2002. Network effects. In Cave, M. et al. (eds). *Handbook of Telecommunication Economics*. Amsterdam: Elsevier.

Lindgren, M., Jedbratt, J. and Svensson, E. 2002. *Beyond Mobile: People, Communications and Marketing in a Mobilized World*. Basingstoke: Palgrave.

Maitland, C., Bauer, J. and Westerveld, R. 2002. The European market for mobile data: Evolving value chains and industry structures. *Telecommunications Policy*, 26(9/10), 485–504.

Nam, C., Kim, S. and Lee, H. 2008. The role of WiBro: Filling the gaps in mobile broadband technologies. *Technological Forecasting and Social Change*, 75(3), 438–48.

Olleros, X. 2008. The lean core in digital platforms. *Technovation*, 28(5), 266–76.

Olsson, A. 2003. *Understanding Changing Telecommunications – Building a Successful Telecom Business*. Chichester: John Wiley & Sons Ltd.

Osterwalder, A. and Pigneur, Y. 2004. An ontology for e-business models. In Currie, W. (ed.). *Value Creation from e-Business Models*. Amsterdam: Elsevier

Porter, M. 1985. *Competitive Advantage: Creating and Sustaining Superior Performance*. New York: The Free Press.

Raghaven, S., Golden, B. and Wasil, E. 2008. *Telecommunications Modeling, Policy, and Technology*. New York: Springer.

de Reuver, M. and Bouwman, H. 2008. The dynamic STOPF model in practice. In Bouwman, H. et al. (eds). *Mobile Service Innovation and Business Models*. Heidelberg: Springer.

de Reuver, M. and Haaker, T. 2009. Designing viable business models for context-aware mobile services. *Telematics and Informatics*, 26(3), 240–48.

Rogers, E. 2003. *Diffusion of Innovations*. 5th edn. New York: Free Press

Rohlfs, J.H. 2001. *Bandwagon Effects in High-Technology Industries*. Cambridge, Mass: The MIT Press.

Rouvinen, P. 2006. Diffusion of digital mobile telephony: Are developing countries different? *Telecommunications Policy*, 30(1), 46–63.

Sainio, L-M. and Puumalainen, K. 2007. Evaluating technology disruptions in a strategic corporate context: A case study. *Technological Forecasting and Social Change*, 74(8), 1315–33.

Shapiro, C. and Varian, H. 1999. *Information Rules – A Strategic Guide to the Network Economy*. Boston, Mass: Harvard Business School Press.

Shin, D. 2008. The assessment of 3rd generation mobile policy in Korea: A web of stakeholder analysis. *Technological Forecasting and Social Change*, 75(9), 1406–15.

Singh, S. 2008. The diffusion of mobile phones in India. *Telecommunications Policy*, 32(9/10), 642–51.

Sohn, S. and Lee, J. 2008. Competing risk model for mobile phone service. *Technological Forecasting and Social Change*, 74(9), 1416–22.

Steinbock, D. 2002. What happened to Europe's wireless advantage? *info*, 4(5), 4–11.

----- 2003. Globalisation of wireless value system: from geographic to strategic advantages. *Telecommunications Policy*, 27(3/4), 207–35.

Tassey, G. 2000. Standardization in technology-based markets. *Research Policy*, 29(4/5), 587–602.

Timmers, P. 1999. *Electronic Commerce – Strategies and Models for Business-to-Business Trading*. Chichester: John Wiley & Sons Ltd.

Wei, R. 2008. Motivations for using the mobile phone for mass communications and entertainment. *Telematics and Informatics*, 25(1), 36–46.

Williams, H., Whalley, J. and Li, F. 2003. Interoperability and electronic commerce: A new policy framework for evaluating strategic options. In Steinfeld, C. (ed.). *New Directions in Research and Electronic Commerce*. West Lafayette, Indiana: Purdue University Press.

Windrum, P. 2004. Leveraging technological externalities in complex technologies: Microsoft's exploitation of standards in the browser war. *Research Policy*, 33(3), 385–94.

CHAPTER 6 Issues in Mobile Industry Structure – Vendors

Introduction

Although operators necessarily have close relationships with equipment vendors, at the end of the day equipment manufacturing is a competitive business and, as is the case with all equipment-making, it is vendors in the emerging countries that have emerged during the past decade to challenge the often cosy relationships between operators and vendors in developed countries.

The introduction of new technology provides an obvious opportunity for ambitious vendors to challenge the existing order, normally by offering comparable equipment at reduced prices, but also on occasion by providing a superior product at comparable prices. If this coincides with a period of retrenchment among operators, during which equipment orders are anyway depressed, then something has to give. This will probably take the form of either M&A (merger and acquisition) activity or bankruptcies or both.

The hallmark of all capital equipment industries is that they are highly cyclical. Once equipment is bought it has a natural lifetime before wearing out or becoming inefficient, at which point there is a need to replace it. Hence, if a new technology dictates that there is a surge in orders from customers, activity at vendors will spike before falling back sharply, only to spike again when replacement is required. But this cycle may not have a natural lifetime because technology may change too quickly, necessitating the writing off of equipment well before the end of its natural life. This process is itself affected by the economic environment. When the economy is buoyant, the equipment cycle will shorten as customers can afford to write down existing equipment and reinvest in more recent technology, whereas this is highly problematic when the economy is depressed.

The above factors are often significant in most industries, and do indeed periodically lead to restructuring. But it is very rare indeed that events occur that are comparable to what happened from the mid-1990s onwards once it became clear that the mobile telecommunications sector was about to emerge into a high-speed world. The purpose of what follows is to shed light on what happened, both in a sector-wide context and also from the point of view of individual vendors. The focus will be upon vendors that supply mobile networks (infrastructure) and vendors that supply equipment such as handsets, although these are clearly not the only vendors that contribute to the provision of high-speed mobile services.

Background

The second half of the 1990s was a golden era for virtually all technology companies and was particularly marked by a plethora of alliances as well as merger and acquisition (M&A) activity (Amesse, Latour, Rebolledo and Séguin-Dulude 2004). However, it was particularly golden for those associated with mobile telecommunications. As a result of the introduction of the initial digital licenses during the first half of the decade, 2G rapidly established itself with the result that 2G was spreading rapidly around the world with highly beneficial consequences for vendors. However, the associated discussion about the provision of high-speed data networks meant that vendors (and operators) were being valued not only according to their existing balance sheets, but according to the balance sheets that they were expected to produce in the future.

As is now well known, one could at this time start up something in a garage and sell it on for tens, if not hundreds, of millions within months without actually needing to produce a working prototype. Our concern here is with considerably larger companies. Nevertheless, it is salutary to remember that Qualcomm, for example, was a relative newcomer at this point, with a market value of only $3.4 billion at the end of 1998. One year later it was worth $115 billion. Nokia was another company that burst upon the scene during the late 1990s: its value rose from $47 billion at the end of 1998 to $208 billion one year later, Longer-established vendors could not be expected to match this performance, but virtually all at least doubled in value. Lucent Technologies could only manage an equivalent rise in value from $91 billion to $235 billion, but a more than tripling in value was more common, as in the case of Alcatel, Ericsson and Motorola.

But it was not to last – indeed, how could it? The past two years 2008 and 2009, may not have been pleasurable ones for the great majority of vendors, but consider the following. During 2000, the market value of illustrative equipment vendors fell as follows: Lucent Technologies by 78 per cent, Motorola by 44 per cent and Ericsson by 20 per cent. Yet Nokia only fell by 5 per cent and Alcatel actually rose in value by 50 per cent. In practice, this was the lull before the real storm hit. Between January 2001 and March 2002, the market value of Nortel Networks fell by 87 per cent, Alcatel by 75 per cent, Lucent Technologies by 69 per cent, Ericsson by 62 per cent, Nokia by 48 per cent and Motorola by 37 per cent. Yet the year March 2002 to March 2003 brought little respite for most vendors, with, for example, further falls for Ericsson (by 68 per cent), Lucent (by 64 per cent), Nortel (by 42 per cent) and Motorola (by 40 per cent).

The overall effect was astonishing. By March 2003, for example, the market value of Lucent had fallen from its peak of $235 billion in December 1999 to a mere $6 billion, that of Nortel from $136 billion to $8 billion, that of Ericsson from $115 billion to $11 billion and that of Motorola from $90 billion to $19 billion. At its peak, Lucent was one of the 10 most valuable companies in the world while Nortel was ranked in the top 30, as was Ericsson.

True, things then improved significantly in most cases, but the damage was deep-rooted. Even in mid-2007, with equity markets booming, vendors were mostly just about treading water and Lucent, for example, could barely reach a valuation of $20 billion, less than one-tenth of its peak value. One year on, and another recession had hit the telecommunications sector. Which raises an interesting question: if the vendor sector has still not fully recovered from the previous recession, how will it manage this time around?

Restructuring Stage 1

The first point to bear in mind is that the sector does not look exactly the same because there was a good deal of restructuring as a consequence of the 2000 to 2002 meltdown. Most notably, Alcatel sought a tie-up with Lucent in May 2001, offering $23 billon, which was rejected (*The Economist* 2001). Ostensibly, Lucent, which would provide most of the expertise in wireless, was concerned that what was billed as a 'merger of equals' was to all practical purposes a takeover by Alcatel which would have ended up with 58 per cent of the new entity. However, the economic logic behind the tie-up was not in dispute, so when Alcatel returned to the table in March 2006, offering $13.5 billion in another 'merger of equals' designed, this time around, to leave Alcatel shareholders with an unequal 60 per cent stake in the proposed Alcatel-Lucent (Sovich and Cheng 2006), the offer was crowned with success. More commonly, divisions of vendors, rather than entire companies, were spun off and conjoined. One such entity is Sony Ericsson Mobile Communications, created in 2001 from the mobile handset divisions of Sony and Ericsson that were struggling to survive as independent units. In addition, Ericsson announced in October 2006 that it would be acquiring some of Marconi's telecommunications operations for $2.1 billion, while the $25 billion merger of Nokia Networks and Siemens to form Nokia Siemens Networks (NSN), announced at much the same time, was consummated the following year.

This restructuring could in part be ascribed to disappointment at the performance of high-speed networks of all kinds. Certainly, the licensing of high-speed mobile networks, discussed in Chapter 3, was an unexpectedly long drawn-out process even in Europe where it was mandated, while the launching of networks was held back by the realization that subscribers were less ready than had been predicted to pay for high-speed services. But by 2007 the outlook had improved significantly. In particular, the handsets that had become available to take advantage of such services were no longer clunky, but rather sleek items fit for fashionistas to flash around as symbols of their high-tech lifestyle.

With economic indicators once again highly positive, vendors should have prospered. However, the numbers speak for themselves. Commencing in November 2008, and looking back one year to happier times, the data show, for example, that the share prices of Alcatel-Lucent (down roughly from $6 to $2), Ericsson (from $90 to $55), Motorola (from $17 to $5) and Nokia (from $27 to $13) severely downgraded to varying degrees. Indeed, reductions on this scale were very similar to those witnessed at the beginning of the previous recession, as noted above.

But were share prices falling for the same reasons? Clearly, there were some common factors. As noted above, the vendor sector is ultimately dependent upon orders for equipment from operators. Most of these also suffered sharp share price reductions at the time, although some were much worse affected than others. For example, whereas it had become conventional wisdom that emerging markets were the place to be, Millicom International's assertion in October 2008 that its customers were suffering from rising food prices, and hence had less to spend on telecoms, caused a rapid reappraisal of its prospects. However, the likes of Deutsche Telekom and DoCoMo insisted that their balance sheets were in far better shape than in 2000–2001. In addition, whereas, especially in the USA, large numbers of people were abandoning their fixed-wire connections, this suggested by implication that the mobile sector was in relatively good shape.

Nevertheless, with customers likely to delay replacing their handsets, even vendors strongly placed in the handset sector could be expected to be squeezed. Furthermore, as evidenced by the behaviour of Vodafone in acquiring fixed-wire licensees, fixed-mobile convergence appeared to be a sensible strategic option for vendors during the second half of the decade – and, indeed, underpinned the creation of Nokia Siemens Networks – but would the commitment of scarce management resources to the integration of divergent organizations prove costly? In relation to NSN and Alcatel-Lucent this was arguably the case since both were struggling with their restructuring. NSN had clearly underestimated the challenge inherent in combining two weak companies – two weaks do not necessarily make one strong – and with technology moving on remorsefully through 3G and on towards 4G, even relatively strong positions in legacy products could rapidly lose their value.

Overview of Markets

Before moving on to look at some vendor case studies, it is necessary to make some distinctions between the various vendors cited. To begin with, it is useful to examine data on the number of mobile networks supplied by vendors. So far as GSM is concerned, the market is predictably dominated by Europe-based Ericsson and NSN, whereas CDMA (code division multiple access) networks are largely the preserve of North-America-based Motorola and Nortel, with Chinese vendors Huawei and ZTE an emerging force. However, Motorola is far less prominent in relation to cdma2000 1xRTT and 1xEV-DO. The dominance of Ericsson and NSN becomes increasingly marked as technology progress from GSM (global system for mobile) through EDGE (enhanced data rates for GSM evolution) and W-CDMA (wideband code division multiple access) to HSDPA (high-speed downlink packet access) and HSUPA (high-speed uplink packet access), although in the process Huawei has taken over the mantle of third-placed vendor from Alcatel-Lucent. Overall, given the dominance of GSM over CDMA, Ericsson and NSN remain by far the largest providers of mobile networks, but the Chinese challenge to their position is the most notable feature of 2008 and 2009. Opinion within the sector appears (without any real scientific basis) to favour three eventual survivors over the medium-to-long term – predictably Ericsson, NSN and Huawei – with ZTE as an also-ran and the others possibly being taken over by these or going bankrupt.

The position in relation to vendors of handsets has recently become clouded by the advent of the iPhone and its imitators (see Chapter 7). Although the touch-screen market remains small in relation to total handset sales, its growth has served to propel Apple into the top 10 by volume among handset vendors, and has accounted for significant variability in the quarterly share taken by Nokia. Not that Nokia's dominance has been seriously challenged – it remains much the largest handset vendor with a market share fluctuating close to the 40 per cent mark. Samsung of South Korea has established itself as the runner-up, albeit with a share fluctuating around the 15 per cent mark, and is followed by LG Electronics (also of South Korea), Sony Ericsson, and Motorola whose fall from grace is discussed below (Telecoms.com 2008).

Vendor Case Studies

ALCATEL-LUCENT

The effective takeover of Lucent Technologies of the USA by Alcatel of France has already been outlined above, as has the severe effect of the 2000–2002 recession upon both parties. In respect of Alcatel, a loss of $4.3 billion was reported for the calendar year 2001, compared to a profit in excess of $1 billion for the calendar year 2000. On top of this loss, restructuring charges of nearly $2 billion were incurred, so it is hardly surprising that a strategy of selling off non-core assets (unrelated to its limited wireless activities) was subsequently pursued in conjunction with significant reductions in the workforce.

The initial approach to Lucent was seen as a means to offset weak growth in corporate technology spending and delays in the roll out of next generation networks (NGNs). The irony was that although Lucent appeared to be a perfect partner, given that Alcatel's USA sales accounted for only 22 per cent of its total revenues in 2000, the proposed 'merger of equals' would of necessity have to be a merger of unequals – in effect a French takeover – because Alcatel could hardly turn a blind eye to matters such as its relatively high value and larger work force. But in the light of the fallout from the prior 'merger of equals' between Daimler and Chrysler, the perfect partnership was likely to be marred by an inevitable clash of corporate cultures, and there was likely to be a difference of opinion as to where the inevitable post-merger job losses would be located (*The Economist* 2001).

The failure to proceed with the merger looked to have more serious implications for Lucent than for Alcatel; in trying to keep pace with its major rivals, Cisco Systems and Nortel Networks, it had badly overstretched itself. This was particularly worrying because the days of the monolithic vendor, able to supply the full range of products required by its customers, appeared to be numbered. Rather than use a single source for equipment, in the process acquiring some that was distinctly not state-of-the-art, operators were increasingly buying the best equipment available from a variety of sources and integrating it in-house.

The failure of the merger led to various other vendors casting their eye over Lucent. But what they really wanted were pieces rather than the whole, and whereas selling the pieces would have raised more money than selling Lucent as it stood, shareholders would subsequently have faced a massive tax bill. Furthermore, whereas Alcatel-Lucent would have jumped regulatory hurdles with relative ease, that would not have been the case were, say, Ericsson to have picked up the division of Lucent where Ericsson was already a market leader.

With its chosen route to redemption blocked, Alcatel responded in September 2002 by announcing a restructuring involving its separation into fixed, mobile and private communications divisions and the reduction of its workforce to 60,000 from its peak of 110,000 in 2000. It subsequently embarked upon a series of joint ventures, among which one concerned its mobile division. This consisted of a merger between its loss-making handsets business and TCL of China which would take a 55 per cent stake in the new company (Milne 2004). But a mere nine months after its formation, TCL was obliged to take over Alcatel's share of the company, albeit at a cost considerably less than the valuation placed upon its contribution by Alcatel in 2004 (Lau and Guerrera 2005). Evidently, rejuvenating tired Western brands with Chinese prowess in mobile technology was a less than straightforward matter when one's market share was less than one-tenth that of Nokia.

Nevertheless, having spun off its satellite, fibre-optics and mobile handsets businesses, Alcatel felt that a renewed focus on its remaining core infrastructure activities would enable it to prosper. Lex expressed some doubts (Lex 2005) although it considered Alcatel's performance to be 'uninspired' rather than 'disastrous'. Meanwhile, Lucent continued to struggle with its inability to provide a range of products to satisfy operators' moves towards converged fixed/mobile networks, although its wireless business was widely considered to be solid, even if biased strongly towards CDMA rather than GSM.

But all was not as it seemed. Analysts were quick to point out that Lucent had been able to take advantage of an accounting rule which gave it a pension credit, and that whereas this was going to continue through 2006, and hence the short-term outlook was satisfactory, the longer-term outlook was endangered by mounting post-retirement (largely healthcare) costs. Once these were taken into account, Lucent was technically bankrupt (Dow Jones 2006a).

The obvious solution for both parties was to resurrect their aborted merger (Dow Jones 2006b), and in March 2006 it was admitted that this was under discussion (Dow Jones 2006c). Naturally, a 'merger of equals' was proposed even though Alcatel was valued at roughly \$20 billion compared to \$13 billion for Lucent and, in practice, Alcatel made a takeover bid for Lucent priced (at \$13.5 billion in Alcatel shares) so as to leave it with a 60 per cent stake in the proposed Alcatel-Lucent. The new company would be second only to Cisco Systems in the infrastructure sector, and with promised cost savings of \$1.7 billion (partly via a 10 per cent reduction in the combined workforce of 88,000), was expected rapidly to develop into an efficient and effective global competitor (Hollinger and Taylor 2006).

The obvious virtues of the deal were that product gaps would be filled, R&D (research and development) would be better focused and Alcatel would be better able to compete for W-CDMA contracts. However, analysts identified three major challenges: integrating the respective operations in time to supply the next generation of equipment, retaining key workers and avoiding a clash of cultures. Unsurprisingly, each vendor claimed to be entirely in tune with its opposite number, but external misgivings were reflected in the continued decline in both vendors' share prices (Sovich and Patrick 2006). While regulatory and shareholder approvals were pending, Alcatel went ahead with the purchase of Nortel's GSM/W-CDMA assets, but this was seen as playing catch-up in relation to Ericsson and NSN.

The USA government approved the merger in December 2006. In January 2007, the result of adding Alcatel's revenues for November and December 2006 to those of Alcatel-Lucent for December was so poor that a profit warning became necessary and the latter's share price fell by over 10 per cent. Analysts noted that whereas Alcatel had paid out €12 billion, it was expected to achieve total synergies worth €11 billion, so there was barely any need for Lucent to show a profit. But if these synergies did not appear – and the initial results cited were none too promising – then future prospects looked grim (Lex 2007a). It was noted in particular that Lucent's dependency upon Sprint Nextel and Verizon Wireless for its CDMA revenues made it vulnerable to problems at these carriers, and that Sprint's own merger with Nextel was fast becoming a poisoned chalice. A second profit warning was largely ascribed to problems in North America, but with costs remaining constant while prices were on a downward spiral, a third warning became necessary in September (Cellular-news 2007).

Advice from analysts was varied, but included additional layoffs (which were duly promised in October), a move into markets with better growth potential, possibly via an acquisition, and a restructuring of the wireless division, to include plant closures, in the face of a price war with Ericsson (Mullen 2007).

A fifth consecutive quarter of net loss was reported in May 2008 and, in July, the pre-merger CEOs of Alcatel (Serge Tchuruk – now Chairman of Alcatel-Lucent) and Lucent (Pat Russo – now CEO of Alcatel-Lucent) both announced their resignations. Opinions varied as to their culpability for writing off 60 per cent of the combined equity of the two vendors post-merger, although they undoubtedly under-estimated the clash of cultures (Lex 2008a). In October, Alcatel-Lucent announced a further loss for 2008Q3 and admitted that a change in strategic direction was needed that would treat customers more as partners and require the company to focus on services and applications as well as hardware (Morris 2008b). By now the share price had fallen to a mere €2.

In December, the new CEO, Ben Verwaayen, denied that Alcatel-Lucent would withdraw from the wireless market and committed the company to invest heavily in LTE (long-term evolution) even though he had only recently scaled back the LTE joint venture with Japan's NEC. Analysts responded that this was unwise as they would never be able to compete with Ericsson (Hollinger 2008). A further quarterly net loss announced in February 2009, bringing the total for 2008 to €5.2 billion, reduced the share price to €1.50. However, Alcatel-Lucent continues to insist that it is on track to break even in 2009 and make a modest profit in 2010. If it fails to clear the first hurdle, the markets will not be forgiving.

MOTOROLA

Motorola provides an interesting case study because it demonstrates just how easy it is to lose focus in a fast-moving technologically advanced environment – in this case, wireless handsets. Motorola never looked as if it would catch up Nokia, but for a lengthy period it was regarded as the clear number two in the handset pecking order. Along with the other vendors, it suffered badly during the 2000–02 recession, but salvation was on hand in the form of the RAZR handset, a slim and stylish device for calling and texting launched in mid-2004. After the introduction of the RAZR, Motorola looked set to be a much stronger challenger to Nokia among handset vendors, with an ongoing market share above 20 per cent. Nevertheless, despite the enormous demand for the RAZR in its various versions, it was plagued with faults because quality control was inadequate, and this provided the opportunity for other handset vendors to encroach into Motorola's markets.

In addition, Motorola adopted the strategically obvious, and less stressful, route of developing sister handsets known, for example, as the KRZR and the ROKR, but these were perceived as markedly inferior to equivalent handsets from other vendors – for example, the Sony Ericsson Walkman series of music phones easily outsold the ROKR. Unfortunately, because it was so sure that the KRZR in particular would take over from the RAZR as its flagship high-margin device, Motorola allowed the price of the RAZR to fall significantly in order to maintain sales volume.

In effect, Motorola was suffering from some delusions of grandeur. It believed that the RAZR and its spin-offs would allow it to reduce the market-share gap between itself and

Nokia even though Nokia was clearly going to defend that gap in whatever way it could. Both vendors were likely to suffer reduced profitability in such circumstances, given the need to slash prices as the primary method for acquiring market share, especially since the likes of Samsung and Sony Ericsson were simultaneously determined to close the market-share gap with Motorola. This might, perhaps, have eventually turned out reasonably well had the handset market been subject to high barriers to entry, but in practice a vendor with a new handset made attractive either by its superior technology (as was true of the iPhone) or its low price (because it was manufactured in China and used the operator's brand) could appear at any time and undermine the market leaders (Lex 2007a).

Indeed, the trend that Motorola largely failed to respond to was that customers were moving in two divergent directions; towards cheap low-end devices and towards expensive high-end devices such as the iPhone. Whereas the ROKR was initially perceived as a fashionable high-end handset, with its buyers correspondingly fairly indifferent to the quality of the user experience, it had developed into a mass-market mid-range product by 2006 with an accompanying decline in its profit margin (Global Insight 2007a). Sitting on huge cash reserves – amounting to $11 billion by the end of 2006 – Motorola's management might understandably have been expected to behave conservatively, although it was certainly aware that its handset portfolio was missing some important ingredients such as a decent range of 3G devices. Hence, in March 2007 the CEO, Ed Zander, announced that the company intended to improve the performance of its Mobile Devices division, albeit not in the immediate future (Taylor 2007a).

Certainly, the new handsets unveiled in May failed to impress, although the RAZR2 held some promise (Taylor 2007b). By mid-year, Motorola's market share had halved from its peak of roughly 22 per cent and had for the time being been overtaken by Samsung. Motorola would accordingly have to increasingly rely for profits on its business devices and home entertainment gadgets. Furthermore, the prospects for 2008 were less than promising as the North American market, where Motorola was market leader in handsets and generated nearly half of its total sales, was likely to slip into a consumer-led recession

In March 2008, under pressure from activist investor Carl Icahn, Motorola announced that it would be spinning off the troubled Mobile Devices division, which at the time accounted for roughly 50 per cent of total revenue, into a separately listed company (Cellular-news 2008a). Shortly thereafter, its market share fell below the 10 per cent level. However, despite some attempts in the interim to project its operations in a more positive light, Motorola announced in October that it would be cutting 3,000 jobs to add to the 9,000 lost during the previous year, and that it would be delaying the spin-off due to a dramatic collapse of handset sales. In 2007Q3, Motorola had sold 37.2 million handsets, whereas in 2008Q3 it sold only 25.4 million (Taylor 2008c), leaving it in fourth place in terms of market share, one below Sony Ericsson. The lack of a range of 3G handsets suitable for the European market and parts of Asia left a major hole in its product range, even if it could claim to be doing better in the USA and China (which was preparing for the launch of TD-SCDMA (time division synchronous code division multiple access)).

Thus, on the one hand, there was no great expectation that any spin-offs or disposals would actually take place, at least not until well into 2009. Meanwhile, on the other hand, it was estimated that if Motorola was to remain independent, it would need to raise roughly $4 billion to tide it over during the ensuing two-year period (Taylor 2008b). The financial markets, foreseeing ongoing losses in the Mobile Devices division, responded negatively. For example, Standard & Poor's downgraded Motorola's long-term credit rating to junk status in December 2008. Motorola responded by slashing its pension benefits and announcing cuts of up to 50 per cent of its workforce in the Mobile Devices division together with a much slimmed-down range of new handsets to be launched during 2009 (Wood 2009).

With a mere 6 per cent of the handset market, and no touch-screen smartphone on the horizon, doubts began to be expressed whether Motorola would survive as a handset vendor (Lex 2009a). A further 4,000 job losses, primarily in the Mobile Devices division, were announced in January 2009, but the real saving graces were the large amount of cash on the balance sheet and the profitable operations of Motorola's other two divisions (Taylor 2009c).

In conclusion, Motorola fundamentally failed to supply the market with what it wanted, namely cheap low-end handsets and high-end, preferably touch-screen, smartphones. Other vendors, some of them relatively new entrants into the sector, were quick to take advantage. Motorola, understandably, claimed that the Mobile Devices division was being turned around, in part due to the launch of Android-based smartphones, and there is some evidence that it may be proved right. In early November, Verizon Wireless launched the Motorola Droid – a super-slim (like the RAZR) touch-screen smartphone with a slide-out QWERTY keyboard (known in Europe as the Milestone) – which was the first handset to market incorporating the Android 2.0 OS (Gabriel 2009b). With its emphasis upon search activities, the Droid could be contrasted with the almost simultaneously launched Cliq/Dext, a touch-screen smartphone with an emphasis upon social networking (Metz 2009; Reedy 2009).

Motorola's market share in 2009Q3 was a mere 4.7 per cent. The new handsets will undoubtedly cause this to rise, but the market is becoming increasingly competitive, so for now Motorola's prospects must remain cloudy at best.

NORTEL NETWORKS

As the data at the beginning of this chapter reveal, Nortel Networks, a network vendor somewhat surprisingly based in Toronto, Canada, had been reduced to a shadow of its former self during the period 2000–2003, at the beginning of which it employed over 90,000 workers. However, within two years this number was down below 35,000 in response to massive write-offs and allegations of financial irregularities (Hunter 2003). During 2004, Nortel announced that 2003 net profits had been overstated by 50 per cent, leading to the removal of the CEO (Frank Dunn) and CFO. However, the replacement CEO himself only lasted for 18 months, and his successor (Mike Zafirovski) had to oversee a $2.5 billion settlement of a class action lawsuit brought by disgruntled shareholders.

A third restatement of 2003–2005 earnings followed in 2006, predictably leading to a plethora of additional lawsuits and a fourth restatement in 2007 (Masters and Taylor 2007). Nevertheless, Nortel shared in the much improved climate for technology stocks at the time, and its share price climbed back to roughly $30. Buoyed by this apparent improvement in its fortunes, Nortel was even involved in the bidding for a former unit of Lucent Technologies, Avaya, when it was put up for sale in June 2007. At the time, Avaya was Cisco's main competitor in the market for IP (Internet protocol) telephone systems, which rely on data networks to route calls within companies, but there were concerns that it lacked the requisite scale to remain competitive. In the event, Nortel was outbid (Politi 2007). However, Nortel did manage to raise $320 million by selling its unsuccessful W-CDMA business to Alcatel-Lucent (BWCS 2006).

In June 2008, Nortel announced that it would be switching the focus of its wireless operations in favour of LTE – a strategy received very positively by the financial markets (Taylor 2008a). It also made fairly positive noises about its future sales and profit margins (Dummett 2008). But there could ultimately be no escape from the fact that Nortel had become too small to compete effectively in many of its markets, and that its reliance upon CDMA rather than GSM equipment was at best unhelpful, LTE plans notwithstanding, since existing CDMA equipment would be gradually allowed to wear out over roughly a 10-year period.

In the light of the progressive decline in the share price which had reached $5 at the end of 2007, the prevailing view by mid-2008 was nevertheless that Nortel's prospects were now being heavily discounted (Lex 2008b). But Nortel soon showed how foolish this was by issuing a sales and profit warning in conjunction with a proposal to put one of its fastest-growing fixed-wire businesses, Metro Ethernet Networks, up for sale in September (Parker 2008a) – although potential buyers were conspicuous by their absence. Then, in November 2008 with the 'credit crunch' firmly in place, Nortel announced a 2008Q3 loss of $3.4 billion (the worst for seven years) leading, *inter alia*, to job losses and a salary freeze (Cellular-news 2008b). These new job losses started from a base point of only 30,000. Still, with its share price down to just over $1, the only real strategic issue was how to survive through 2009.

In practice, it had become clear even before 2008 was out that survival would not be possible unless a radical restructuring was to occur. The only real issue was the method (Silver and Lublin 2008). The answer came when Nortel filed for Chapter 11 bankruptcy in both the USA and Canada on 14 January 2009 (Simon and Taylor 2009). The Latin American and Asian operations were excluded from the bankruptcy proceedings even though they accounted for nearly one-half of the contracts. The filing was triggered by an impending demand for a bond interest payment of $107 million, despite the fact that Nortel held $2.3 billion in cash at the time. However, given the long-term debt of $4.5 billion, it appeared to be sensible to use the remaining cash to see the company through the bankruptcy process with minimum disruption to customers.

By early 2009 it had become clear that Nortel would not be able to emerge from bankruptcy in anything like its existing form, and it was restructured internally in order to expedite the disposal of a wide range of business divisions. In May, the 50 per cent plus one share stake in the LG-Nortel joint venture was offered for sale (Matsumoto 2009). In June, Nortel announced the sale of its core (profitable) CDMA and LTE wireless infrastructure business to NSN (Taylor 2009b) although rumours swirled that a major creditor, MatlinPatterson, was trying to put together a consortium to recapitalize the entire company. In the event, it chose merely to outbid NSN in July, only to be outbid in turn by the eventual buyer, Ericsson, which paid $1.13 billion in cash (Brickley 2009). Also in July, Avaya (see above) agreed to buy the assets of the Enterprise division together with the shares of Nortel Government Solutions and DiamondWare (Taylor 2009c), only to find itself competing with Siemens Enterprise Communications. Avaya eventually won out with a bid of $900 million and completed in December. In September, Nortel announced the planned sale of the Packet Core assets of its Carrier Networks business (Telecompaper 2009b) – Hitachi offered $10 million for the Evolved Packet Core (essentially LTE) assets in October – followed in November by the proposed auction of its GSM and GSM-R (railways) assets (Middleton 2009) won by Ericsson and Austria's Kapsch Carrier Com (Le Maistre 2009). In October, a provisional bid was made by Ciena for the Networking and Carrier Ethernet businesses (Cellular-news 2009) which it won in November at the expense of NSN for $769 million. What remained were the successful Carrier VoIP and Application Solutions, the legacy mobile packet core assets, the LG-Nortel majority stake and some intellectual property (Le Maistre 2009).

In retrospect, the core problem, which Mike Zafirovski arguably addressed too late and with insufficient vigour, was that Nortel was a player in too many markets simultaneously, holding a position of no better than third in any of its four main markets (of which one was wireless infrastructure). Given its lack of scale, it should have actively promoted mergers or joint ventures. In addition, its balance sheet was too weak to provide the investment funds needed to support these varied businesses. Once the most valuable company in Canada, Nortel had become so insignificant by 2008 that it is even doubtful whether its bankruptcy will create noticeably improved market conditions for its competitors, although Ericsson must hope to the contrary (at least in respect of CDMA and LTE), and hence that it will prevent the need for further sectoral restructuring.

Recent Experience at Other Vendors

It is clearly not possible to provide extensive case studies of all relevant vendors. What follows is therefore a more truncated analysis of some the other companies that play a more than passing role in this and other chapters.

Nokia launched its first GSM handset in 1992. With GSM mandated as the technology for 2G in Europe, Nokia took its first-mover advantage and combined it with aggressive branding – customers were more likely to specify a Nokia handset at a retailer than specify their choice of operator – and reliable, user-friendly and feature-rich devices to build up an unassailable lead in the handset market. It was so successful that it had to radically restructure its operations in order to cope, which proved to be a blessing in disguise when it had to respond to the 2000–2002 recession.

Its only weakness was in the market for CDMA equipment where it failed to connect with North American operators. But where it chose to compete, its sheer scale was its mainstay – when competitors did well it generally did better, and when it was suffering its competitors generally did worse. Furthermore, it was quick on its feet – a feature of the likes of Samsung during the mid-2000s but hardly of the likes of the much larger Motorola.

In its glory days of the 1990s, Nokia could boast of achieving operating margins of nearly 25 per cent. In late 2006, it could do little better than half that amount, yet one year later the margin was back to 22 per cent and its handset market share had risen from roughly 36 per cent to 40 per cent (a record level, helped in no small way by Motorola's precipitous decline – see above). What Nokia was able to achieve was a shift from saturated developed markets to less-developed markets with much greater growth potential and to introduce low-end handsets there at cheap prices (in the region of $50) while generating decent profit margins not much different from those on much pricier handsets (Lex 2007d).

A combination of an improved market share and operating margin, combined with a stable average selling price was likely to prove difficult to maintain – as investors appeared to believe, judging by the less than buoyant share price – but for Nokia's competitors it was too much to hope for at virtually any point in time. Mindful of the dangers, and having hived off its network assets into NSN (see below), the residual handset operation was pointed firmly in the direction of the convergence of the Internet and mobility (*Financial Times* 2007; Global Insight 2007b).

Nokia was not immune to the market downturn of 2007–2008. In September 2008, it issued a profit warning, having just previously announced that its market share was holding up satisfactorily. It admitted that its mid-range product line had become tired, but identified the major problem as aggressive price reductions initiated by competitors that it was not willing to match on the grounds that the benefit to sales would be outweighed by the damage to profitability (*Financial Times* 2008). A notable absentee from the Nokia handset range at this time was an attractive touch-screen smartphone, but Nokia remained unperturbed, merely stating that its scale, brand and global presence would ultimately leave it in better shape than its rivals.

It was obliged to reiterate this when it issued a second profit warning in November, driven by a forecast that 2009 would see global handset sales fall on an annualized basis for the first time since 2001 (Parker 2008b). However, in April 2009 it stated that the severe destocking by retailers that had been manifest during the previous six months appeared to have eased, and in June launched its touch-screen smartphone, the N97. Its prospects give cause for optimism in the medium term, but whether it will be able to restore its operating margin, which fell to 12 per cent during the first half of 2009, back above 20 per cent on an ongoing basis is a moot point (Lex 2009b; Ward 2009).

Nokia Siemens Networks (NSN) was announced in June 2006 as a 50/50 joint venture between Siemens of Germany's COM division (minus its Enterprise Business unit) and Nokia's Network Business Group. Unlike other divisions of Nokia, the latter had been struggling in the post-2002 recovery period and a previous attempt to create a joint venture with Sanyo in February 2006 to create a scaled-up CDMA business had failed to reach fruition. What Nokia needed was a means to plug the fixed-wire gap in its product portfolio so that it could meet the need expressed by operators for a fixed-mobile convergence product range from the same vendor, and COM certainly appeared to fit the bill.

NSN was a follower, as opposed to a leader, in terms of restructuring driven by convergence in that it followed on from the creation of Alcatel-Lucent and Ericsson's acquisition of parts of Marconi. However, it managed in one fell swoop to create the third-largest vendor in the overall infrastructure market and the second-largest in mobile infrastructure with a near 25 per cent market share. Although fears were expressed that NSN would face difficulties, these were largely directed at the fixed-wire business and convergence, whereas the mobile infrastructure business was expected to prosper from the off. Overall, however, NSN did take some time to settle down after it was officially set up on 1 April 2007 – it was delayed by a corruption scandal at Siemens – and its results for the rest of that year were disappointing. Indeed, in October, Siemens was alleged to be willing to sell its stake in NSN to its partner even though Nokia would have struggled to pay cash for this.

The need to streamline NSN was paramount, but was expected to take until 2010 to be achieved and to include 9,000 lay-offs (of which 6,000 were announced in November

2009). Meanwhile, NSN felt able in October 2008 to pronounce itself the market leader in 3G with 130 W-CDMA/HSPA contracts, and it was the first vendor to announce that it would be able (by the year end) to supply trial equipment for LTE (Morris 2008a). But its share of the mobile infrastructure market was no larger in mid-2009 than it had been in April 2007, thereby indicating how fierce competition was in this part of its business (Fitchard 2009). The core issue was straightforward: NSN needed to shun low-margin contracts in order to restore its finances, yet this provided an opportunity for the likes of Ericsson and Huawei to grab market share at the expense of short-term profitability. And in this market, scale ultimately matters a great deal.

It is now clear that Siemens, having in other respects exited the telecommunications sector, wants to sell its stake in NSN – it wrote off $2.5 billion of its value in December 2009 – and Nokia has indicated a willingness to follow suit. But of prospective buyers there is no sign.

Sony Ericsson, which was created in 2001, is important both as an example of an international venture – in this case between the mobile phone divisions of Ericsson of Sweden, Ericsson itself a provider of mobile infrastructure, and Sony of Japan, each with a half share – and as a handset vendor that throughout the mid-2000s looked to be one of the success stories of the sector. As noted above, its Walkman handsets were highly regarded by those who valued their handsets as providers of music on the go, and as Motorola lost its way so Sony Ericsson aspired to leap-frog it into third place, as measured by market share, among handset vendors.

By March 2008, however, it had become clear that, if anything, Sony Ericsson might end up with a share below that of LG Electronics as it ran into severe difficulties in its core European markets. This was a particular issue since the vendor did not produce cheap handsets suitable for emerging markets such as China and India. In Europe, the maturity of the 2G market, the trend towards longer operator contracts and a reluctance to buy expensive handsets as recessionary conditions began to take hold, combined with the need for Sony Ericsson to invest heavily in R&D in order to improve the competitiveness of its product range, provided a challenging environment for the vendor. By August, in scenes reminiscent of the collapse of Motorola, the first job losses were announced and amounted to 17 per cent of the workforce. Sony Ericsson was seen as over-dependent upon the Walkman and Cybershot brands, and as Motorola had shown, more of the same was not a sensible strategy.

Hence, much depended upon the forthcoming launch of the Xperia X1, a prospective touch-screen competitor to the iPhone. Ultimately, it was no longer a question of where Sony Ericsson stood in the market share pecking order, but whether it was profitable (Minto 2008). But in October, it posted its first quarterly loss in five years – three more have since followed – and by March 2009 rumours had begun to circulate that Ericsson wanted to sell out to Sony although Sony was struggling to raise the necessary funds (Telecompaper 2009a). Sony Ericsson then issued a profit warning which reflected the expectation that the handset market was about to suffer a worse year even than 2001, with sales falling by as much as 15 per cent on an annualized basis.

Ultimately, Sony Ericsson suffered from its over-reliance upon its music player/camera feature phones, which rivals were able to emulate, while making inadequate plans to create competitive products in the smartphone market. Even when it eventually appeared, the X1 was never going to be a market leader, and its replacements were based upon more than one OS (operating system), which was inefficient (Parker 2009b). But the

Walkman and Cybershot brands (prefixed by 'Sony') remained its most valuable assets, and the new Walkman X-series was doing well. Which makes it somewhat difficult to understand why, in May 2009, it emerged that Sony Ericsson intended to drop the brands in developed markets in favour of a new unified entertainment offering, Entertainment Unlimited, with associated handsets branded as 'idou', a strategy which had been unveiled in February (Gabriel 2009a). The vendor claims that customers want handsets providing a full range of entertainment, not ones with specialist appeal. Not all are convinced.

As for parent Ericsson, its market-leading performance as a vendor of W-CDMA infrastructure during the mid-2000s appeared to leave it with few concerns. It played the role of predator, acquiring Marconi in 2005 and Redback Networks in 2006, each for roughly $2 billion. Furthermore, it strongly resisted the blandishments of potential joint venture suitors, preferring to opt for organic growth combined with bolt-ons. However, it proved no more immune to the deteriorating economic environment of 2007 than other vendors mentioned above, and issued a profit warning in October citing deteriorating conditions in emerging markets, Europe and the USA (Ibsen 2007). Its share price fell by 25 per cent in one day, the largest fall in the company's history, leading it to switch strategy from the pursuit of market share to that of improved operating margins (Ibsen and Parker 2007). Overall, given Ericsson's relatively high dependence upon shorter-term contracts such as network upgrades and extensions, it is expected to be the main beneficiary of improved conditions, and its sound financial situation compared to its rivals ensures that it will be a long-term survivor.

Japan

It is also worth making some comments about the situation in Japan. Japan is renowned for its electronics industry, but in respect of mobile handsets its history is unique because domestic vendors concentrated almost exclusively upon the home market. This was relatively large (it reached 100 million subscribers in early 2007), but it had approached the limits of market penetration at an early date and hence it was dependent upon advances in technology inducing customers to upgrade their handsets at regular intervals. This, in turn, required massive handset subsidies to be offered by operators in return for customers signing a two-year, or possibly longer, contract.

During the early to mid-2000s, no single vendor could truly claim to be the market leader – Sharp led the roster of roughly 10 vendors that accounted for 90 per cent of total sales with a market share fluctuating between 20 per cent and 25 per cent. But despite the absence of scale economies, the difficulties faced by non-Japanese vendors in entering the market – Nokia had managed to gain a market share of less than one per cent despite 20 years of effort – combined with operators' contractual arrangements, provided little pressure for the long-overdue restructuring. Then, in early 2008, under pressure from the government, operators switched to a new model whereby, in return for lowered monthly usage fees, subscribers effectively repaid the subsidy – which would typically amount to several hundred dollars because handsets were stuffed with state-of-the-art technology – in instalments over a two-year period.

Sales promptly collapsed as the replacement cycle rose to roughly two years, and restructuring finally began in earnest, if somewhat half-heartedly. In January 2008, Kyocera had announced that it had agreed to buy Sanyo's loss-making mobile phone division despite the absence of synergies, and Lenovo had quit the handset sector. In March, Mitsubishi Electric followed suit. More significantly, in September 2009, NEC, Casio Computer and Hitachi agreed to merge their respective handset operations, thereby (in April 2010) creating NEC Casio Mobile Communications with their respective initial ownership set at 66 per cent, 17.3 per cent and 16.7 per cent. However, this will only serve to create an entity with a market share of roughly 20 per cent, just behind Sharp and ahead of Panasonic (Yamaguchi 2009), so the sector's structure will clearly remain suboptimal.

The iPhone is doing well in Japan, so there are signs that overseas-sourced handsets may now be able to gain a foothold. Nokia quit the Japanese market in November 2008, but it is interesting to speculate whether the new arrangements will induce it to return. As for Japanese vendors, they predictably see a need to export to restore their fortunes – for example, Sharp is targeting China – but their ability to build the requisite handsets is quite anther matter.

There is an interesting contrast with China where virtually the opposite trend could be observed during the mid-2000s. Previously the preserve of local vendors which relied more upon design than upon functionality, the market was steadily taken over by foreign vendors. Nokia even managed to reach roughly a 35 per cent market share in line with its performance in other markets, while Samsung and even Motorola made significant inroads. However, Asian vendors such as China's state-owned ZTE that are certain to survive well into the next decade are anxious to reverse the above trend. ZTE, a vendor with both mobile network and handset divisions, is both smaller and less internationally minded than Huawei, but can be expected to leverage its access to vendor financing from the China Development Bank to good effect (Parker 2009a).

Joint Ventures

Joint ventures are common but they are not particularly successful – possibly only half can be so described. As demonstrated above, bringing together two weak operations in the hope that the combined entity will now succeed simply because it is bigger is unlikely to pay off. One key factor underpinning success is undoubtedly the existence of complementarities. A good example is Sony Ericsson where Sony, whose mobile handset division largely produced operator-branded phones in Japan, combined with Ericsson which had scale in Europe but was loss-making. Sony could contribute R&D expertise while Ericsson could provide marketing know-how. In addition, it is important that a joint venture is properly financed. Here again, Sony Ericsson got it right by investing roughly $1 billion during the first two (loss-making) years of its operation.

Thirdly, there is a need for transparency. Not surprisingly, parent companies are happy to offload unprofitable operations into joint ventures or to skew prices in favour of the parent. Full public disclosure of the joint venture's finances is necessary to control this. But, arguably, the most crucial aspect of all is the need for a joint venture to develop its own culture even though those running it are likely in practice to be transferring in and out of the respective parent companies.

In all of the above respects, Sony Ericsson largely got it right, whereas rivals such as Alcatel-Lucent assuredly did not (Lex 2007c). But, as noted above, the ability to build on a good start during a market upturn may not survive a subsequent downturn of any severity if the joint venture has come to rest upon its laurels.

Still, dumping unloved divisions into joint ventures does at least have the virtue of allowing a company to focus on what it sees as its core activities. A particularly telling example is Siemens which eventually succeeded in offloading the last of its telecoms-related activities in August 2008 (Preuschat 2008).

Conclusion

The cyclical nature of capital goods industries was referred to at the beginning of this chapter. In principle, the spending sequence in a recession is well established. General consumer spending falls first, followed three to six months later by spending on telecoms services, which nevertheless remains relatively resilient. Operators respond by cutting back on investment and vendors take a hit. With data generation rising sharply in response to the launch of high-speed data networks, but competition and regulatory oversight on the increase, leading operators tend to send out mixed messages, as they did between June 2008 and June 2009. Because at that time there was widespread acceptance that the world was in recession, analysts tended to emphasize any negative aspect of operator announcements, which in turn had a depressive effect upon vendor share prices. Vendors, in turn, had little option but to respond with a universal message of 'very limited visibility', which simply served to depress their share prices even further.

Ironically, perhaps, given the mayhem caused during the 2000–02 recession among vendors of network equipment, expectations about their performance during the subsequent period tended to remain low whereas expectations about the performance of handset vendors tended to rise sharply, and hence their share prices were far more vulnerable when recession returned during the so-called credit crunch. Whereas the 2000–02 recession was driven by equipment cutbacks at operators, reacting in part to the slow growth of demand for high-speed data services, the problem more recently has been to do with a reluctance by retail customers in emerging markets to buy new handsets and those in developed markets to replace their handsets (although spending on telecoms has become one of the last things to be cut back during hard times). In 2007, the replacement cycle averaged around three years whereas it is currently much closer to four years (Parker, Taylor and Harding 2008).

It has been noted above that the mobile infrastructure market is increasingly a battleground between three or four vendors. Accurate figures for handset sales are notoriously difficult to pin down – for example, vendor shipments are not the same thing as actual sales. Chapter 7 will go into considerable detail about the smartphone market, but that remains for now of modest importance overall – the big five vendors accounted for almost 900 million of the roughly 1.1 billion handsets shipped during 2008, so the likes of Apple and RIM do not (as yet, at any rate) rate as heavy hitters in the overall market. These five provide contrasting pictures, with Nokia so far in the lead as to be unassailable while Samsung no longer appears to have anything to fear from Motorola which is struggling to remain in touch with Sony Ericsson and LG.

Nokia has stated that it prefers to maintain margins and lose market share as a consequence, rather than vice versa. But it starts from a position of much the largest market share and will no doubt be happy to see some of its competitors go under, following the example of BenQ Mobile which had brought together the mobile handset divisions of Siemens of Germany and BenQ of Taiwan, but filed for bankruptcy in September 2006 when BenQ refused to make any further investment (Lennighan 2007). Some other minor players in the handset market such as VK Corp. of South Korea have also gone bankrupt.

Vendors were able to survive last time around, with the cushion of the late 1990s fat years to fall back on. This time around there is less fat and hence there will be insolvencies. There are also going to be mooted joint ventures and mergers, although these take a lot of time to come to fruition, so much depends upon how long the recession will last – optimists claim that it is already over, but not everyone agrees. In conclusion, one thing is certain, which is that by the end of 2012, the vendor sector is going to look very different to how it looked in 2000.

To what extent this is going to reflect the roll-out of 4G remains for now a moot point. Because LTE is a new technology it has the potential to be disruptive, and to present opportunities for ambitious vendors to gain market share at the expense of rivals with large investments in legacy systems. However, although there is pressure in parts of Asia to push on as rapidly as possible to LTE, the position of network operators in Europe is relatively conservative. In essence, their view is that there is considerable mileage left in HSPA+ (high-speed packet access +), and in the aftermath of the credit-crunch their shareholders will not react positively to plans to scale up capital investment (Parker and Taylor 2009).

References

Amesse, F., Latour, R., Rebolledo, C. and Séguin-Dulude, L. 2004. The telecommunications equipment industry in the 1990s: From alliances to mergers and acquisitions. *Technovation*, 24(11), 885–897.

Brickley, P. 2009. US, Canadian courts OK Ericsson $1.13bn buy of Nortel unit. 29 July. Available at http://www.totaltele.com (accessed 30 July 2009).

BWCS. 2006. Alcatel-Lucent irons out deal to buy Nortel's UMTS section. 4 December. Available at http://www.bwcs.com (accessed 5 December 2006).

Cellular-news. 2007. Shares in Alcatel-Lucent tumble after 3rd profit warning this year. 13 September. Available at http://www.cellular-news.com (accessed 14 September 2007).

----- 2008a. Motorola to split off handset business. 26 March. Available at http://www.cellular-news.com (accessed 27 March 2008).

----- 2008b. Nortel reports loss of $3.4 billion – cuts 1,300 jobs. 10 November. Available at http://www.cellular-news.com (accessed 11 November 2008).

----- 2009. Nortel to sell optical networking and carrier Ethernet businesses. 8 October. Available at http://www.cellular-news.com (accessed 9 October 2009).

Dow Jones. 2006a. Lucent, Alcatel better off together. 28 March. Available at http://www.cellular-news.com (accessed 1 April 2006).

----- 2006b. Lucent, Alcatel discussing $33 billion merger. 28 March. Available at http://www.cellular-news.com (accessed 1 April 2006).

----- 2006c. Lucent, Alcatel joined better able to face telcos. 28 March. Available at http://www.cellular-news.com (accessed 1 April 2006).

Dummett, B. 2008. Nortel posts Q2 loss; reiterates outlook. 1 August. Available at http://www.totaltele.com (accessed 2 August 2008).

Financial Times. 2007. Nokia changes its tune on mobile strategy. 5 December. Available at http://www.ft.com (accessed 6 December 2007).

----- 2008. Nokia stuns market with strong warning. 5 September. Available at http://www.ft.com (accessed 6 September 2008).

Fitchard, K. 2009. NSN loses market share as services outshine infrastructure. 16 July. Available at http://www.printthis.clickability.com (accessed 21 July 2009).

Gabriel, C. 2009a. Sony Ericsson could ditch best asset, the Walkman brand. 21 May. Available at http://www.rethink-wireless.com (accessed 22 May 2009).

----- 2009b. Verizon promises biggest ever marketing splash for Motorola Droid. 29 October. Available at http://www.rethink-wireless.com (accessed 30 October 2009).

Global Insight. 2007a. Motorola's resurgent handset profits falter as RAZR loses its edge. 22 January. Available at http://www.comunicationsdirectnews.com (accessed 23 January 2007).

----- 2007b. Nokia unveils new partnerships and acquisitions; outlines plans for the future. 4 December. Available at http://www.comunicationsdirectnews.com (accessed 5 December 2007).

Hollinger, P. 2008. Alcatel to remain in wireless market. 5 December. Available at http://www.ft.com (accessed 6 December 2008).

Hollinger, P. and Taylor, P. 2006 Alcatel and Lucent agree €30bn merger. 2 April. Available at http://news.ft.com (accessed 3 April 2006).

Hunter, D. 2003. *The Bubble and the Bear: How Nortel Burst the Canadian Dream*. Canada: Doubleday.

Ibsen, D. 2007. Ericsson slumps after new warning. 20 November. Available at http://www.ft.com (accessed 2 November 2007).

Ibsen, D. and Parker, A. 2007. Ericsson plunges on warning. 16 October. Available at http://www.ft.com (accessed 17 October 2007).

Lau, J. and Guerrera, F. 2005. TCL and Alcatel unravel joint venture. 16 May. Available at http://www.ft.com (accessed 17 May 2005).

Le Maistre, R. 2009. Nortel: What's left on the shelf? 25 November. Available at http://www.lightreading.com (accessed 26 November 2009).

Lennighan, M. 2007. End of the line for BenQ. 27 February. Available at http://www.totaltele.com (accessed 1 March 2007).

Lex. 2005. Alcatel. 29 April. Available at http://www.ft.com (accessed 1 May 2005).

----- 2007a. Handsets. 5 January. Available at http://www.ft.com (accessed 6 January 2007).

----- 2007b. Alcatel-Lucent in trouble. 23 January. Available at http://www.ft.com (accessed 23 January 2007).

----- 2007c. Sony Ericsson. 2 July. Available at http://www.ft.com (accessed 3 July 2007).

----- 2007d. Sony Ericsson. 2 August. Available at http://www.ft.com (accessed 3 August 2007).

----- 2008a. Alcatel-Lucent. 29 July. Available at http://www.ft.com (accessed 30 July 2008).

----- 2008b. Mortal Nortel. 3 August. Available at http://www.ft.com (accessed 4 August 2008).

----- 2009a. Motorola. 15 January. Available at http://www.ft.com (accessed 16 January 2009).

----- 2009b. Nokia. 30 September. Available at http://www.ft.com (accessed 1 October 2009).

Masters, B. and Taylor, P. 2007. Former Nortel executives charged. 12 March. Available at http://www.ft.com (accessed 13 March 2007).

Matsumoto, C. 2009. Nortel offers up its LG-Nortel stake. 27 May. Available at http://www.lightreading.com (accessed 28 May 2009).

Metz, R. 2009. Review: Motorola's Cliq is a snappy smart phone. 21 October. Available at http:/www.cellular-news.com (accessed 22 October 2009).

Middleton, J. 2009. Nortel to offload GSM assets. 1 October. Available at http://www.telecoms.com (accessed 2 October 2009).

Milne, R. 2004. Alcatel in handset merger deal with TCL. 26 April. Available at http://www.ft.com (accessed 27 April 2004).

Minto, R. 2008. Sony Ericsson looks for a new lease of life. 4 August. Available at http://www.ft.com (accessed 5 August 2008).

Morris, A. 2008a. Nokia Siemens Networks stresses scale is key to survival. 29 October. Available at http://totaltele.com (accessed 1 December 2008).

----- 2008b. Running faster is no longer enough for Alcatel-Lucent – Verwaayen. 30 October. Available at http://totaltele.com (accessed 1 December 2008).

Mullen, J. 2007. Wireless unit seen critical for Alcatel-Lucent. 25 October. Available at http://www.totaltele.com (accessed 26 October 2007).

Parker, A. 2008a. Nortel plunges after profit warning. 23 April. Available at http://www.ft.com (accessed 24 April 2008).

----- 2008b. Nokia warns handset sales to shrink in 2009. 14 November. Available at http://www.ft.com (accessed 15 November 2008).

----- 2009a. Confident ZTE sees double-digit growth. 25 March. Available at http://www.ft.com (accessed 26 March 2009).

----- 2009b. Sony Ericsson needs to get smart, and fast. 6 May. Available at http://www.ft.com (accessed 7 May 2009).

Parker, A. and Taylor, P. 2009. Delays to 4G hurt handset makers. 15 February. Available at http://www.ft.com (accessed 16 February 2009).

Parker, A., Taylor, P. and Harding, R. 2008. Handset makers face reversal of fortune. 26 October. Available at http://www.ft.com (accessed 27 October 2008).

Politi, J. 2007. Nortel edged out of Avaya auction. 5 June. Available at http://www.ft.com (accessed 6 June 2007).

Preuschat, A. 2008. Siemens to part with last telecoms operations. 1 August. Available at http://www.totaltelecom (accessed 2 August 2008).

Reedy, S. 2009. Motorola on socializing the mobile experience. 20 November. Available at http://www.printthis.clickability.com (accessed 21 November).

Silver, S. and Lublin, J. 2008. Nortel taps legal advice on filing for bankruptcy. 10 December. Available at http://www.totaltele.co (accessed 11 December 2008).

Simon, B. and Taylor, P. 2009. Nortel files for bankruptcy protection. 14 January. Available at http://www.ft.com (accessed 15 January 2009).

Sovich, N. and Cheng, R. 2006. Alcatel to buy Lucent for $13.4bn in stock. 3 April. Available at http://www.totaltele.com (accessed 4 April 2006).

Sovich, N. and Patrick, M. 2006. Lucent's Alcatel merger stirs doubts. 18 August. Available at http://www.totaltele.com (accessed 19 August 2006).

Taylor, P. 2007a. Motorola slashes outlook, CFO retires. 21 March. Available at http://www.ft.com (accessed 22 March 2007).

----- 2007b. Analysts hint at further Motorola job cuts. 16 May. Available at http://www.ft.com (accessed 17 May 2007).

----- 2008a. Nortel shares bounce 13 per cent on shift in strategy. 12 June. Available at http://www.ft.com (accessed 13 June 2008).

----- 2008b. Motorola falls on analyst update. 24 June. Available at http://www.ft.com (accessed 25 June 2008).

----- 2008c. Motorola cuts 3,000 jobs and delays spin-off. 30 October. Available at http://www.ft.com (accessed 1 November 2008).

----- 2009a. Motorola calls on non-mobile divisions. 12 February. Available at http://www.ft.com (accessed 13 February 2009).

----- 2009b. Nortel to liquidate assets. 21 June. Available at http://www.ft.com (accessed 22 June 2009).

----- 2009c. Nortel sells unit to Avaya for $475 million. 21 July. Available at http://www.ft.com (accessed 22 July 2009).

Telecompaper. 2009a. Ericsson looks to sell packet Sony Ericsson stake to Sony – report. 18 March. Available at http://www.telecompaper.com (accessed 19 March 2009).

----- 2009b. Nortel to sell packet core network assets. 22 September. Available at http://www.telecompaper.com (accessed 23 September 2009).

Telecoms.com. 2008. Three handset vendors duke it out for third place. November. Available at http//www.telecoms.com (accessed 28 November 2008.

The Economist. 2001. The runaway bride. 2 June, 17.

Ward, A. 2009. Nokia weighed down by networks venture. 15 October. Available at http://www.ft.com (accessed 16 October 2009).

Wood, N. 2009. Motorola to cut 50 per cent of handset workforce. 13 January. Available at http://www.totaltele.com (accessed 15 January 2009).

Yamaguchi, Y. 2009. NEC, Casio, Hitachi to merge mobile phone ops. 14 September. Available at http://www.totaltele.com (accessed 15 September 2009).

CHAPTER 7 The iPhone: A Case Study of Innovation in the Mobile Handset Market

Introduction

On 26 June 2007, the share price of Apple fell by 2 per cent – on the face of it a fairly trivial event except that it was about to launch (exclusively in the USA in conjunction with AT&T) the arguably most-hyped mobile device in history known as the iPhone (Allison 2007b). The iPhone is a variant of what are generally categorized as smartphones, although your average handset is now so smart that the term is becoming outdated.[1]

The iPhone had initially surfaced at the MacWorld trade show in January 2007 when Steve Jobs announced that, having previously changed the music industry via the launch of the iPod, Apple now intended to do the same for the mobile communications industry via the iPhone (Allison 2007a). By June, shareholders had become edgy because the value of Apple had risen by $34 billion since January and they stood to lose much of their gains if the iPhone proved to be a one-week wonder. The curious aspect of this was that the device in question was, by the standards of the time, fairly expensive, relatively large, on the heavy side, lacking a keyboard, short on memory and would initially at least be launched over a so-called 2.5G network rather than one capable of 3G speeds – the sort of attributes that would leave the reputation of any 'normal' handset vendor in tatters.

But not that of Apple, because it had set out to address the fundamental flaw in existing handset design, namely the inherently user-unfriendly nature of the beast. Anyone working their way through a lengthy series of drop-down menus, or trying to type accurately on a tiny keyboard, finds the experience frustrating, and may also be unable to access parts of the Internet. What Apple set out to do was to give the user unrestricted access to the Internet via an operating system (OS) comparable to that found on most desktop computers, as well as to simplify the process of making voice calls, listening to music and managing contact lists. That was achieved by replacing the keyboard with a touch-screen that could immediately replace one set of button functions with another according to whether the user wanted to make a voice call, send an email or whatever.

1 Compared even to three years ago, a standard 3G handset is incredibly smart. Just how smart a smartphone can be is constantly being redefined. Currently, it is reasonable to expect it, *inter alia*, to send and receive email, provide calendar operations, access multimedia applications, surf the Web and synchronize with a computer. A recent (June 2009) handset, the Samsung Jét (S8000), was being marketed as 'smarter than a smartphone' (Cellular-news 2009f).

Competing Handsets

So why were shareholders edgy? In the first place, with the iPod market reaching maturity, Apple effectively had a large chunk of its future profitability resting on the success of the iPhone. However, it was not taking on the manufacturers of MP3 players, as it did with the iPod, but the giants of the handset market such as Nokia, Sony Ericsson and Samsung, all of whom had already introduced models that provided fairly effective competition for the iPhone even if they lacked its Mac OSX and Safari web browser. Apple chose to make direct comparisons with three QWERTY keyboard smartphones, the BlackBerry Curve 8300, Palm Treo 750 and Samsung Blackjack, which was arguably counterproductive as it accentuated the fact that the lack of a QWERTY keyboard meant that the iPhone was somewhat unsuitable for anyone wanting to send lots of mobile e-mails or do lots of texting – that is, businessmen and youths. Apple also made comparisons with the Nokia N75 (Cellular-news 2007a).

In addition to the N75, Nokia's N95 with its far superior five mega-pixel camera and video capability could not be ignored. Like the iPhone, the N95 was marketed as a 'multimedia computer' – after all, $700 seemed a touch pricey for something merely called a cell phone (Svensson 2007). The HTC Touch was expected to exist as a niche product, but was followed up in October by the Touch Dual, capable of operating over a HSDPA (high-speed downlink packet access) network (Thomas 2007). However, if buyers wanted a sophisticated camera, the iPhone was not state-of-the-art and, significantly, it had drawbacks as a music machine. The iPhone was designed to work exclusively with iTunes software, so users were obliged to set up an online account with the latter. This may have seemed to be a price worth paying, but if a user simply wanted to replace his or her MP3 player with a handset incorporating similar capabilities, then, for example, the Sony Ericsson Walkman series appeared better able to fit the bill.

Subsequently, operators hit back with an attractive range of alternative handsets, almost all provided free with new contracts costing the equivalent of $35 a month and a new range of iPhone alternatives was launched by Verizon in November 2007 (Malykhina 2007). Apple responded by adding new facilities to the iPhone including (initially only in the USA) maps with location awareness in January 2008 and SMS (short message) to multiple recipients service. In June 2008, Sprint Nextel introduced the Samsung Instinct which, like the iPhone, lacked a keypad. This was the latest in a sequence of iPhone imitators such as the LG Voyager from Verizon which had an exterior touch-screen and folded out to reveal a non-touch-screen paired with a keyboard.

Among the strongest competitors in late 2008 were arguably the Sony Ericsson Xperia X1 and Samsung Soul, although the 3G version of the popular BlackBerry, the Bold, launched in May over the AT&T network, could certainly not be discounted (Gallagher 2008). But with 1.5 million pre-sale orders and a late-October 2008 launch date fixed for the T-Mobile USA network, the HTC G1 (Google) handset, designed to run on the Android OS, looked likely to become the iPhone's main threat during the first half of 2009. When the G1 was first announced, Android handsets were seen as addressing a different market segment to the iPhone in that they were expected to be of the relatively cheap kind that are normally given away free to contract customers (Charny 2007b). At the time the G1 was launched, it was still seen in many quarters as complementary to the iPhone and subject to significant drawbacks such as excessive weight, smallness of screen and lack of memory (Mossberg 2008).

T-Mobile managed to sell 100,000 G1 handsets within six months from launch (Cellular-news 2009e), but its shelf life, like that of so many previous smartphones, was short. The updated G2, also known as the Google Ion (Googleion 2009), was due to be launched in mid-2009 and was expected to prove to be more competitive than its predecessor, especially as the G2 was being marketed by Vodafone, branded as the Magic. T-Mobile USA sold is own version of the G2 in the USA branded as myTouch 3G. A variant based on a new user-interface called the Sense, also made by HTC and branded the Hero, followed shortly thereafter (Talkandroid 2009). The first non-HTC Android handset, the I7500 made by Samsung, was scheduled for a launch in Europe in June 2009 (Cellular-news 2009d) followed by the Huawei U8230 (Cellular-news 2009h).

Meanwhile, the latest BlackBerry, the Storm, made its mark in early 2009, with Verizon Wireless selling one million Storms in roughly two months despite mixed reviews and many teething problems (Berg 2009). An alternative to the Storm, if a QWERTY keyboard was *de rigueur*, was the Tour, launched in June 2009 and capable of connecting to both W-CDMA (wideband code division multiple access)/UMTS (universal mobile communications system) and cdma2000 1xEV-DO networks (TechCentral 2009).

Table 7.1 provides a sample of the handsets that have at various times been put forward as viable alternatives to the iPhone even though some of the earlier ones did not have touch-screens – a full review of each handset can also be found, for example, at www.3g.co.uk and http://reviews.cnet.com and see also 3G.co.uk (2008a). It should be borne in mind that handsets are normally identified by a code, for example the Nokia N97, but often come with a nickname designed to appeal to prospective users and which can vary according to which operator is launching the handset or the country/region in which it is being launched. Furthermore, upgraded versions of an initial design may be introduced, such as the Samsung Omnia II, HTC Touch Pro2 and Blackberry Storm 2.

Table 7.1 Illustrative iPhone rival handsets

Manufacturer	Name	Date launched	Type[1]	Details
HTC	Touch	June 2007	T	Cellular-news (2007a)
HTC	Touch Dual	October 2007	T	Thomas (2007)
HTC	Touch Diamond	June 2008	T	BWCS (2008c)
HTC	G1 [Google]	October 2008	Q, T	Cellular-news (2008h)
HTC	G2 [Google]	April 2009	TC	BWCS (2009a)
HTC	Hero	June 2009	TC	Talkandroid (2009)
HTC	Tattoo	October 2009	T	Middleton (2009d)
Huawei	C8000	2009H2	Q, T	Cellular-news (2009h)
Huawei	U8230	2009H2	Q, T	Cellular-news (2009h)
LG Electronics	Prada	March 2007	T	Cellular-news (2007a)
LG Electronics	KG-600	February 2008	T	Taylor (2008a)
LG Electronics	Vu	April 2008	T	Cheng (2008)
LG Electronics	Dare	July 2008	T	Dow Jones (2008d)
LG Electronics	Renoir	October 2008	TC	Cellular-news (2008d)
LG Electronics	Cookie	October 2008	TC	Cellular-news (2009c)
LG Electronics	Arena	February 2009	T	Cellular-news (2009a)
LG Electronics	GM-730	June 2009	T	Cellular-news (2009g)
Motorola	Krave	October 2008	T	Telecoms.com (2008d)
Motorola	Dext/Cliq	October 2009	Q, T	Wood (2009b)
Motorola	Droid/Milestone	November2009	Q, TC	Gabriel (2009f)
Nokia	N95	April 2007	Q	Cellular-news (2007a)
Nokia	5800 XpressMusic	October 2008	TM	Cellular-news (2008e)
Nokia	N97	June 2009	Q, TC	Taylor (2008b)
Nokia	N900	October 2009	Q, TC	Telecompaper (2009d)
Palm	Prē	June 2009	Q, T	Svensson (2009)
Palm	Pixi	December 2009	Q, T	Unstrung (2009)
Research in Motion	BlackBerry Storm	November 2008	T	Svensson (2008b)
Research in Motion	BlackBerry Tour	June 2009	Q	TechCentral (2009)
Samsung	Soul	June 2008	T	Taylor (2008a)
Samsung	Instinct	February 2008	T	Svensson (2008a)
Samsung	Omnia	June 2008	T	Dow Jones (2008b)
Samsung	Pixon	November 2008	T	3G.co.uk (2008c)
Samsung	Memoir	February 2009	TC	3G.co.uk (2009a)
Samsung	S5600	May 2009	T	3G.co.uk (2009b)
Samsung	Star	May 2009	Q, T	3G.co.uk (2009b)
Samsung	Galaxy	September 2009	T	Cellular-news (2009d)
Samsung	Jét	June 2009	TC	Cellular-news (2009f)
Sony Ericsson	Xperia X1	November 2007	Q, T	Parker and Taylor (2008)
Sony Ericsson	Xperia X2	October 2009	Q, TC	Cellular-news (2009k)
Sony Ericsson	Satio	2009Q4	TC	Telecompaper (2009b)
Toshiba	TG01/T-01A	July 2009	T	Gabriel (2009a)

Source: Compiled by authors from a variety of sources

Note: [1] Q = Qwerty keyboard; T = touch-screen; TC = touch-screen with 5 mega-pixel camera; TM = touch-screen with high-end music capability

Apple Connects with AT&T

It could be argued that the iPhone had something that its touch-screen competitors lacked, namely the support of a major carrier, in this case AT&T – important because it meant that the iPhone could be subsidized by the operator. However, this was not the case if the competing models were defined more broadly, as noted above, and it was possible that there would be unintended consequences (Lex 2007). In the first place, AT&T was selling the iPhone contract in conjunction with its EDGE (enhanced data (rates) for GSM evolution) network – although the iPhone automatically switched from EDGE to much faster Wi-Fi networks if available. Now EDGE was a major step up from GSM (global system for mobile)/cdmaONE, but it was a carthorse compared to HSDPA and cdma2000 1xEV-DO Rev. A, so users might prefer a combination of more speed combined with one of the handsets previously mentioned. Whether the availability of Wi-Fi connectivity was a plus or minus for the iPhone was a moot point, in the first place because Wi-Fi was usually more expensive than downloading on a 3G network and, secondly, because the iPhone was largely directed towards the consumer rather than the business market, and it was the latter that was making most use of Wi-Fi (Mossberg and Boehret 2007).

A further drawback for potential corporate buyers was that whereas their BlackBerrys were compatible with their office systems, the iPhone might prove difficult to integrate (TelecomWeb 2007), while, for their part, personal buyers might need to switch network operators at some initial cost to themselves (Cellular-news 2007b).

In addition, the arrangement between Apple and AT&T had altered the nature of the commercial strategies pursued by both companies in the USA (Jones 2007). Because the great majority of all handsets in the USA were bought initially by operators, equipment vendors had been obliged to adopt the specifications that operators imposed, which, it was argued, explained in good part why handsets had remained user-unfriendly. In contrast, Apple had always controlled rigorously what had gone into its products, but had now exposed itself to any shortcomings in the AT&T network. Apple, not surprisingly, had been able to insist on far more control over the iPhone's features than AT&T normally allowed to other vendors, and these vendors might be unable, in practice, to ride on Apple's coat-tails, but AT&T might not have properly taken into account the fact that iPhone owners would be able to synchronize their devices with iTunes over a PC connection – so-called 'side-loading' – and thereby avoid downloading (at a price) over the air.

Downloads were – and indeed remain – critical for operators given the billions of dollars spent on upgrading their networks essentially to counteract the fact that not enough downloading was going on. The ring tone and mobile gaming markets were no longer expected to provide long-term growth potential, and, somewhat surprisingly, most handset owners simply stored their pictures in their handsets without sending them on to others. As a consequence, texting remained the main source of revenue-generating data flows while push email was increasingly popular, yet neither required a 3G network to work satisfactorily.

By much improving the Internet browsing experience, Apple stood to gain a decent share of the worldwide market for smartphones. However, it was doubtful that Apple would be able to replicate the marketing plan implemented for the iPod which resulted in market segmentation and progressively cheaper variants of the device. The costs of re-engineering the iPhone would almost certainly not be economic given the likely size of the potential market. And if the iPhone remained expensive, it needed to deliver a superior performance or the initial euphoria would soon fade, especially since prospective purchasers were concerned about high handset costs and the two-year contracts required by AT&T.

It should also be borne in mind that when the AT&T network came to be upgraded to W-CDMA and HSDPA, the initial versions of the iPhone would not be compatible. Nevertheless, reflecting the hype, almost the entire initial stock – AT&T would not reveal the numbers, but 700,000 was the best estimate at the time – was allegedly sold on the opening two days, although many buyers appeared intent upon an immediate, profitable, resale on eBay.

In practice, the initial sales figures appeared to have been heavily overestimated. On 24 July, AT&T announced that only 146,000 iPhones had been activated during their first day and a half on sale (the final days of 2008Q2), and it was reported elsewhere that demand had slowed dramatically with store inventories at healthy levels. The Apple share price, which had peaked at $144, slid back to $139. Apple immediately countered that 270,000 iPhones had been sold and remained positive about its forecast of 10 million sales in 2008. Subsequently, the share price suffered because of adverse comments online about the $79-plus-shipping replacement cost of a malfunctioning battery (Cellular-news 2007c), and on 30 July it fell to $132 on an unconfirmed rumour that Apple was cutting back on production.

A Strategic Course Change

In Europe, the intention was to launch the iPhone during 2007 only in France, Germany and the UK – the markets where Apple first launched iTunes in 2004 – with the rest of Europe plus Asia launching during 2008. In the UK, Telefónica was predicted to be the contracted operator in preference to Vodafone, but in either case the expectation was for the launch to be over EDGE rather than 3G. This was unlikely to impress potential buyers, so much attention was paid to the possibility that Apple would subsequently launch with a 3G-compatible version of the iPhone.

In late August, the three operators alleged to be ready to sign up proved, predictably, to be Orange in France, T-Mobile in Germany and Telefónica in the UK. All were said to have agreed to hand over to Apple 10 per cent of the revenues accrued from voice calls and data transfers using the iPhone. However, AT&T is now believed to have paid close to 15 per cent and the failed bidders in the UK claim to have offered up to 25 per cent, implying that Telefónica offered more (PriMetrica 2007a). In practice, negotiations with operators appeared to be stickier than expected because on 6 September 2007, Apple announced a bombshell: the 4GB model would be discontinued and the $599 price of the 8GB iPhone would be reduced by one-third to $399 with immediate effect. This reduction was the quickest ever for a headline Apple product.

Perhaps surprisingly, many analysts were supportive of the price reduction, arguing that the profits made in the first two months after the iPhone launch were now available to help establish Apple as a serious player in the handset market. However, those who had paid the initial high price were outraged, even though they were offered compensation – albeit only of a kind, in that the $100 voucher could only be redeemed in an Apple store where nothing of interest to them would cost so little, and as a result they would have to spend more of their money to use the voucher.

'We want to do the right thing for our valued iPhone customers. We apologise for disappointing some of you, and we are doing our best to live up to your high expectations' was the view taken by CEO Steve Job. Bloggers, for some reason, failed to agree. Furthermore, a GMI (Global Market Insite) Poll conducted at the end of October indicated that the price remained too high to tempt the majority of potential buyers.

So just how much money was Apple making from the iPhone? According to Bernstein Research, AT&T could end up paying Apple up to $360 over the course of every two-year contract, suggesting that there was still plenty of room left for reducing the price of the handset. This calculation assumed that the AT&T ARPU (average revenue per user) rose from $58 a month to between $70 and $80 when an iPhone was acquired.

The one-millionth iPhone was sold on 9 September 2007 after 74 days on sale (Associated Press 2007). By this point, business users had begun to grumble about the iPhone's failings in relation to Internet-based work and appeared to be using both an iPhone and a BlackBerry, indicating that the iPhone's success would depend upon how it shaped up against the better-quality consumer devices such as the LG Chocolate and LG Voyager, introduced by Verizon in October 2007 to tackle the iPhone head-to-head. Although the Voyager came with the additional advantage of both a touch-screen and a QWERTY keyboard, not to mention connectivity with Verizon's high-speed cdma2000 1xEV-DO network, it needed the insertion of a memory card as built-in memory was limited and its screen was smaller than that of the iPhone.

In late October, Apple reported that it had sold 1.12 million iPhones during the third quarter and 1.38 million all told. It was claimed that 250,000 customers intended to unlock their handsets despite Apple's attempts to forestall this behaviour, but AT&T still appeared to be happy with its deal, claiming that 40 per cent of iPhone customers were new subscribers. According to the latest ChangeWave Research handset survey, 82 per cent of iPhone owners were very satisfied with their purchase – the next best score was 51 per cent for Research In Motion (RIM) – while 16 per cent of those expecting to acquire a new handset within six months nominated the iPhone as their preferred choice. In mid-January 2008, worldwide sales were said to have risen to roughly four million.

International Launches

In late September 2007, Deutsche Telekom announced that it had taken up the rights to market the iPhone in Germany, Croatia, Hungary, the Netherlands and Austria (thereby pushing out Telekom Austria). In Germany, T-Mobile already had an EDGE network suitable for iPhone use and could also offer connection to 8,600 Wi-Fi hotspots. In late October, three packages were put on offer. On the iPhone's first day on sale in early November, more than 10,000 sales were recorded (Preuschat 2008a). Shortly thereafter, Vodafone obtained an injunction barring T-Mobile from selling the iPhone if provided with a 'SIM-lock' and/or if offered only 'in combination with a mobile phone contract ... running at least 24 months'. T-Mobile agreed to comply with the injunction with immediate effect, but raised the price of an unlocked handset to €999 ($1,480) (BWCS 2007). However, on appeal, T-Mobile's exclusive contracts were reinstated. In April, with the 3G version pending, T-Mobile cut the price to €99 in conjunction with an XL contract costing at least €89 a month, and to €249 for subscribers paying €29 a month. Nevertheless, sales were reported to be only 75,000 at the end of August.

Table 7.2 Representative iPhone launches

Country	Operator	Launch
USA	AT&T	29/06/07
USA	AT&T	11/07/08 [3G]
USA	AT&T	19/06/09 [3G S]
Puerto Rico	AT&T	17/10/08
US Virgin Isles	AT&T	17/10/08
UK	Telefónica	09/11/07
UK	Telefónica	11/07/08 [3G]
UK	Telefónica	19/06/09 [3G S]
Germany	T-Mobile	09/11/07
Germany	T-Mobile	11/07/08 [3G]
France	Orange	25/11/07
France	Orange	18/07/08 [3G]
France	Orange	19/06/09 [3G S]
Ireland	Telefónica	14/03/08
Austria	T-Mobile	15/03/08
Austria	T-Mobile	11/07/08 [3G]
Czech Rep.	T-Mobile	22/08/08 [3G]
Netherlands	T-Mobile	11/07/08 [3G]
Hungary	T-Mobile	22/07/08 [3G]
Poland	T-Mobile	22/07/08 [3G]
Australia	Vodafone	11/07/08 [3G]
Czech Rep.	Vodafone	22/08/08 [3G]
India	Vodafone	22/08/08 [3G]
Italy	Vodafone	11/07/08 [3G]
New Zealand	Vodafone	11/07/08 [3G]
New Zealand	Vodafone	10/07/09 [3G S]
Portugal	Vodafone	11/07/08 [3G]
Italy	Telecom Italia	11/07/08 [3G]
Australia	Optus	11/07/08 [3G]
Australia	Optus	26/06/09 [3G S]
Australia	Telstra	11/07/08 [3G]
Singapore	SingTel	22/08/08 [3G]
Singapore	SingTel	10/07/09 [3G S]
India	Bharti AirTel	22/08/08 [3G]
Philippines	Globe	22/07/08 [3G]
Austria	ONE	11/07/08 [3G]
Belgium	Mobistar	11/07/08 [3G]
Kenya	Orange	15/12/08 [3G]
Madagascar	Orange	12/12/08 [3G]
Poland	Orange	22/07/08 [3G]
Portugal	Optimus	11/07/08 [3G]
Romania	Orange	22/07/08 [3G]
Senegal	Orange	29/12/08 [3G]
Switzerland	Orange	11/07/08 [3G]
France	SFR	08/04/09 [3G]
France	Bouygues	29/04/09 [3G]
Malaysia	Maxis	20/03/09 [3G]
Malaysia	Maxis	31/07/09 [3G S]
Indonesia	Telkomsel	20/03/09 [3G]
Greece	CosmOTE	23/03/09 [3G]

Table 7.2 ***Concluded***

Country	Operator	Launch
Denmark	TeliaSonera	11/07/08 [3G]
Estonia	TeliaSonera	11/07/08 [3G]
Finland	TeliaSonera	11/07/08 [3G]
Latvia	TeliaSonera	11/07/08 [3G]
Lithuania	TeliaSonera	11/07/08 [3G]
Norway	TeliaSonera	11/07/08 [3G]
Sweden	TeliaSonera	22/08/08 [3G]
USA	América Móvil	11/07/08 [3G]
Brazil	América Móvil	15/06/07
Mexico	América Móvil	20/09/08 [3G]
Mexico	América Móvil	31/07/09 [3G S]
Argentina	América Móvil	11/07/08 [3G]
Chile	América Móvil	22/08/08 [3G]
Colombia	América Móvil	22/08/08 [3G]
Colombia	América Móvil	31/07/09 [3G S]
Ecuador	América Móvil	22/08/08 [3G]
El Salvador	América Móvil	22/08/08 [3G]
El Salvador	América Móvil	31/07/09 [3G S]
Guatemala	América Móvil	22/08/08 [3G]
Guatemala	América Móvil	31/07/09 [3G S]
Honduras	América Móvil	22/08/08 [3G]
Paraguay	América Móvil	22/08/08 [3G]
Peru	América Móvil	22/08/08 [3G]
Uruguay	América Móvil	22/08/08 [3G]
Uruguay	América Móvil	31/07/09 [3G S]
Hong Kong	Hutchison	11/07/08 [3G]
Hong Kong	Hutchison	10/07/09 [3G S]
Japan	Softbank	11/07/08 [3G]
Thailand	TrueMove	18/01/09 [3G]
Thailand	TrueMove	04/08/09 [3G S]
Spain	Telefónica	11/07/08 [3G]
Czech Rep.	Telefónica	22/08/08 [3G]
Argentina	Telefónica	11/07/08 [3G]
Brazil	Telefónica	11/07/08 [3G]
Chile	Telefónica	11/07/08 [3G]
Colombia	Telefónica	11/07/08 [3G]
Ecuador	Telefónica	11/07/08 [3G]
El Salvador	Telefónica	11/07/08 [3G]
Guatemala	Telefónica	11/07/08 [3G]
Nicaragua	Telefónica	11/07/08 [3G]
Panama	Telefónica	11/07/08 [3G]
Peru	Telefónica	11/07/08 [3G]
Uruguay	Telefónica	11/07/08 [3G]
Venezuela	Telefónica	11/07/08 [3G]
Canada	Rogers	11/07/08 [3G]
Russia	MegaFon	10/09/08 [3G]
Saudi Arabia	Etisalat	21/02/09 [3G]
UAE	Etisalat	21/02/09 [3G]
Taiwan	Chunghwa	13/12/08 [3G]
Bermuda	M3 Wireless	26/03/09 [3G]

Source: Compiled by authors from a variety of sources

Note: For a list of countries where the iPhone is available, see http://www.apple.com/iphone/countries

In France, Orange won distribution rights in mid-September and stated that it would not be subsidizing the handset. However, there was still no agreement over the specific terms of the French deal by early October, allegedly because under French law – as in Belgium (Telecoms.com 2008c) – the iPhone would also have to be made available unlocked to non-contract customers (PriMetrica 2007b). Subsequently, it was reported that a 30 per cent revenue payment to Apple had been agreed with the 8GB handset retailing at €399 including taxes with a two-year contract, but otherwise €649 to discourage anyone from buying an unlocked handset. The launch date was set at 29 November. France Télécom announced that it had sold 30,000 handsets during the first five days in the shops, rising to 70,000 after one month (PriMetrica 2008a), and that one-half of the buyers had opened new subscriptions with Orange. However, sales had risen only to 96,000 by the end of March, suggesting that the 2008 target of 400,000 to 500,000 sales might not be achievable.

By May 2008, France Télécom was also in discussions to launch the 3G iPhone in Poland, Spain and elsewhere, and those discussions rapidly came to fruition (Le Maistre 2008). Given that Apple claimed to have a 'cultural affinity' with Telefónica, analysts were given to wonder why this did not extend to Telefónica Germany. The UK launch took place on 9 November with the 8GB model retailing at £269 ($566) in a market where 3G handsets normally came free with new contracts and a lot more inclusive minutes were normally offered compared to the iPhone package (Global Insight 2007a). However, the handset still did not work on the 3G network because this would impose too great a drain on the battery.

In January, it was revealed that only 190,000 iPhones had been sold during the first two months of release. When subsequently introduced, the 16GB iPhone retailed at £329, but in April 2008 the 8GB version was reduced to £169 if acquired from Telefónica even though Apple itself still retained the original price. The previous failure to subsidize the handset was felt to account for the relatively poor sales (Bradshaw 2008). Telefónica announced that it would be the exclusive distributor for the iPhone in Ireland as from March 2008.

TeliaSonera and Telenor were among the next batch of European operators to engage in talks with a view to launching the iPhone, albeit not until 2008. Their decision was likely to depend on developments over Christmas 2007, and these were such as to persuade TeliaSonera to proceed in its seven European markets towards the end of 2008. Swisscom also expressed an interest while Telefónica was said to be negotiating in Spain, where it duly signing up in June 2008 (Sinclair 2008) only when the 3G version was forthcoming. However, mobilkom Austria rejected the iPhone because of its lack of 3G connectivity, and in February 2008, the CEO of KPN claimed that the iPhone was 'pretty useless' and that its arrival in Germany had not affected sales at KPN subsidiary E-Plus (Steen 2008).

It was rumoured that Vodafone was negotiating the exclusive right to sell a 3G version of the iPhone in certain European markets. Such an arrangement would not affect existing deals involving GPRS (general packet radio service)/EDGE. In May, this became a more concrete proposal to launch the 3G iPhone in 10 countries later in the year (BWCS 2008b). Among these was to be found the first country, Italy, where Telecom Italia was already licensed, to be permitted a second licensee (Lennighan 2008a).

Elsewhere in the world, agreements had yet to be formalized, and the situation was less clear-cut. For example, the regulator in the UAE announced in October that it would not allow Apple to establish any exclusive network agreements – unlocked iPhones were selling for up to $1,100 on the black market (Global Insight 2007b). Meanwhile, in Australia, Telstra appeared to be the likely partner for the iPhone, although its HSDPA network operated in the 850 MHz band which meant that a new version of the iPhone would be required (Dow Jones 2008c). An Australian operator, in this case SingTel Optus, together with SingTel itself, was collaborating with Advanced Info Services to agree a launch in Thailand where 60,000 iPhones had allegedly been smuggled in by January 2008 (PriMetrica 2008b), but it faced competition from TrueMove which launched in January 2009. Meanwhile, in China, both China Mobile and China Unicom expressed a general interest, but with the operators less than keen on a revenue-sharing agreement and the typical handset selling at $135 in China, there did not seem to be much of a market for an expensive iPhone even setting aside the problem of creating a Chinese-language interface. Hence it was no surprise when discussions with China Mobile were terminated in January 2008, although unofficial discussions continued until February 2009 when China Mobile balked at Apple's insistence on selling software direct to customers (Cellular-news 2009b). Armed with a new W-CDMA licence, China Unicom then struck a non-exclusive deal to market the iPhone which was signed in March 2009 (Telecoms.com 2009a) and the launch (using handsets lacking Wi-Fi, which grey-market models came equipped with, and more cheaply) took place at the end of October (McDonald 2009).

Talks with DoCoMo in Japan were also under way, although nothing was expected to transpire until 2008 after a 3G version of the iPhone became available. Whether DoCoMo would accept a revenue-sharing model, and on what terms, remained a matter of conjecture, but in any event the sophisticated Japanese subscriber was unlikely to want a one-size-fits-all handset plus services package (Phromchanya 2008). Meanwhile, Vodafone Essar was expected to launch in India in September with the 8GB handset priced around $680 – the 16GB handset would follow in 2009.

In North America, Rogers Communications became an early licensee in Canada. In Latin America, América Móvil took up the running with a proposal to introduce the 3G iPhone across its numerous networks towards the end of 2008, immediately followed by Singapore Telecom, Globe (Philippines), Optus (Australia) and Bharti AirTel (India), indicating further permissions for multiple licences in the latter two countries (Telecoms.com 2008a). This development appeared to reflect the fact that the marketing of the iPhone through a single operator and on a contract basis made little sense in countries where pre-paid was the norm. Hence, it was no great surprise when Softbank was licensed in Japan even though DoCoMo was still reluctant to sign up, and multiple licences were anticipated in Hong Kong and Macau after Hutchison agreed to proceed, as well as in Russia where all three national operators agreed contracts in August. Multiple licensing effectively became established practice when Telefónica signed up in June 2008 to launch the 3G iPhone in the Czech Republic and 12 Latin American countries towards the year end. Nevertheless, T-Mobile was unable to launch in the UK and USA, despite signing an agreement covering its other networks, because of existing exclusive arrangements in those countries (Preuschat 2008b).

The Unlocking Issue

Unlocking the iPhone has been an issue ever since the handset was first launched. Not surprisingly, hackers immediately set out to find ways to unlock the iPhone so that it could be used on any network (Charny 2007a). Apple, equally unsurprisingly, attempted to make this as difficult as possible. AT&T, for example, admitted that a 'sizeable' number of iPhones that had initially been sold by Apple were not connected to its network – estimates in January 2008 of the proportion of iPhones sold unlocked ranged between 20 per cent and 30 per cent (Dow Jones 2008a), and this was confirmed in February when AT&T reported that it had sold two million iPhones compared to the 3.7 million released by Apple (Penton Media 2008). Operators generally tried to prevent anyone from acquiring more than a couple of iPhones on the assumption that any larger number would be bought with a view to resale to someone wanting an unlocked handset. It may be noted that the likes of Wirelessimports.com contributed to the emergence of the problem in the USA by offering either an unlocking solution for $179 or the unlocked version of the 8GB iPhone for $599.

There is evidence that 400,000 customers signed up with China Mobile by the end of 2007 using unlocked iPhones (BWCS 2008a). This would indicate that all of these had been shipped from Europe and the USA – hence the disparity in numbers there. The Chinese market is especially desirable for trans-shipments since the iPhone's focus on games, music and the Internet marries well with the Chinese desire to be entertained on their handsets. The Chinese are also typically less price-sensitive.

In Europe there was an added twist since iPhones in Belgium and France had to be sold unlocked (Dow Jones 2007a), and an unlocked version had initially to be made available in Germany. As a result of the particular wording of the legal framework in countries such as these, exclusive sales partnerships were unlikely to be as easily enforceable across Europe as they were in English-speaking countries, the legal systems of which are based upon the common law. Inevitably, the iPhone would lose some of its cachet if it could be obtained from any outlet, although the price would ensure that it retained some exclusivity – assuming the rest of the marketing model stayed in place. However, it had to be questioned why an operator would continue to be willing to hand over a significant share of its revenue from the sale of a handset that was available elsewhere. Equally, it remained unclear whether legal challenges would be forthcoming on the grounds that the very high price of the unlocked model simply constituted a deliberate attempt to penalize those not buying contracts with relatively cheap handsets.

That Apple did not appear to have given any consideration to the above indicated that it was not quite as smart as it was purported to be. However, it was Orange that was left with egg on its face when, in December 2008, the Competition Council in France struck down, with immediate effect, Orange's exclusive arrangement with Apple. The Council argued that the five-year exclusivity was far too long in the context of the relatively uncompetitive French mobile market (containing only three licensees) – and in that respect left the position rather ambiguous with respect to whether or not a precedent was being set that would subsequently be adhered to in Germany, Spain, the UK and the USA. According to the Council, Orange had only invested €17 million to launch the iPhone but had achieved sales worth €222 million (Hall 2008). Orange immediately appealed, only to lose its case before the Paris Court of Appeals in January. A further appeal to the Cour de Cassation was scheduled to come next (Dow Jones 2009b), but SFR was quick to respond with a launch pencilled in for April.

The iPhone Needs to Evolve

Apple shareholders had cause to celebrate at Christmas 2007 because, on 26 December, the share price rose above $200 for the first time (compared to a year-low of $76). They were less happy at a report early in the New Year which involved the Chinese-made X-SIM II (or Turbo SIM), a device that combined with a normal SIM card so as to bypass the normal identification process involved in opening a handset (Cellular-news 2008a). Because no actual unlocking was required, not merely could an iPhone be used on any network, but it would be highly problematic for Apple to claim that warranties no longer applied, and there would not be any issues with software upgrades. They were also less than enamoured to learn that two million HTC Touch handsets had been sold – roughly half as many as the iPhone over the same period. But it was, curiously, Apple's normal practice of conservative forecasts, which it then comfortably exceeded, that cost it dearly in late January 2008, because in the volatile equity markets at that time investors chose to take it at its (pessimistic) word and the share price plunged below $160 (Allison 2008a). In any event, scepticism was rapidly building concerning Apple's prospects of selling 10 million iPhones before the end of the year (Charny 2008a).

A critical issue was that the iPhone was still largely the preserve of young tech-savvy owners – known as iCultists (Shrimsley 2007) – who had previously used Apple products, but who nevertheless had reservations about iPhone features such as the screen size. Getting mainstream buyers to adopt the iPhone was a challenge yet to be fully addressed. In February 2008, Apple announced that it would be unveiling its Software Development Kit (SDK) in early March (Cellular-news 2008b). Previously, software developers could only write Web-based applications, but henceforth they would be able to build native applications. However, Web-based videos for the iPhone had to be specially formatted to a file type that Apple endorsed, and since only YouTube had so far been willing to do this, only a small number of videos were available. Elsewhere, Adobe's Flash Player was in common usage, but Apple claimed that it was not good enough – it was optimized for laptops and hence was allegedly not fast enough for handhelds (Charny 2008d), although some considered this to be disingenuous (Charny 2008b). The new Flash Lite released in February 2009 also appeared to be incompatible with the iPhone (Charny 2009).

Also in February, Apple added to the iPhone range by offering a 16GB version for $499. On the whole, however, a 3G-compatible version would have been preferred by potential subscribers, and it was assumed that this would shortly be unveiled – the CEO of AT&T had let slip in November 2007 that this would be forthcoming during 2008, but not precisely when. However, there were those who claimed that Wi-Fi and EDGE were perfectly satisfactory and that 3G would serve to reduce battery life and raise the cost of the handset (Lennighan 2007a).

The SDK and Open-Source Applications

The imminent arrival of the SDK focused minds. Many developers had already successfully written and published downloadable software for the iPhone despite Apple's attempts to thwart them. They tended to expect Apple to seize control of the price of applications (which would be delivered via the iTunes online store) and the size of the revenue share, and to require pre-approval of their programmes. However, if Apple failed to permit open-sourcing, this could potentially trigger a mass exodus of frustrated developers into the arms of Google's open-source Android platform (Charny 2008c).

In the event, Apple released what it claimed was the same set of developer tools used by its own engineers, although it intended to take a 30 per cent cut of money earned from downloads via the iTunes store. It also unveiled a set of enterprise capabilities such as Microsoft Exchange (*Financial Times* 2008a).

The 3G iPhone

At the end of April 2008, it was reported that Apple had placed an order with its Asian suppliers to provide 200,000 3G handsets by the end of May and 500,000 a week during June. However, this was not altogether good news for European licensees – despite their urgent desire to have the 3G version – because most of them were sitting on large unsold stocks of the 2G version (Cellular-news 2008c).

In the USA, the worst collateral damage directly attributable to the iPhone was, predictably, suffered by domestic vendor Motorola. Motorola had previously basked in the success of its best-selling RAZR, but significant numbers of RAZR owners were upgrading to the iPhone (Lennighan 2008b). Sales of the BlackBerry also appeared to have been quite badly affected, although it was no small matter that operators were subsidizing the BlackBerry to the tune of $150, less than half of AT&T's subsidy per iPhone (Weinberg 2008). In October 2008, Steve Jobs boasted that the 3G iPhone had outsold the BlackBerry during 2008Q3, although it was noted that whereas it was true that the 6.9 million iPhones sold during the quarter did indeed exceed the 6.1 million BlackBerry handsets sold during 2008Q2, this was hardly a fair comparison pending RIM's figures for 2008Q3.

So what was reality, and what hype? In March 2008, Steve Jobs introduced the 3G iPhone as potentially 'the best mobile device ever in enterprise', given its push email capability, advanced calendar programmes and synchronization capabilities with Microsoft products (Telecoms.com 2008b). But whereas businessmen certainly loved the iPhone for personal use, they cavilled at its lack of a physical keyboard given the need for high-volume texting and e-mailing (Hay 2008). However, the crucial factor was that businesses themselves saw important advantages in the way that RIM handled its relationship with its business customers, allowing the latter in effect to personalize their BlackBerrys to suit the specific needs of their workforces. Hence it was evident that there was no mass desertion going on from BlackBerrys to iPhones in the business world. The mass desertion, if that was an apt phrase – was from handsets, such as the RAZR, primarily intended for personal use and this in turn reflected in no small way the change in the iPhone pricing model.

IPHONE PRICING

Some reference has already been made above to the fact that the price of the original 2G iPhone was cut significantly as competition increased and, more particularly, in the face of rumours that a 3G version was soon to appear. However, the approach to the pricing of the 3G version was quite different, mainly due to recognition on Apple's part that the marketing strategy used for the 2G version was not sustainable. Sales had not been as good as had been hoped, and this could be attributed in good part to the attempt by Apple to overturn the traditional industry practice of handset subsidization. To expand sales aggressively, iPhone prices would have to be reduced. Since the iPhone was expensive to manufacture, that meant subsidies were now to be the order of the day (*Financial Times* 2008b).

For example, T-Mobile announced in June 2008 that in Germany, where it remained the exclusive agent, it would be selling the new 8GB version for between €1 and €170 and the new 16GB version for between €20 and €250, depending on the rate plan (which still required a two-year contract). Meanwhile, in the USA, AT&T would be subsidizing the 8GB handset down to $200 and the 16GB handset down to $300. In July, Telefónica also insisted that a two-year contract had to be signed in Spain, with subscribers required to agree to a flat-fee data plan costing €15–25 a month in addition to a minimum voice plan costing €9–90 a month. However, anyone signing up for an 8GB iPhone, a €15 data plan and a €9 voice plan, for example, would pay only €299 for the handset. In the UK, Telefónica went so far as to offer the iPhone free with its most expensive packages.

It is worth noting, however, that this strategic switch would prove damaging to operators' bottom lines for some time to come. For example, despite extolling the benefits of the iPhone in terms of much higher net income (Cellular-news 2008f), AT&T predicted that earnings per share would be reduced by between 10 and 12 cents during both 2008 and 2009, while Optus reported a 'drag on margins' (McFarland 2008). Given the calculation that AT&T had been paying Apple on average more than $600 per 2G iPhone, this was hardly surprising. An updated calculation by iSuppli put the manufacturing cost per unit of a 3G iPhone at only $173[2] – significantly less than for the 2G version despite its enhanced features. On the basis that the average price being paid to Apple was $525 – it is interesting in this respect that AT&T launched a pre-paid service where the 8GB version retailed at $599 and the 16GB version at $699 – this meant that Apple was making a profit of up to $352 per iPhone. On the whole, this seemed to indicate that the balance of power was in its favour *vis-à-vis* AT&T (Charny 2008e) although AT&T, in its turn, was able to attract large numbers of churning customers away from other operators by virtue of its monopoly control over the iPhone in the USA.

OTHER FACTORS

Naturally, reduced prices were not the only factor explaining the relative success of the 3G iPhone. It should be borne in mind, first of all, that the reduction had two significant side-effects, namely a much reduced incentive to sell unlocked iPhones on the black market and much improved prospects for sales in developing markets. So far as Europe was concerned, the simple fact of being able to access the pervasive high-speed data networks was arguably the most crucial factor as all competing devices could do so.

But what precisely were new subscribers getting? In essence, they were getting the 2G iPhone with added benefits in the form of cheaper prices, faster downloads and extra applications (Hutchinson 2008). Not everyone was convinced that this would suffice. In the first place, markets dominated by pre-pay, such as Italy, China and India, would be only modestly affected by the switch to a subsidization model for contract customers. Secondly, the customary two-year contract was much longer than the norm of one year

2 The equivalent cost for the 16 GB 3G S was $179 (Cellular-news 2009j).

for competing devices and, thirdly, competition would rise sharply as the likes of Nokia launched increasing numbers of premium touch-screen handsets.

The likes of Charny and Cheng (2008) and Global Insight (2008a) were not persuaded. Charny and Cheng pointed out that price cuts were unlikely to prove a market driver for long, as evidenced by the history of the RAZR, which started out as a high-end device but ended up being given away free to contract customers, and that the 3G iPhone could suffer the same fate given that it did not appear to be technologically much superior to its 2G predecessor. Global Insight, for its part, argued that the iPhone needed to adopt the BlackBerry model, which used its email capabilities as a unique service to attract customers. But what unique service could the iPhone provide?

Perhaps it did not need to, was the implicit reply contained in both initial sales data and the 2008Q3 shipment data for smartphones. When it released its 2G iPhone, Apple had to wait 74 days to reach sales of one million units, whereas it sold the same number of 3G iPhones in just three days – albeit in 21 countries rather than only in the USA, mostly without exclusive deals and at rather cheaper prices, so the numbers were not directly comparable (Wood 2008).

Historically, smartphones did not constitute a particularly big market segment, and hence had largely been left in the capable hands of Nokia, RIM and Motorola. But the widespread introduction of high-speed data communications had made smartphones relatively attractive, especially given the comparably high prices being charged for high-end consumer devices. Hence, although Apple's 6.9 million shipments in 2008Q3 were indeed of significance, representing as they did a 17.3 per cent market share compared to its 3.6 per cent market share one year previously, the year-on-year growth of 27.9 per cent in the overall market to 39.9 million units offered protection to all participants. As it transpired, Steve Job's boast (see above) had turned out to be true, since RIM sold only 6.1 million BlackBerrys during the quarter. However, because this represented a year-on-year growth of 83 per cent – the BlackBerry Bold had come late to market and the Storm was about to arrive – it raised a question mark over whether the iPhone would sell better than BlackBerrys during 2008Q4, even if the iPhone beat RIM into second place in the J. D. Power and Associates customer satisfaction report (Cellular-news 2008g).

2008 in Retrospect

No handset can be all things to all men, and especially not when first launched. But it can work near perfectly, as in the case of the 2G iPhone. On the whole, the 3G iPhone was very well received – see, for example, 3G.co.uk (2008b) – although certain shortcomings were identified. That the camera was only two mega-pixel, as was that in the 2G version, was something of a surprise given that this was at the bottom end of the range for high-end handsets, and the virtual keyboard was not considered to be as reliable as a mechanical keyboard. The handset was also prone to overheating. However, the main cause for complaint was the tendency for signal strength to fluctuate, leading to dropped calls and lengthy download times (Allison 2008b). Apple rapidly issued a patch and sought to place the blame on faulty chips manufactured by Infineon Technologies, but this being the USA the court cases rapidly began to mount up, commencing with a claim for damages arising from Apple's claim that the 3G iPhone could download data twice as fast as its predecessor (Charny 2008f).

Overall, 2008 was not particularly kind to Apple, although clearly it would be misguided to correlate its share price exclusively with the performance of the iPhone – it also suffered, *inter alia*, from declining prices for the iPod. In November 2008, the 12-month share price drop amounted to 40 per cent and the six-month price drop amounted to 47 per cent. However, it is of interest that, from a high of $200 around the turn of the year, the share price had dropped sharply to $120 before rallying again to hit $180 in May. It again hit $180 in mid-August, after which it fell consistently to its low point of $85 on 10 October, subsequently struggling to recover.

It is easy enough to ascribe the October 2008 situation to the so-called 'credit crunch', and to point out that it has also affected Apple's competitors. Furthermore, despite proclaiming its total inability to forecast more than a week ahead at a time, Apple claimed to be armed with the strongest product line in its history, not to mention no debts and '$25 billion of cash safely [sic] in the bank' (Allison 2008c). But the credit crunch was bound to unleash cut-throat competition among handset vendors, so a particularly acute question was whether Apple could meet its 10 million iPhone sales target by the year end. If it failed, then this was probably due in part to the Android G1 handset – the build cost of which was estimated by iSuppli to be $144, somewhat less than the 3G iPhone (Cellular-news 2008h) – for which 1.5 million pre-orders were placed with T-Mobile USA by mid-October, a figure easily comparable to 3G iPhone sales.

All Change in 2009

2009 was inevitably going to bring with it a surge in the launch of competitive devices. As noted above, for example, the Android G2 handset, with a 1.5 software upgrade (known as the Donut) on the G1, was scheduled to launch in mid-2009. It did not have a physical keyboard, unlike the G1, but retained the trackball and was HSDPA-enabled. Initially, its main advocates were T-Mobile, which followed the UK launch of the G1 with launches in Austria, the Czech Republic and the Netherlands in January 2009 and in Germany and Poland in February. In practice, a variant of the G2, branded as myTouch, was first launched by T-Mobile USA in July for $200 in conjunction with a two-year contract. The G2 was also popular with Vodafone which used the Magic brand, first launched in the UK in April 2009 (Android GII 2009), and it was expected to sell well on the grey market in China as had the G1. The Android-based HTC Hero followed.

In the USA, in particular, the BlackBerry remained an ongoing threat to the iPhone. As a result of a buy-one-get-one-free promotion by Verizon Wireless, the Curve became the best-selling smartphone in the USA in 2009Q1 with the Storm and Pearl also doing well at the expense of the iPhone and Android G1 (Telecompaper 2009b). A quad-band 3G model, the Tour, was forthcoming. For its part, Sprint Nextel pinned its colours to the Palm Prē, launched over its 3G network on 6 June for $200 (after a $100 mail-in rebate) with a two-year contract and a $70 to $90 a month messaging and data plan. At least 50,000 Prēs were sold during its first weekend, and the five-fold increase in the share price of Palm commencing in November 2008 was testimony to the belief that it would be a serious challenger to the iPhone (Cheng 2009).

Both the date and price were fairly crucial elements of the package because of the imminent launch of the new version of the iPhone. It had increasingly become clear that Apple would need to adapt its iPhone marketing strategy as it went along. It had, for example, announced at Christmas 2008 that the 3G iPhone would be made available at Walmart's 2,500 stores, with the 8GB version retailing at $197 and the 16GB version at $297 in conjunction with a relevant two-year AT&T contract. The odd features of this arrangement were, first, the juxtaposition of a high-end brand with a discount retailer and, secondly, the fact that the discount was a mere $2 or roughly 1 per cent. With the Best Buy consumer electronics chain (as well as AT&T's own outlets) already selling iPhones, Apple's strategy appeared to be to move during 2009 to establish market share, although it was highly debatable whether the typical Walmart customer could afford a $70 a month contract.

But improvements to the marketing strategy were never going to suffice, and Apple's strategy contained three other elements. In the first place, Apple set out to retain the iPhone's uniqueness by filing for protective patents. For example, in January 2009, Patent 7,479,949 awarded to 'Steve Jobs et al' a patent for a method of 'detecting one or more finger contacts with the touch screen display' to command computing devices (Dow Jones 2009a).

However, the two crucial elements in the battle plan concerned hardware and software, or alternatively, devices and applications. Priced at $199 for the 16GB version and $299 for the new 32GB version in conjunction with a two-year AT&T contract, the third version of the iPhone, called the 3G S (or 3GS where 'S' stands for 'speed'), certainly started off on the right foot with AT&T reporting that all pre-orders could be picked up on 19 June, but that there would be no further handsets available for two weeks (BWCS 2009c). However, it was generally agreed that the 3G S was not significantly different from the 3G – basically it was faster and had more memory and an improved camera capable of handling video (Telecompaper 2009c).

Android fought back with its version 2.0 OS (known as the Éclair). Somewhat surprisingly, it was Motorola that launched the first handset based on Android 2.0 – known as the Droid in the USA and the Milestone in Europe. A follow-up to the social network oriented Cliq/Dext (Metz 2009), and marketed by Verizon Wireless in November, the Droid – the slimmest-ever handset with a QWERTY keyboard – specialized in location services, arguably the most trendy application for smartphones at that time (Gabriel 2009f). Twice as fast as the original Android handsets, the Droid appeared to present a real challenge to the iPhone, especially given its launch over the superior Verizon network in the USA together with applications downloaded from Android Market (Telephonyonline 2009).

But the iPhone prices certainly appeared to be attractive, and the 3G version could now be picked up for a mere $99 if so desired – provided you were a new subscriber. However, much to the annoyance of loyal subscribers, an upgrade from the 3G to the 3G S would cost $200 (Huffman 2009) although this was subsequently made more elastic. Furthermore, an unlimited voice, texting and data package worked out at roughly $150 a month for the iPhone compared to Sprint Nextel's 'Simply Everything' package for the Prē at $100 (Kane 2009).

The other crucial battle was being fought over the range of applications being made available on different platforms. The iPhone Mac OS 3.0 beta release in March 2009 included an updated SDK with more than 1,000 new application programming interfaces (APIs), and by the end of April some 35,000 applications were available and had been downloaded one billion times in total, a doubling or more in both cases compared to mid-January (Wireless Week 2009). By mid-July, 65,000 applications were available and more than 1.5 billon downloads had been racked up (Middleton 2009c) – these figures had risen to 85,000 and two billion respectively by end-September (Reedy 2009a). Both Google, with the Android handsets, and Palm with the Prē, were only too well aware that they needed to provide sufficient attractive applications to prevent the iPhone winning the war hands-down. One interesting side-effect was that network operators such as AT&T were obliged to upgrade download speeds to 7.2 Mbps in order to cope with the tidal flow of applications (BWCS 2009b).

The Palm Prē (and its successor, the Pixi) uses a Linux-based operating system called the WebOS which is claimed to be superior to that of the iPhone (Deagon 2009). A particular feature is the ability to keep multiple applications open at the same time (Svensson 2009). However, the range of applications available via the App Catalog (numbering only a few hundred) cannot compete with the Apple SDK – which is arguably why only Sprint Nextel has so far backed the WebOS – and as yet the Google Android Market (which appeared as version 1.6 in September 2009 and claimed to contain more than 10,000 applications in October) and RIM's BlackBerry App World have made little impression (Middleton 2009b). For its part, Microsoft provides the Windows Mobile (WinMo) OS for smartphones, and in May 2009 it added the My Phone beta platform to its range, initially without charge and including 200GB of free storage. The strategy was to launch its own applications store in 2009 with an initial portfolio of 20,000 applications (Middleton 2009a). Windows Marketplace for Mobile (aka Skymarket) was scheduled to accompany the launch of WinMo 6.5 (Gabriel 2009b), but the latter's appearance in October was generally seen as an attempt to hold the fort until an improved version, WinMo 7.0, with comparable features to the iPhone, could be marketed in 2010 (Waters 2009).

Interestingly, the issue as to whether applications are the primary responsibility of hardware provider, OS provider or network provider or any combination thereof remains in the balance. Nokia evidently believes that it can leverage its leading share of the device market by launching its Ovi Store which even AT&T in the USA is happy to make available to its Nokia-owning subscribers (Telecompaper 2009a) despite having its own Media Mall.[3] The Ovi store launched with 1,000 applications and is currently adding more than 100 each day (Gabriel 2009c); it is due to be rebuilt from scratch during 2010. For its part, LG Electronics announced the launch of its Applications Store with 1,400 applications in 15 languages in July 2009 (Wood 2009a), while Orange launched its App Shop in December 2009. In contrast, Samsung (which already had an Application Store) appeared to have slipped up in leaving the launch of its bada ('Ocean' in Korea) platform, essentially an adaptation of an existing proprietary OS (Telecoms.com 2009b) until 2010H1 – was there really the need for yet another platform (Parker 2009)? In contrast, Sun Microsystems believes that since its Java is installed in two million devices, it should do well by providing applications via its Java Store to owners of handsets with JavaFX Mobile installed.

---

3 The Ovi Store is the *de facto* store for Symbian applications given Nokia's control of Symbian and the reluctance of the Symbian Foundation to compete with its members to avoid fragmentation (Middleton 2009e). Fragmentation has also become an issue with Android since each operator is producing a slightly different version in its handsets, and apps developers cannot afford to test on every one (Wood 2009c).

Conclusion

During the first three days of its launch, over three million of the iPhone 3G S handsets were sold, and a further six million subscribers downloaded the new 3.0 software, causing Steve Jobs to declare that 'Customers are voting and the iPhone is winning' (Cellular-news.com 2009i). Google, Nokia, Palm, RIM and a good many other parties sincerely hope he is wrong. By July 2009, the primary struggle for ascendancy appeared to be between the iPhone, Palm Prē and Nokia's N97 (Taylor 2009), but with new touch-screen smartphones emerging rapidly, it was likely that the iPhone's main rivals would be in a constant state of flux.[4]

Clearly, Apple was on something of a roll at the end of 2009.[5] It benefits from a symbiotic relationship between hardware and software that no other company can match. The applications do not generate huge sums, but they make a big difference to total sales of handsets (Peers 2009). With its established enterprise customer base, RIM enjoys similar hardware/software benefits, but it is struggling badly on the touch-screen front and recent financial results have been poor (Lex 2009). During 2008, these two companies sold only 3 per cent of all the word's handsets, but they made 35 per cent of the aggregate operating profits (Silver 2009). Apple is expected to improve on both measures during 2009 since it continues to benefit from large subsidies provided by operators that are tying customers to relatively long contracts. For everyone else, the struggle is to acquire or retain the economies of scale that are the alternative route to riches. This is where Nokia has the advantage while other vendors struggle to stay afloat. Nevertheless, it is suffering in the short term and the N97 and N900 may not suffice (Hibberd 2009a).

For operators, being linked to a successful smartphone also has considerable advantages in principle, in terms both of attracting churning customers and enabling the extension of contract times. It was possible initially to sign up for exclusivity with the iPhone, and exclusivity in relation to this and other touch-screen smartphones continues to be possible although it no longer appears to be the preferred model. It was revealed in September 2009 that the standard exclusivity period in the original iPhone deals had been two years. This meant that many of the deals came up for renegotiation towards the end of 2009, and in the case of the UK, for example, both Orange and Vodafone signed up to market the iPhone when the exclusive Telefónica deal expired in September (Lennighan 2009). However, AT&T was expected to renew its exclusive arrangement in June 2010 – the other major operators also have exclusive deals but with competing vendors, all of which the FCC (Federal Communications Commission) is keen to strike down. Exclusivity is a two-edged sword since the current price per handset paid, for example, by AT&T to Apple only makes sense if it results in defections from competing networks and keeps new customers loyal. Its benefits are also much reduced when the great majority of those likely to want an iPhone have already bought one. Its benefits to Apple are also in question since the evidence from France is that the iPhone's market share rose significantly once the exclusive arrangement with Orange was struck down (Gabriel 2009h).

---

4 10 million units of the Samsung Star, for example, were sold during the first six months post-launch, making it the fastest-selling handset in Samsung's history. Altogether, Samsung sold in excess of 40 million touch-screen handsets during 2009 (Cellular-news 2009l).

5 As reflected in its share price which had risen from $90 on 1 January 2009 to $209 on 24 December.

A combination of first-mover advantage in an expanding high-end market combined with tight hardware and software integration underpins Apple's success in breaking into a new market in the face of entrenched and extensive competition. There is good reason to expect Apple to continue to be the front-runner during 2010 and to make disproportionate profits, but competition has been swift to appear and there is everything to play for. What is fairly certain is that competitors cannot simply emulate the Apple model[6] although most are for now doing more or less that. In the medium term, Google, for one, purports to believe that the application-store model will be overtaken by a browser-based model,[7] but then, perhaps, it would and there is evidence that it does not really believe its own claim (Gabriel 2009d). Meanwhile Vodafone has abandoned its proprietary Live! Portal to form the Joint Innovation Lab (JIL) initiative with China Mobile, Verizon Wireless and Softbank in the hope that this will create a distinctive user experience with huge potential for developers. This is based upon a version of open-source Linux known as LiMo that looked to have been consigned to history by its direct competitor Android until Vodafone stepped in, but with Motorola concentrating on Android, Samsung is left as the only significant vendor of LiMo-based devices (Gabriel 2009e).

Forecasting the touch-screen market beyond 2010 is very difficult because it involves new relationships between vendors and operators with the ever-expanding applications market as the glue binding them together. These relationships, in turn, are increasingly being affected by the roll-out of high-speed networks, up to and including HSPA+. The iPhone itself is no longer the 'must have' handset that it was in 2007–08, at least in its existing markets, and now that evidence is leaking out that operators have not necessarily done all that well financially from their initial contracts with Apple, there is likely to be a concerted effort to market rival handsets (Lagorce 2009). Indeed, if all concerned read a recent 'myth-busting' report by Strand Consult (2009) and accept its analysis, there will be a good many executives at operators who will be anxious to put their iPhone experiences behind them – the report is blunt in referring (at page 101) to operators handling the iPhone as 'naïve'. What the report is particularly keen to demonstrate is that sales of the 3G iPhone were much better than those of the original model simply because it was sold in a lot more countries, and that this phenomenon could not be repeated on anything like the same scale with the 3G S. It also contends, *inter alia*, that the iPhone has not helped operators to acquire more customers compared to the rest of the market, possibly because they took their eye off the ball in their other markets, and that operators have found it much easier to sell the iPhone to existing customers than to use it as a weapon to attract new and churning customers. There is a great deal of statistical data to back up these contentions, and there can be no doubt that, overall, the iPhone has been over-hyped in the media. At the end of the day, the prospects of an innovative product such as the iPhone in an increasing high-speed world will be limited unless it can continue to evolve in the face of ever-increasing competition.

Which leads on to what is probably the core issue for 2010 and beyond, namely open-source or not? The first point to make is that, in terms of the USA market, the

6 In the litigious USA, emulation is never a good idea. Apple, for example, has never been backward in coming forward to protect its patents, as in the threatened action against the Palm Prē in relation to its multi-touch interface. But litigation is a two-way street, and in October 2009, for example, Nokia launched a legal challenge relating to an infringement of 10 patents in respect of the iPhone (Ward and Gelles 2009).

7 The debate between adherents of embedded applications and browser-based applications was the key issue addressed at the Open Mobile Summit in San Francisco in November 2009 – see Fitchard (2009).

two most financially successful smartphone vendors are Apple and RIM. Their success is largely ascribed to their tight control over their operations – viewed by some as a form of dictatorship, but defended by the vendors as benign. Yet they rely upon outside developers for most of their applications, even if they ultimately decide whether those applications are officially sanctioned, so their systems are not altogether closed. Furthermore, the Internet itself would have developed much more slowly had it not been open to anyone to put forward innovations.

The Symbian Foundation, among other bodies, has decided to go in the opposite direction, promoting openness – but not full openness. One difficulty with full openness is that all ideas must be considered to be equal, yet it is evident that some peoples' contributions ultimately need to be given priority. Openness is most readily applied to applications, but in principle can apply to platforms. The more people work on improving platforms, the faster the upgrades will appear (Hibberd 2009b).

If this argument is sustainable, then Apple and RIM[8] have probably reached the high-point of their influence, and while they will remain substantive players in the ever-growing smartphone market, the more open the platform the more influential it will become in the medium term. However, one thing is certain, namely that the proliferation of platforms and applications is a classic illustration of excess competition. Sooner or later, the number of winners must shrink to match the available economies of scale. Unlike in previous examples, such as those discussed in Chapter 5, it is highly probable that the winners on this occasion will indeed win on merit. Surprisingly, this spells rather bad news for Microsoft since Windows Mobile is neither open-source nor can it get away with a proprietary platform like Apple – rather it charges fees. HTC, for one, has recently switched in a big way from Windows Mobile to Android, and Sony Ericsson, which had ignored Android 1.5, announced that it would be bringing out the Xperia X3 (aka Rachael) based on Android 2.0 in 2010Q1. But Google has its own problems with Android in that Android is marketed by vendors rather than Google itself despite Google's strong brand image.

As to the nature of the devices, there remains much to play for. The first clear conclusion is that smartphones will increasingly displace basic handsets, if only because they will increasingly come free with contracts. Touch-screen smartphones are very much in the news and appear to constitute the most desirable form of device,[9] but they are mostly high-end and expensive for now[10] and, perhaps surprisingly, consumer surveys do not indicate that a touch-screen facility is a priority. The addition of a QWERTY keyboard is another point of interest. Apple, of course, eschewed a keyboard whereas RIM retained one because its devices were aimed primarily at business users.

Interestingly, Nokia has attempted to find a way around this issue via the N900 launched in August 2009. The N900 is intended to be more like a mini-computer than a handset, and is based on Maemo, the Linux (open-source) OS to be used in all future Nokia high-end devices targeted at users who want computer features such as a full browser and storage (1GB on the N900) on a small device. Meanwhile Symbian (Symbian

8 Although RIM argues to the contrary (Nuttall 2009).

9 In early November 2009, for example, the number of touch-screen handsets in the USA stood at 24 million, an annual growth rate of 160 per cent. The iPhone accounted for one-third of all sales. In contrast, the number of smartphones as a whole had grown only 60 per cent.

10 However, models such as the Samsung Star are categorized as mid-range, so it is evident that a touch-screen facility is by no means confined to high-end handsets.

2009), the core open-source Nokia platform, will be retained for high-end phone-centric devices (Gabriel 2009g). This segmentation is designed to avoid cannibalization, but it does add yet another OS to the existing oversupply, so the response to the N900 will be of much interest (Morris 2009). This is also because the N900, like the N97, has a slide-out QWERTY keyboard, and there will always be those who claim that the absence of a virtual keyboard is a sign of an inadequate interface (Beaumont 2009). However, the N900's ability to multitask will give Apple's engineers some sleepless nights.

Finally, it is worth asking whether there is such a thing as a 'killer app'. History suggests strongly that, in practice, killer apps come and go without leaving much of a mark. As indicated, mapping/navigation – that is, being able to identify where you are and the full range of services in the vicinity, not to mention the ability to be directed to your destination – is currently a top-selling application.[11] Money transfer is also highly popular (3G.co.uk 2009c). However, the Sony Ericsson X10, to be launched in 2010Q1, features an application called Sony Ericsson Timescape which integrates status updates on Facebook and Twitter in one place, alongside email and text messages, and it will be no surprise if social networking applications, which are also available on the likes of the Samsung Genio and Motorola Dext (Reedy 2009b), dominate 2010.

References

Allison, K. 2007a. Apple takes risks in bid to shake up the mobile market. 25 June. Available at http://www.ft.com (accessed 26 June).

-----2007b. iPhone costs hit Apple shares. 26 June. Available at http://www.ft.com (accessed 27 June).

----- 2008a. Apple plunges as outlook disappoints. 22 January. Available at http://www.ft.com (accessed 23 January 2008).

----- 2008b. Signal gripes mount over new iPhone. 17 August. Available at http://www.ft.com (accessed 18 August 2008).

----- 2008c. Apple cautious in spite of profit rise. 21 October. Available at http://www.ft.com (accessed 22 October 2008).

Android GII. 2009. HTC Magic Vodafone Android phone revealed. 23 June. Available at http://www.androidg2.org (accessed 24 June 2009).

Associated Press. 2007. Apple to launch iPhone in UK on Nov. 9. 18 September. Available at http://communicationsdirectnews.com (accessed 19 September 2007).

Beaumont, C. 2009. Nokia N900: A computer in every pocket. 4 November. Available at http://www.telegraph.co.uk (accessed 5 November 2009).

Berg, A. 2009. Will Blackberry users take Pre in the Palms of their hands? 18 June. Available at http://www.wirelessweeek.com (accessed 19 June 2009).

Bradshaw, T. 2008. O2, Carphone Warehouse, slash iPhone service. 15 April. Available at http://www.ft.com (accessed 16 April 2008).

11 This development, inevitably, is not without its controversies. Google Maps Navigation is loaded free of charge on the Motorola Droid as a feature designed to differentiate the handset from other Android-based models – a precursor of fragmentation even given a shared platform? But third-party developers charge for their navigation applications, and some are available on other Android-based handsets sold by the Droid's own networks (such as Verizon Wireless).

BWCS. 2007. Unlocked iPhones go on sale in Germany, as Vodafone linked to bigger prize. 22 November. Available at http://www.bwcs.com (accessed 23 November 2007).

----- 2008a. Unlocked iPhones flood China. 18 February. Available at http://www.bwcs.com (accessed 19 February 2008).

----- 2008b. Vodafone to sell iPhone around the world. 6 May. Available at http://www.bwcs.com (accessed 7 May 2008).

----- 2008c. Diamond aims to cut up Apple. 7 May. Available at http://www.bwcs.com (accessed 8 May 2008).

----- 2009a. Google chooses Vodafone for G2. 18 February. Available at http://www.bwcs.com (accessed 19 February 2009).

----- 2009b. AT&T doubling download speed as iPhone users complain. 21 April. Available at http://www.bwcs.com (accessed 22 April 2009).

----- 2009c. Customers keen on new Apples. 16 June. Available at http://www.bwcs.com (accessed 17 June 2009).

Cellular-news. 2007a. Are there any iPhone alternatives? Here are some rivals. 25 June. Available at http://www.cellular-news.com (accessed 26 June 2007).

----- 2007b. How big a threat is the iPhone to the Blackberry? 1 July. Available at http://www.cellular-news.com (accessed 2 July 2007).

----- 2007c. Apple issues battery replacement program for iPhone. 10 July. Available at http://www.cellular-news.com (accessed 11 July 2007).

----- 2008a. SIM adaptor forces the iPhone to skip SIM lock verification. 5 January. Available at http://www.cellular-news.com (accessed 6 January 2008).

----- 2008b. Apple announces updates to the iPhone. 16 January. Available at http://www.cellular-news.com (accessed 17 January 2008).

----- 2008c. European operators facing losses on iPhone sales. 30 April. Available at http://www.cellular-news.com (accessed 2 May 2008).

----- 2008d. LG shows off 8-megapixel touch-screen phone. 4 September. Available at http://www.cellular-news.com (accessed 5 September 2008).

----- 2008e. Nokia launches upgraded S60 smartphone OS. 2 October. Available at http://www.cellular-news.com (accessed 3 October 2008).

----- 2008f. AT&T hit by high iPhone subsidy costs. 22 October. Available at http://www.cellular-news.com (accessed 23 October 2008).

----- 2008g. Apple ranks highest in business wireless smartphone customer satisfaction. 7 November. Available at: http://www.cellular-news.com (accessed 8 November 2008).

----- 2008h. Report reveals manufacturing cost of T-Mobile's Google phone. 11 November. Available at http://www.cellular-news.com (accessed 12 November 2008).

----- 2009a. LG previews new touch-screen phone. 8 February. Available at http://www.cellular-news.com (accessed 9 February 2009).

-----. 2009b. China Mobile iPhone talks break down – again. 10 February. Available at http://www.cellular-news.com (accessed 11 February 2009).

----- 2009c. LG claims 2 million touch-screen phone sales. 10 April. Available at http://www.cellular-news.com (accessed 11 April 2009).

----- 2009d. Samsung shows off its first Android phone. 27 April. Available at http://www.cellular-news.com (accessed 3 May 2009).

----- 2009e. T-Mobile UK – G1 sales to hit 100,000 as new features launch. 29 April. Available at http://www.cellular-news.com (accessed 3 May 2009).

----- 2009f. Samsung shows off 'fastest ever' smartphone. 16 June. Available at http://www.cellular-news.com (accessed 17 June 2009).
----- 2009g. LG shows off Windows mobile phone with 3D menus. 16 June. Available at http://www.cellular-news.com (accessed 17 June 2009).
----- 2009h. Huawei shows off Android and Windows Mobile based smartphones. 17 June. Available at http://www.cellular-news.com (accessed 18 June 2009).
----- 2009i. Apple sells over one million iPhone 3GS in three days. 22 June. Available at http://www.cellular-news.com (accessed 23 June 2009).
----- 2009j. Apple's iPhone 3GS costs $178.96 to manufacture. 25 June. Available at http://www.cellular-news.com (accessed 26 June 2009).
----- 2009k. Sony Ericsson shows off updated windows mobile smartphone. 15 September. Available at http://www.cellular-news.com (accessed 16 September 2009).
----- 2009l. Samsung sells 50 millionth full touch screen mobile phone. 1 December. Available at http://www.cellular-news.com (accessed 2 December 2009).
Charny, B. 2007a. Apple braces for unlocked iPhones to hit Europe. 12 November. Available at http://www.cellular-news.com (accessed 13 November 2007).
----- 2007b. Google phone OS seen not competing with iPhone. 15 November. Available at http://www.totaltele.com (accessed 16 November 2007).
----- 2008a. Doubts arise about Apple selling 10m iPhones. 12 February. Available at http://www.totaltele.com (accessed 13 February 2008).
----- 2008b. iPhones have sizzle, but Adobe asks 'where's the Flash?' 20 February. Available at http://www.totaltele.com (accessed 21 February 2008).
----- 2008c. Apple seeks control of iPhone software. 20 February. Available at http://www.totaltele.com (accessed 21 February 2008).
----- 2008d. Adobe's Flash Player not suited for iPhone, Jobs says. 5 March. Available at http://www.totaltele.com (accessed 6 March 2008).
----- 2008e. New iPhone rules: More Apple revenue, fewer unlocked phones. 3 July. Available at http://www.totaltele.com (accessed 4 July 2008).
----- 2008f. Apple faces suit over iPhone glitches. 22 August. Available at http://www.totaltele.com (accessed 23 August 2008).
----- 2009. Adobe sends mixed signals about Flash Lite for iPhone. 18 February. Available at http://www.totaltele.com (accessed 19 February 2009).
Charny, B. and Cheng, R. 2008. Apple's iPhone must avoid the RAZR trap. 11 June. Available at http://www.totaltele.com (accessed 12 June 2008).
Cheng, R. 2008. iPhone influence apparent in latest wireless handset lineup. 4 April. Available at http://www.totaltele.com (accessed 5 April 2008).
----- 2009. Sprint, Palm face challenges in maintaining Pre momentum. 8 June. Available at http://www.totaltele.com (accessed 9 June 2009).
Deagon, B. 2009. Can Pre save Palm? Some critics rave, but field crowded. 19 May. Available at http://www.cellular-news.com (accessed 20 May 2009).
Dow Jones. 2007a. Orange sees Christmas without iPhone in France – report. 5 October. Available at http://www.totaltele.com (accessed 6 October 2007).
----- 2008a. Apple lost up to $400 million in revenues from unlocked iPhones. 29 January. Available at http://www.cellular-news.com (accessed 1 February 2008).
----- 2008b. Samsung launches Omnia, rival to iPhone. 9 June. Available at http://www.totaltele.com (accessed 10 June 2008).

----- 2008c. Telstra belatedly joins iPhone sellers in Australia. 20 June. Available at http://www.totaltele.com (accessed 21 June 2008).

----- 2008d. LG to release iPhone rival in US. 3 July. Available at http://www.totaltele.com (accessed 4 July 2008).

----- 2009a. Apple wins patent for iPhone touch-screen controls. 27 January. Available at http://www.totaltele.com (accessed 28 January 2009).

----- 2009b. French court upholds ban on iPhone-Orange deal. 4 February. Available at http://www.totaltele.com (accessed 5 February 2009).

Financial Times. 2008a. Apple unveils iPhone grand plan. 9 March. Available at http://www.ft.com (accessed 10 March 2008).

----- 2008b. Apple set to slash iPhone prices to lift sales. 9 June. Available at http://www.ft.com (accessed 10 June 2008).

Fitchard, K. 2009. Debating the future of open mobility. 5 November. Available at http://www.printthis.clickability.com (accessed 6 November 2009).

Gabriel, C. 2009a. DoCoMo and Telefonica in handset pact that could pressurize Vodafone. 20 May. Available at http://www.rethink-wireless.com (accessed 21 May 2009).

----- 2009b. Operators play with open models for mobile apps stores. 1 July. Available at http://www.rethink-wireless.com (accessed 2 July 2009).

----- 2009c. Vodafone drops Live! Brand as European cellcos regroup. 16 July. Available at http://www.rethink-wireless.com (accessed 17 July 2009).

----- 2009d. Verizon and Google form axis against AT&T/Apple – but how open is it? 7 October. Available at http://www.rethink-wireless.com (accessed 8 October 2009).

----- 2009e. Motorola backs away from LiMo to focus on Android. 13 October. Available at http://www.rethink-wireless.com (accessed 14 October).

----- 2009f. Verizon promises biggest ever marketing splash for Motorola Droid. 29 October. Available at http://www.rethink-wireless.com (accessed 30 October 2009).

----- 2009g. Nokia won't demote Symbian, but Linux OSs rule in North America. 19 November. Available at http://www.rethink-wireless.com (accessed 20 November 2009).

----- 2009h. End of exclusives will benefit Apple as it defends store model. 24 November. Available at http://www.rethink-wireless.com (accessed 25 November 2009).

Gallagher, D. 2008. RIM shares hit all-time high on launch of 3G BlackBerry. 13 May. Available at http://www.totaltele.com (accessed 14 May 2008).

Global Insight. 2007a. United Kingdom: dissecting the details of the U.K. iPhone deal. 19 September. Available at http://communicationsdirectnews.com (accessed 20 September 2007).

----- 2007b. Regulator blocks exclusivity deals with U.A.E. network operators. 17 October. Available at: http://communicationsdirectnews.com (accessed 18 October 2007).

----- 2008a. The iPhone 2.0 – cheaper, faster, better; but is it enough? 10 June. Available at http://communicationsdirectnews.com (accessed 11 June 2008).

Googleion. 2009. Google Ion. 25 June. Available at http://googleion.com (accessed 26 June 2009).

Hall, B. 2008. French watchdog cancels iPhone contract. 17 December. Available at http://www.ft.com (accessed 18 December 2008).

Hay, T. 2008. iPhone plays catch-up to BlackBerry in enterprise. 4 November. Available at http://www.totaltele.com (accessed 5 November 2008).

Hibberd, M. 2009a. Nokia lead threatened by smartphone market growth. 12 August. Available at http://www.telecoms.com (accessed 13 August 2009).

Hibberd, M. 2009b. An experiment in openness. 26 October. Available at http://www.telecoms.com (accessed 27 October 2009).

Huffman, M. 2009. iPhone users bristle at AT&T upgrade policy. 9 June. Available at http://www.cellular-news.com (accessed 10 June 2009).

Hutchinson, J. 2008. First impressions: iPhone 3G. 11 July. Available at http://www.infoworld.com (accessed 12 July 2008).

Jones, D. 2007. Peeling back Apple's iPhone revenue. 17 September. Available at http://www.unstrung.com (accessed 18 September 2007).

Kane, Y. 2009. Pricing plans aplenty for new iPhone 3G S in US. 10 June. Available at http://www.totaltele.com (accessed 11 June 2009).

Lagorce, A. 2009. European operators face dilemma with iPhone deals. 10 September. Available at http://www.totaltele.com (accessed 11 September 2009).

Le Maistre, R. 2008. Orange expands its iPhone empire. 16 May. Available at http://www.unstrung.com (accessed 17 May 2008).

Lennighan, M. 2007a. 3G, who needs it? 11 October. Available at http://www.totaltele.com (accessed 12 October).

----- 2008a. Voda announces 10 iPhone deals, no exclusivity in Italy. 6 May. Available at http://www.totaltele.com (accessed 7 May 2008).

----- 2008b. iPhone steals market share from Moto, others. 29 May. Available at http://www.totaltele.com (accessed 3 June 2008).

----- 2009. Vodafone joins U.K. iPhone club. 29 September. Available at http://www.totaltele.com (accessed 30 September 2009).

Lex. 2007. Apple's iPhone. 28 June. Available at http://www.ft.com (accessed 3 July 2007).

Lex. 2009. Off the Rim. 25 September. Available at http://www.ft.com (accessed 26 September 2009).

Malykhina, E. 2007. Verizon Wireless rolls out fresh PDAs and smartphones. 18 October. Available at http://www.informationweek.com (accessed 19 October 2007).

McDonald, J. 2009. iPhone come to China without key feature. 30 October. Available at http://www.cellular-news.com (accessed 1 November 2009).

McFarland, L. 2008. Optus CEO says iPhone to weigh on margins. 12 November. Available at http://www.totaltele.com (accessed 13 November 2008).

Metz, R. 2009. Review: Motorola's Cliq is a snappy smart phone. 21 October. Available at http:/www.cellular-news.com (accessed 22 October 2009).

Middleton, J. 2009a. Microsoft My Phone opens to public. 20 May. Available at http://www.telecoms.com (accessed 21 May 2009).

----- 2009b. Apple takes 12% of mobile apps market. 22 May. Available at http://www.telecoms.com (accessed 23 May 2009).

----- 2009c. iPhone app downloads top 1.5 billion. 14 July. Available at http://www.telecoms.com (accessed 15 July 2009).

----- 2009d. Android gets Tattoo. 8 September. Available at http://www.telecoms.com (accessed 9 September 2009).

----- 2009e. If they build it, will they come? 9 November. Available at http://www.telecoms.com (accessed 10 November 2009).

Morris, A. 2009. Nokia's N900 Memo device starts shipping. 10 November. Available at http://www.totaltele.com (accessed 11 November 2009).

Mossberg, W. 2008. Google answers the iPhone. 16 October. Available at http://www.totaltele.com (accessed 17 October 2008).

Mossberg, W. and Boehret, K. 2007. Testing out the iPhone. 27 June. Available at http://www.totaltele.com (accessed 3 July 2007).

Nuttall, C. 2009. BlackBerry maker takes fight to rivals. 12 November. Available at http://www.ft.com (accessed 13 November).

Parker, A. 2009. Samsung to allow independent 'apps'. 10 November. Available at http://www.ft.com (accessed 11 November 2009).

Parker, A. and Taylor, P. 2008. Challenge to iPhone heats up. 22 September. Available at http://www.ft.com (accessed 23 September 2008).

Peers, M. 2009. iPhones App side for Apple. 13 July. Available at http://www.totaltele.com (accessed 14 July 2009).

Penton Media. 2008. Unlocked iPhone market in shades of gray. 5 February. Available at http://communicationsdirectnews.com (accessed 6 February 2008).

Phromchanya, P. 2008. DoCoMo has not given up on iPhone. 3 September. Available at http://www.totaltele.com (accessed 4 September 2008).

Preuschat, A. 2008a. T-Mobile Germany sold 70,000 iPhone contracts. 28 January. Available at http://www.totaltele.com (accessed 3 February 2008).

----- 2008b. T-Mobile to sell iPhone in most of its markets. 10 June. Available at http://www.totaltele.com (accessed 11 June 2008).

PriMetrica. 2007a. O2 UK agrees 40% revenue sharing deal for iPhone – report. 18 September. Available at http://www.telegeography.com (accessed 19 September 2007).

----- 2007b Has Apple's Orange deal already gone rotten? 1 October. Available at http://www.telegeography.com (accessed 2 October 2007).

----- 2008a. Orange iPhone sales top 70,000 in first month, broadly in line with target. 4 January. Available at http://www.telegeography.com (accessed 5 January 2008).

-----. 2008b. AIS to offer iPhone. 21 January. Available at http://www.telegeography.com (accessed 22 January 2008).

Reedy, S. 2009a. What Apple's milestone means to app stores. 28 September. Available at: http://telephonyonline.com (accessed 29 September 2009).

Reedy, S. 2009b. Motorola on socializing the mobile experience. 20 November. Available at http://www.printthis.clickability.com (accessed 21 November 2009).

Shrimsley, R. 2007. Cometh the iPhone, cometh the iCult. 28 June. Available at http://www.ft.com (accessed 29 June 2007).

Silver, S. 2009. Apple, RIM outsmart phone market. 20 July. Available at http://www.totaltele.com (accessed 21 July 2009).

Sinclair, J. 2008. Telefonica to sell iPhone this year in Spain. 4 June. Available at http://www.totaltele.com (accessed 5 June 2008).

Steen, M. 2008. KPN unmoved by iPhone. 6 February. Available at http://www.ft.com (accessed 7 February 2008).

Strand Consult. 2009. *The Moment of Truth – A Portrait of the iPhone.* Copenhagen: Strand Consult.

Svensson, P. 2007. Review: Rival phones no match for iPhone. 24 August. Available at http://communicationsdirectnews.com (accessed 25 August 2007).

----- 2008a. Sprint to make big push for new touch-screen phone; looks vaguely familiar. 1 April. Available at http://www.cellular-news.com (accessed 2 April 2008).

----- 2008b. RIM challenges Apple with Storm, a touch-screen phone you can feel. 8 October. Available at http://www.cellular-news.com (accessed 9 October 2008).

----- 2009. Review: Dazzling Palm software beats the iPhone. 3 June. Available at http://www.cellular-news.com (accessed 4 June 2009).

Symbian. 2009. Symbian FAQs. 5 November. Available at http://www.symbian.org (accessed 7 November 2009).

Talkandroid. 2009. HTC Hero Android phone details. 25 June. Available at http://www.talkandroid.com (accessed 26 June 2009).

Taylor, P. 2008a. A strong flavour of Apple. 15 February. Available at http://www.ft.com (accessed 16 February 2008).

----- 2008b. Nokia releases its answer to Apple's iPhone. 2 December. Available at http://www.ft.com (accessed 3 December 2008).

----- 2009. Smarter ways to talk. 3 July. Available at http://www.ft.com (accessed 4 July 2009).

TechCentral. 2009. RIM launches BlackBerry Tour 'world phone'. 22 June. Available at http://www.techcentral.ie (accessed 23 June 2009).

Telecompaper. 2009a. Nokia launches Ovi Store globally. 26 May. Available at http://www.telecompaper.com (accessed 27 May 2009).

----- 2009b. SonyEricsson unveils 3 multimedia phones. 29 May. Available at http://www.telecompaper.com (accessed 30 May 2009).

----- 2009c. Apple launches iPhone 3G S, offers iPhone 3G for USD 99. 9 June. Available at http://www.telecompaper.com (accessed 10 June 2009).

----- 2009d. Nokia unveils high-power N900 Maemo handset. 27 August. Available at http://www.telecompaper.com (accessed 28 August 2009).

Telecoms.com. 2008a. More iPhone carriers revealed. 12 May. Available at http://www.telecoms.com (accessed 13 May 2008).

----- 2008b. Apple faithful not disappointed; 3G iPhone lands in July. 10 June. Available at http://www.telecoms.com (accessed 11 June 2008).

----- 2008c. SIMlock free iPhone comes to Belgium. 10 July. Available at http://www.telecoms.com (accessed 11 July 2008).

----- 2008d. Verizon launches Moto touchscreen gadget. 15 October. Available at http://www.telecoms.com (accessed 16 October 2008).

----- 2009a. China Unicom adds iPhone to portfolio. 25 March. Available at http://www.telecoms.com (accessed 26 March 2009).

----- 2009b. Bada bing Bada boom. 8 December. Available at http://www.telecoms.com (accessed 9 December 2009).

TelecomWeb. 2007. Analysis: Why the iPhone won't replace the BlackBerry. 10 September. Available at http://communicationsdirectnews.com (accessed 11 September 2007).

Telephonyonline. 2009. Defining open mobile. 24 November. Available at http://www.telephonyonline.com (accessed 25 November).

Thomas, D. 2007. HTC launches 4 devices, sells 800,000 touch phones. 2 October. Available at http://www.cellular-news.com (accessed 3 October 2007).

Unstrung. 2009. Palm shows off second WebOS handset. 11 September. Available at http://communicationsdirectnews.com (accessed 12 September 2009).

Ward, A. and Gelles, D. 2009. Nokia takes on iPhone in legal fight. 22 October. Available at http://www.ft.com (accessed 23 October).

Waters, R. 2009. Microsoft aims to close gap in mobile race. 4 October. Available at http://www.ft.com (accessed 5 October 2009).

Weinberg, S. 2008. Could BlackBerry get last laugh on iPhone? 23 October. Available at http://www.totaltele.com (accessed 23 October 2008).

Wireless Week. 2009. Apple hits 1 billion downloads. 27 April. Available at http://www.communicationsdirectnews.com (accessed 3 May 2009).

Wood, N. 2008. 3G iPhone sales top 1m in first weekend. 14 July. Available at http://www.totaltele.com (accessed 15 July 2008).

----- 2009a. AT&T, Apple likely to extend iPhone deal – iSuppli. 11 September. Available at http://www.totaltele.com (accessed 12 September 2009).

----- 2009b. Orange to sell Motorola's make or break Android phone in Europe. 16 September. Available at http://www.totaltele.com (accessed 17 September 2009).

----- 2009c. App developers struggle to profit from Android. 1 December. Available at http://www.totaltele.com (accessed 2 December 2009).

3G.co.uk. 2008a. Top ten alternatives to the iPhone. 16 July. Available at http://www.3g.co.uk (accessed 17 July 2008).

----- 2008b. Apple iPhone 3G review by 3G.co.uk. 8 August. Available at http://www.3g.co.uk (accessed 9 August 2008).

----- 2008c. Samsung Pixon M8800 review by 3G.co.uk. 24 November. Available at http://www.3g.co.uk (accessed 25 November 2008).

----- 2009a. Samsung Memoir 8 MP camera phone launched. 6 February. Available at http://www.3g.co.uk (accessed 7 February 2009).

----- 2009b. Samsung unveils new touch lineup. 10 March. Available at http://www.3g.co.uk (accessed 11 March 2009).

----- 2009c. The best 3G mobile applications for 2012. 19 November. Available at http://www.3g.co.uk (accessed 20 November 2009).

CHAPTER

8 *Microsoft, Google and Yahoo!: A Case Study in Internet Evolution*

Introduction

Although high-speed data services downloaded to mobile devices have been slow to take off, there is every reason to believe that the launch of hundreds of high-speed networks and their continuous upgrading, combined with touch-screen smartphones such as the iPhone, currently represents a totally different landscape even compared to that two or three years ago. However, although the expectation of a massive growth in high-speed subscribers obviously presents structural and strategic issues for operators and equipment vendors, it is equally the case that it presents structural and strategic issues for the main providers of software. In particular, this means the likes of Microsoft, Google and Yahoo! which had previously tended to be treated in isolation. This chapter examines why this is no longer the case, and explores how the relationships between these three major new economy companies have had to change direction.

Background

On 1 February 2008, Microsoft launched an unsolicited takeover bid for Yahoo! worth $44.6 billion, offering to pay either entirely in cash or via a combination of cash and Microsoft shares. The bid priced each Yahoo! share at $31, a 62 per cent premium on the price ruling the previous day. Microsoft justified the bid in terms of the need to take on a resurgent Google in the Internet search market and claimed that it would lead to synergies worth $1 billion. According to Steve Balmer, the Microsoft CEO, the merged company would 'deliver superior value to our respective shareholders and better choice and innovation to our customers and industry partners' (Parker, Edgecliffe-Johnson and Palmer 2008).

Mr Balmer went on to point out that as part of previous correspondence between the protagonists, the chairman of Yahoo! had written to him in February 2007 claiming that it was inappropriate even to consider the possibility of a merger at that time because Yahoo! expected its reformulated strategy to deliver significant benefits to shareholders. However, these benefits had yet to appear and Wall St was becoming disillusioned. But Microsoft was also becoming disillusioned because Yahoo! had agreed to cooperate with Microsoft in fighting back against Google's dominance of the search market, yet had not kept its side of the bargain. Furthermore, Yahoo!'s weakening financial position indicated that it would provide an increasingly less competitive counterweight to Google as time went on. Given that MSN Search had only a 3 per cent share of the market while Yahoo! had nearly 13 per cent, any further decline on Yahoo's part would effectively leave MSN as a permanent niche player in a market increasingly dominated by Google.

Although the $31 per share represented a substantial premium over the market price at the time, it is worth noting that the Yahoo! share price had previously slid by 45 per cent from its October 2007 high. Given Microsoft's expectation in respect of potential cost savings, improved innovation and a sharpened responsiveness to opportunities in the video and mobile markets, it could be argued that Microsoft was paying over the odds. On the other hand, Yahoo! had stakes in Yahoo! Japan Corp. and Alibaba Group Holding Ltd which were valued at more than $12 per Yahoo! share. However, primarily because of the stake in Yahoo! Japan which stood at only 33 per cent (with Softbank holding a further 40 per cent), Microsoft would need to pay out a further substantial sum – in excess of $20 billion – if it hoped to acquire full ownership over the pieces of Yahoo! that were the most successful.

The fact that the Yahoo! share price only rose to $28 after the offer was announced indicated clearly that Wall St did not expect a bidding war to break out, if only because prospective white knights would fear taking on cash-rich Microsoft, but it was also acknowledged that regulators would not countenance a counter-bid by Google. However, the co-founders of Yahoo! – Jerry Yang and David Filo – held only 10 per cent of the shares so whatever their personal antipathy to a Microsoft takeover of their company, the decision would ultimately be made by others.

Predictably, Jerry Yang, CEO of Yahoo!, did not even deign to reply to the offer. Google's next move was to offer to help Yahoo! avoid the clutches of Microsoft, although it remained unclear how it could do so (Delaney and Karnitschnig 2008). It was suggested that it might help to finance a counter-bid by a third party (Waters and Chaffin 2008a; Waters and Edgecliffe-Johnson 2008a; Global Insight 2008). It was of interest that Yahoo! had recently been negotiating to outsource its Internet search advertising in Europe to Google, thus reflecting the much superior advertising revenue that Google was generating per search. However, it was reluctant to have Google displace its name in the most prominent position on the Web pages.

In essence, part of the difficulty lay in the fact that whereas the three companies had coexisted reasonably peacefully in the past, they were all now jockeying for status in the rapidly evolving world of mobile services. Google, for example, had unveiled plans to become a major player in the mobile operating system market with its Android initiative and had flagged its intention to bid for spectrum in the then ongoing 700 MHz auction in the USA, thereby emphasizing its intention to press for open access conditions. It had also provisionally acquired DoubleClick in December 2007 – confirmed in March 2008. By combining with Yahoo!, Microsoft clearly hoped to create an organization with parallel skills.

Google's Strategy

For most of the past decade, Microsoft had dominated the 'new economy', crushing its rivals while constantly at loggerheads with regulatory bodies around the world. So dominant had it been, that until very recently few had bothered to note the surging value of what started out as a modest search engine founded only in 1998, namely Google. At the end of October 2007, Google's share price exceeded $700 for the first time (Talley 2007), having exceeded $600 only on 8 October, and the company was valued at $220 billion. When it first went public in mid-August 2004, the share price was a mere $85.

Google reached a price/earnings ratio of roughly 35 – the sort of figure generally associated with the 1998–2000 boom, but hardly commonplace in the existing financial environment (although yet-to-become-profitable Facebook was briefly valued at $15 billion). However, it must be borne in mind that Google's share of the global search market is normally in the region of 65 per cent. What is particularly notable about this figure is that whereas Microsoft's software comes built into computers, and is hard to avoid unless one is a die-hard Apple fan, Google users can switch very easily to an alternative search engine even if they are partly discouraged by personalized Google home pages (Lex 2007a).

Companies like Google are not simply measured against their historic records, no matter how successful, but against the viability of their future strategies which must contain new sources of profits. By implication, Google cannot sit on its laurels as the top search engine. So where does its future lie? One major battleground looks set to be social networks. In effect, through its purchase of a small (1.6 per cent) but expensive ($240 million) stake in Facebook, Microsoft had laid down the gauntlet (Lex 2007b). With over 50 million users, Facebook already appeared to be the clear market leader, and had cemented that status by permitting outside developers to produce applications for the site in May 2007. However, Facebook employed its own proprietary technology, so this material could not be used elsewhere on the Web. Noting this Achilles Heel of sorts, Google set out to develop OpenSocial technology (Delaney 2007) which sought to lay down a standard for use on any network. Linking up initially with 12 partners including Oracle and a number of social networks such as Hi5 Networks and LinkedIn Corp., Google subsequently pulled off something of a coup by inducing News Corp's MySpace, Facebook's main rival, as well as a further significant social network, Bebo, to join the OpenSocial alliance (Waters and Allison 2007).

While Google's social network activities have been modest in the USA – hence the need for powerful partners – it does have a significant presence in the likes of Brazil and India where it operates as Orkut. Hence, in addition to the 200 million Internet subscribers to which it now had access in the USA, it also had a superior worldwide network compared to Facebook. While the likes of Yahoo!, Ebay and Amazon also host some social networking activities, they are unlikely to alter the balance of power significantly whichever – if any – side they choose to take. Facebook may or may not now adopt OpenSocial, but history indicates that open standards tend to defeat proprietary standards in the majority of cases.

The other arena in which Google is flexing its muscles is mobile communications. In principle, Google can be seen as a disruptive force because in early November 2007 it launched a Google-designed software and services platform – known as Android – which can be used in a wide range of mobile handsets (Delaney and Sharma 2007; Lex 2007c; Waters and Taylor 2007).[1] In addition to Android, Google announced the formation of

1 In August 2008, T-Mobile USA announced that it would be launching the first Android-based handset, made by HTC, before the end of the year (Dow Jones 2008b). An improved version has recently been launched together with other Android-based smartphones – see Chapter 7.

the Open Handset Alliance (sometimes Consortium) with 34 initial members including Motorola, Samsung, DoCoMo and T-Mobile (Taylor, Waters and Palmer 2007). By implication, it is Microsoft and Nokia that are perceived as the real threats with a desire to lock the market into their proprietary platforms. In January 2008, Yahoo! announced that it intended to produce versions of its services that would be compatible with Android (BWCS 2008).

Google's takeover of DoubleClick was controversial. Whereas Google had a large share of the market in text-based Web advertising, it was not a substantial player in the market for display and graphic advertisements that appear on Web sites. It wanted to acquire DoubleClick because of the latter's role as a broker between Google and its competitors and customers who wish to advertise. DoubleClick places advertisements on Web sites and tracks hits and click-through rates. In that respect, it would face competition from Microsoft/aQuantive and Yahoo!/RightMedia and hence the deal was eventually authorized by the various concerned regulatory authorities (Forelle 2008).

Meanwhile, Microsoft is predictably denigrating its new rival, claiming that companies have core competencies and that they are better off sticking to them and avoiding conflict with other industry players. But it is almost certainly too late to present such an argument because the very nature of the Google model is incompatible with current practice. Whereas mobile operators normally provide a standard package for a monthly fee and charge extra for additional features, Google gives its services away for nothing and makes its money from advertisements on the website (Taafe 2007). Advertising directed at mobile handsets is potentially a multi-billion dollar market, and the Apple model is also groundbreaking in that it obliges partnering operators to hand over a share of their revenues to the handset maker for the first time (see Chapter 7) . The Google CEO has indicated that its widespread adoption of such a model is 'highly likely' (Dow Jones 2007).

The Google share price registered $730 when Android was announced and peaked at $747 in November 2007. However, by the end of February 2008 it had fallen to $460 on the back of disappointing fourth-quarter earnings with revenues rising by 52 per cent compared to analysts' expectations of 55 per cent, as well as a poor growth forecast for the US economy. In addition, news of the Microsoft takeover bid for Yahoo! was seen as unfavourable. Year-on-year growth in searches was growing rapidly, with a 57 per cent share of USA search activity, but growth of click-throughs on advertisements was doing far less well (Nuttall 2008a).

Let Hostilities Commence

Given the paucity of alternative options to the bid, the markets increasingly took the view that Microsoft would prevail and the Yahoo! share price rose to equal that offered by Microsoft. However, since Microsoft's own share price had declined somewhat, the total value of the offer had fallen to $41.5 billion, suggesting that Microsoft would have to make an improved offer if it wanted to prevail (Waters 2008a). Nevertheless, few doubted that it would prevail in the end, although they did have doubts about Microsoft's ability to manage the integration of the companies – thoughts inevitably turned to the sorry tale of AOL and Time Warner (Curwen and Whalley 2004: Chapter 9).

It was argued that Microsoft had become vulnerable because of its reliance upon the slow and expensive upgrade cycle to Windows and Office, and that these had led to regular and costly run-ins with regulators in the USA and Europe whenever Microsoft tried to fold new services into the package (Lex 2008a). The main threat to Microsoft's reliance upon desktop-embedded software accordingly lay in the possibility that more and more software could be delivered on demand from massive centralized servers or simply sent en masse by the likes of Google as a free service – but with advertising attached. By acquiring Yahoo!, Microsoft would gain access to this service thereby giving prominence in advertising to its own software.

On 10 February 2008, Yahoo! formally rejected the Microsoft offer (*Financial Times* 2008a). It was alleged that Yahoo! would only sell at $40 per share although the market price at the time was only $30 and few thought a realistic value to lie above $35 (Lex 2008b). With a 9 per cent stake, Yahoo!'s second-biggest shareholder (after Capital Research) Legg Mason effectively said that it would be agreeable to a deal at around that level (Waters and Authers 2008). But Yahoo!'s board appeared to prefer a strategy of avoiding Microsoft's clutches altogether, and to that end devised a plan whereby News Corporation would inject its online assets including social networking site MySpace plus some cash into Yahoo! in return for a stake of roughly 20 per cent (Chaffin and Nuttall 2008). This was held to be a potential coup for News Corp., but a poor deal for Yahoo! since it was hardly short of Internet traffic whereas both parties were struggling to generate additional advertising dollars from Internet users. However, Rupert Murdoch was quick to point out that News Corp. was not going to get into a fight with Microsoft which had access to a lot more money than him.

Faced by these developments, Microsoft began to flag up the possibility of a hostile bid (Karnitschnig and Delaney 2008; Waters and Allison 2008a). This has rarely proved to be a successful strategy in 'new economy' industries because they are heavily dependent upon the talents of individual staff who can simply up and walk away. However, Microsoft appeared to take the view that it already had much of the necessary engineering know-how and that it would be able to get by even if some of Yahoo!'s engineers did decide to resign (Guth 2008). In effect, what Microsoft was claiming was that the brand name, customer base and infrastructure – which would be comparatively unaffected by a takeover – were more valuable than individual employees.

Clearly, this sparring could not go on indefinitely, so by early March 2008 the battle lines began to be drawn up. On the one hand, some Yahoo! shareholders sued its board for rejecting the Microsoft offer. On the other, Microsoft prepared to nominate a slate of people to sit on the Yahoo! board, thereby instigating a proxy vote fight at the forthcoming AGM (Dow Jones 2008a). Some at Yahoo! began to ponder the desirability of disposing of the Asian assets, but were put off by the difficulties inherent in so doing and the dramatic effect such action would have upon the size and value of Yahoo! (Lex 2008c).

In April, Microsoft began to lose patience and announced that it would be launching a hostile takeover bid before the month end, in the process offering less than had previously been put on the table (Waters 2008b, 2008c). In practice, the slide in Microsoft's share price meant, in any event, that the original offer was now worth $29.4 rather than $31 and upping the offer appeared to some to be a sensible move (Lex 2008d). The Yahoo! board reacted negatively (Waters 2008d), but the views of shareholders were eagerly awaited.

Rumours ran rife. On the one hand, an alliance between Yahoo!, Google and News Corp. was said to be developing. On the other, News Corp. was said to be forming an alliance with Microsoft. One semi-concrete sign was the announcement of a two-week experiment involving Google-supplied relevant advertisements alongside a small sample of Yahoo! search results, indicating to some – not for the first time (see above) – that Yahoo! would be shutting down its search engine and outsourcing the work to Google. It was also known that Time Warner was engaged in talks with a view to injecting its AOL division and some cash into Yahoo! in return for a 20 per cent stake (Waters and Chaffin 2008b; Waters and Nuttall 2008a). However, the likes of Lex continued to believe that the Microsoft offer made most sense (Lex 2008e).

In late April, Yahoo! declared excellent results for the first quarter – Jerry Yang called them 'extraordinary', but they were not sufficiently extraordinary to warrant a marked revaluation of Yahoo! (Lex 2008f; Nuttall 2008b). Accordingly, an improved bid by Microsoft – or even adding cash to raise it back to the original $31 per share – seemed no more or less likely than either a hostile offer or Microsoft walking away (Waters 2008e, 2008f). One factor that influenced many reading the runes was the poor personal relationship between Microsoft CEO Steve Ballmer and Google CEO Eric Schmidt (Pimental 2008). In the event, it was Microsoft that blinked, raising its offer to $33 per share and $46.5 billion in total – Wall St's best guess as to an acceptable price – only for Yahoo! to hold out for $37. So Microsoft walked away (Waters 2008g, 2008h, 2008i) in a move that was seen in some quarters as tactical rather than final.

The Yahoo! share price promptly and predictably nosedived by 20 per cent (Waters 2008j). Both Steve Balmer and Yahoo! CEO (and founder) Jerry Yang were heavily criticized for failing to reach a compromise and – this being America – the lawsuits began to be filed (Nuttall 2008c). Some saw a renewed opportunity for AOL (Chaffin 2008) while others saw Google as the ultimate winner with an opportunity to proceed with its tentative relationship with Yahoo! or go it alone (Waters 2008k).

According to best estimates, Yahoo!'s search operations were now worth $21 billion, its display and communications properties were worth $14 billion and its Asian investments were worth $9 billion. Rumours began to circulate that Microsoft was now interested in buying the display and communications properties and in taking a stake in a residual Yahoo! shorn of its Asian interests (Edgecliffe-Johnson and Waters 2008; Waters and Edgecliffe-Johnson 2008b). Meanwhile, billionaire active investor Carl Icahn decided to try to force Yahoo! to accept the original Microsoft offer by putting up a slate of (largely new) candidates for the Yahoo! board due for election on 3 July even though Microsoft itself had balked at a (hostile) proxy fight. He accordingly bought a stake plus options equivalent to 4.3 per cent of Yahoo! and enlisted the support of shareholders T. Boone Pickens and Paulson & Co. This action served to provide a floor for the Yahoo! share price at $26.50 (Lex 2008g, 2008h). Microsoft then let it be known that it was indeed interested in a friendly acquisition of an unspecified part of Yahoo! (Waters and Nuttall 2008b) – which subsequently turned out to be an $8 billion investment at $35 a share followed by the acquisition of the search business (Hodgson 2008a) – although this was not thought to be as attractive a proposition as a full takeover (Lex 2008i, 2008j) which, in practice, Microsoft did not rule out completely *per se* despite claims to the contrary by Yahoo! (*Financial Times* 2008b).

In late May, Yahoo! postponed its AGM until the end of July and reduced the size of its board to nine from ten (Nuttall 2008d). It also introduced a hugely generous employee severance plan which would be triggered in the event of a change in control such as voting off the majority of the board. In effect, this undermined Carl Icahn's attempt at a successful proxy fight in July although it left the possibility open that electing a few new Microsoft-friendly directors would be sufficient to force Yahoo! back to the negotiating table (Lex 2008k).

On 12 June, Yahoo! and Google announced a new deal whereby Google agreed to display some advertisements originating with Yahoo!. The deal, which was expected to require regulatory scrutiny and hence was not to be implemented for a further three months, was anticipated to generate an additional $800 million a year through improved monetization of certain types of searches, and could lead to further deals related to display advertising (Karnitschnig and Vascellaro 2008). However, the deal was non-exclusive and could be terminated by either party were there to be a change of control in Yahoo! within two years (Waters and Nuttall 2008c). Insofar that it implied that Yahoo! accepted that Google would remain the market leader and that the best strategy was hence to leverage the Yahoo! brand via Google's technology while allowing itself to concentrate on other matters, the deal could not be faulted. However, it was argued that the need to maintain the semblance that Yahoo! remained a competitor to Google for regulatory reasons meant that the deal had to be limited to North America and provided no synergies and hence was too little too late (Lex 2008l).

As for Microsoft, it was, according to Jerry Yang, clearly 'time to move on'. The Yahoo! share price duly moved on – down to $22 – while Microsoft itself moved on with the announcement that it had terminated any further interest in Yahoo!.

In early July 2008, strong rumours began to emerge that, on 17 May, Jerry Yang had offered to sell Yahoo! to Microsoft for $33–34 a share, but that this had been rejected since, by that point, Microsoft was more interested in acquiring only Yahoo!'s search engine business (Karnitschnig and Guth 2008). Subsequently, it was alleged, Microsoft had sought out partners for this deal, but had (so far at least) failed to find any (Vascellaro and Marr 2008). This may possibly explain why, on 6 July, Microsoft once again declared that it was willing to buy the whole of Yahoo! – but only if a new board was appointed first (Waters 2008n). In essence, its argument was that a deal would take at least nine months to close, and if during that time Yahoo! was mismanaged, Microsoft would potentially end up paying much more than Yahoo! was now worth. This placed additional emphasis upon Carl Icahn's proxy fight with Microsoft with which Microsoft claimed that it would not interfere (Kardos 2008).

However, in a further twist, Yahoo! claimed to have received a joint offer from Microsoft and Carl Icahn involving a complex restructuring of Yahoo! that would include the acquisition of its search business by Microsoft, the hiving off of the Asia businesses, and in return would receive $1 billion in cash plus a guaranteed annual payment of $2.3 billion. Claiming that it had been given only one day to respond, Yahoo! promptly rejected the proposal, pointing out that it could restructure without outside help (Nuttall 2008f), and once again offered to sell the entire company for $33 a share. In turn, Carl Icahn claimed that his proposal had been misrepresented, and filed a list of proposed Yahoo! board members with the SEC (Securities and Exchange Commission) (Murti and Kardos 2008).

In an effort to persuade shareholders that the board was just as interested in value creation as Carl Icahn's proposed replacements, Yahoo! announced in mid-July that it was willing to sell off its Asian assets. However, the board and senior management appeared to have survived the proxy fight when it was announced on 20 July that eight of the existing nine members of the board would stand for re-election with the ninth to be replaced by Carl Icahn, and that an additional two members would be appointed from his slate of candidates. This appeared to make little difference to the prospects of a deal involving Yahoo! and Microsoft, and hence the former's share price continued to languish at roughly $22 (Vascellaro, DiColo and Tibken 2008). It was of interest that Legg Mason and some other shareholders appeared to have swung behind Yahoo!, even going so far as to wonder why any concessions at all had been made to Carl Icahn. He, in turn, declined to attend the annual meeting in order to avoid confrontation, but it was clear from comments made by a number of shareholders that there would be moves to reinstigate a proxy fight if Yahoo!'s fortunes failed to revive during the forthcoming year (Nuttall and Waters 2008). Yahoo!'s share price promptly fell below $20, indicating little faith in the prospects of such a revival.[2]

Whither Microsoft?

In late July 2008, Steve Ballmer publicly dismissed Yahoo! as a 'tactic, not a strategy' in the campaign to attract more volume to Microsoft's search function (Lex 2008m). Unfortunately, those – almost everyone, in practice – who remembered how Microsoft's Zune had been seen off in no uncertain manner by the iPod remained sceptical that Microsoft could out-innovate Google, which announced that it had achieved 1,000 billion links to web pages (Nuttall 2008e). The Microsoft strategy appeared largely to consist of spending $2.5 billion a year (Bass and Thomson 2008) and was generally derided by analysts. However, on a more positive note, Microsoft announced that it had agreed a deal to place its search service, along with search-related advertising, on Facebook (Waters and Allison 2008b). In addition, in August, Microsoft released a Web browser that had the potential to affect the advertising model of rivals such as Google. The Internet Explorer 8 browser's InPrivate setting already allowed users to access sites without disclosing their browsing behaviour, thereby preventing the sites from delivering targeted advertisements – the business that Google had newly entered via its acquisition of DoubleClick. By placing the InPrivate button in a much more prominent position, Microsoft sought to reduce the potential for DoubleClick to access customers (Minto 2008).

Google–Yahoo!, Yahoo!–Google or (Google) (Yahoo!)?

On 5 October 2008, the mooted advertising partnership between Google and Yahoo! was put on the back burner because, despite the fact that it was non-exclusive and not a merger, the Justice Department was taking an unexpectedly long time to review the case (Kendal 2008). In the meantime, Mithras Capital, a small Yahoo! shareholder, proposed the sale of Yahoo! to Microsoft at $22 a share, which also involved the sale of Yahoo!'s Asian holdings (Morrison 2008b). Superficially, this might not appear to be a particularly attractive price, but after a brief foray above the $20 mark, the Yahoo! share

2 At the AGM the votes of the Capital Group were initially miscounted. After a recount, Jerry Yang was shown to have received only 66 per cent support from shareholders (Waters 2008o).

price had rapidly dropped below $15 before stabilizing at roughly $13. Not surprisingly, Microsoft CEO Steve Ballmer was supportive, although it was made clear that no specific negotiations to that end were in progress (Hodgson 2008b).

Google appeared to have its own troubles, with its share price being pushed down towards the $300 mark in anticipation of poor third-quarter results. When these turned out to be better than expected, the share price rose sharply to $380, indicating the hugely volatile state of the market (Waters 2008s). Unhappily, Yahoo!'s own results could not be met with comparable relief given that strong growth in the search advertising business had not been matched by display advertising, and a 10 per cent cut in the workforce by the year end was to be the consequence (Nuttall 2008i). Understandably, the proposed joint venture began to buckle under the strains imposed by this volatility, not to mention a reluctance to make concessions to meet concerns raised by the Justice Department (Lex 2008p; Vascellaro 2008a). Carl Icahn predictably recommended that Yahoo! should fall back into the arms of Microsoft (Morrison 2008c), but the protagonists preferred to lay a revised joint venture before the Justice Department (Vascellaro 2008b). This stipulated that the deal would cover a two-year period only, down from the previous 10 years, and that the revenue that Yahoo! could generate should be capped at 25 per cent of Yahoo!'s total search revenue. Furthermore, Google advertisers would be able to opt out of having their advertisements displayed on Yahoo! sites.

This proposal was heavily criticized, essentially on the grounds that Yahoo! thereby reduced the extent of its competition with Google without getting anything in return (Angwin 2008). Angwin went on to argue that Yahoo! had been pursuing a misguided advertising strategy, 'combining its search-and display-advertising sales forces and focusing on technology to try to boost the rates of lower-priced ads across the whole site'. As a consequence, it had neglected its core business of selling premium advertisements on its front page and in its valuable section front pieces such as that of Yahoo! Finance.

On 5 November 2008, the Justice Department informed the protagonists that it would file suit to prevent the revised deal, and, predictably, Google promptly withdrew. Yahoo! expressed its disappointment with Google, but its share price rose to $14.20 on the basis that it would now turn again to Microsoft for its salvation (Nuttall and Waters 2008).

Looking Forward to 2009

It was undeniable that Microsoft had failed to spot the threat posed by Google until it was too late and that it backed itself into something of a corner by failing to buy either the whole of Yahoo! or its search business. For the time being, Microsoft had squashed any rumours that it was planning to pursue other targets such as AOL or Facebook (Waters 2008l). According to Steve Ballmer: 'The most important application for the foreseeable future ... is search ... We don't have to dominate, but we'd better have a darn good chunk of the search market over time.' But precisely how that was to be achieved was far from clear. Historically, Microsoft prospered not by being first to the market with a new technology, but by the speed of its response and superior strategy, yet that had hardly been the hallmark of the pursuit of Yahoo! (Water 2008m).

Not that Yahoo! had won in any meaningful sense – after all, it had failed to push the share price offered by Microsoft up to an acceptable level. It now faced a number of issues. On the one hand, its shareholders were deeply unhappy that it had rejected an offer of $33 a share, given that its share price in June 2008 languished at $23. On the other hand, while the search advertising deal with Google had the potential to boost the share price, it could not make out that the deal was really significant since that would attract the attention of regulators.

But, in truth, regulators could not be expected to avert their eyes, especially in the face of complaints by associations representing consumer advertisers (Waters 2008q). In September 2008, the Department of Justice had duly instigated a preliminary investigation, but Yahoo! and Google continued to claim that that the matter only affected the USA and Canada, and hence should be ignored by the European Commission (Waters and Tait 2008). Unfortunately for Google, the Commission is not famed for ignoring things, and it revealed that it had instigated its own preliminary investigation in July. The Canadian authorities also decided to act independently of the USA (Letzing 2008).

Having voluntarily put the deal on hold – in theory, the agreement had to go live by late-September, no more than 105 days from its signing, but, as seen, practice was quite another matter – Yahoo! and Google needed to return to the offensive. In the event, it was Google which took the initiative, releasing a comic book that introduced Chrome, a new Internet browser designed to compete with Internet Explorer. Based upon open-source Mozilla Firefox, which had taken a fifth of the browser market despite Explorer's advantage of being pre-installed in many computers, Chrome appeared to constitute an attack not merely on Explorer but upon Windows itself given its focus upon managing memory and running applications. It also provided a link to Android with which it shared significant components (Waters 2008p; Lex 2008n).

The Android project was also looking moderately healthy (Taylor and Parker 2008). In the USA, at the end of September 2008, T-Mobile and Google had launched the first Android powered handset, the HTC Dream, designed to operate over T-Mobile's new W-CDMA (wideband code division multiple access)/HSPA (high-speed packet access) network (CommunicationsDirect 2008). However, the reception was muted at best (Lex 2008o; Waters 2008r).

As for Yahoo!, it had Carl Icahn and his allies on its back, and the reality was that they were unlikely to be fully satisfied with anything short of a Microsoft takeover of the entire company. At one point the outcome appeared to rest upon whether dissident shareholders would vote in any or all of Icahn's alternative slate of board directors at the forthcoming AGM, and at the very least Yahoo! was likely to face a number of class action lawsuits brought by disgruntled shareholders (Nuttall 2008g). In the event, however, Mr Icahn and his two nominees were appointed to the expanded 11-member board without fuss, and apart from the odd rumbling from Mr Icahn's general direction, little was expected to come of this until the Yahoo!/Google deal reached fruition one way or the other (Nuttall 2008h).

Yahoo! also faced other difficulties. In August 2008, for example, Fox Interactive Media, owner of MySpace.com, overtook Yahoo! as the top display advertising site in the USA in terms of hits, although the situation was ameliorated by the fact that Yahoo! could charge much higher rates to advertisers because social network users are more interested in searching for friends than for information or products (Morrison 2008a).

Furthermore, the febrile financial environment had hardly been helpful to the various protagonists. When, in late August, the share price of Yahoo! fell below $19, this took the price roughly back to where it had been before the Microsoft saga began. For its part,

Google had fared no better, with its share price falling below $400 in mid-September despite having reached a high of $750 less than a year previously.

So what was to be the likely outcome of this saga? It was always probable that the Google–Yahoo! deal would flounder under regulatory pressure, leaving Yahoo! on its own with a collapsed share price, poor profitability and facing an advertising downturn. Given that Yahoo! would have had to maintain its own search technology and associated cost base in order to satisfy the regulators, it could be agued that no great loss would be suffered. Obviously, the argument that it should have bit the Microsoft bullet would niggle away for a long time to come, but Microsoft now had far less need to return to the negotiating table than it had a year previously. Indeed, should Yahoo! acquire AOL, this would enable Microsoft to mop up two competitors with large online audiences in one fell swoop.

For the moment, Yahoo! appeared to be left with just the one, unsatisfactory, option, namely a deal to buy its AOL unit from Time Warner. However, there was a considerable gap between the valuations indicated by the protagonists (Waters 2008t), and even if Yahoo! succeeded there were those who believed that it would serve only to link together two sinking ships (Angwin 2008). So, perhaps, Microsoft needed to be invited back to the party. Steve Ballmer and Jerry Yang were well aware of this, which is why the latter stated in early November 2008 that '[t]o this day, I would say that the best thing for Microsoft to do would be to buy Yahoo! ... We remain open to everything, but it has to make sense' (Harvey 2008). But there was widespread recognition that a full takeover would unlash a titanic clash of corporate cultures, so an acquisition limited to Yahoo!'s search business seemed to be the better option, and that conclusion was supported by comments made by Steve Ballmer on 7 November 2008 (*Financial Times* 2008c; Wingfield and Vascellaro 2008) and 8 January 2009 (Hodgson 2009).

Ironically, admitting to negotiations to this end would cause the Yahoo! share price to shoot up, so Steve Ballmer was bound to keep his cards close to his chest. But, ultimately, Microsoft had only roughly 10 per cent of the search market while Yahoo! had 17 per cent and Google had 63 per cent. This is a business that benefits from scale, something which Microsoft lacked (Lex 2008q). The fat lady had yet to sing, but when she did, Jerry Yang would not be conducting the orchestra since he fell on his sword in mid-November (pending the appointment of a replacement CEO), returning to his previous role as Chief Yahoo! (Waters and Nuttall 2008d). His departure meant some delay in negotiations with AOL, but was expected to be helpful in bringing Microsoft back to the negotiating table (Li 2008).

A New Brush in 2009 Brings Closure

It all now rather appeared to depend on the new CEO Carol Bartz who was brought in from outside Yahoo!. With Jerry Yang and Yahoo! President Susan Decker standing down, she was expected to make conciliatory overtures to Microsoft, but instead announced that her gut feeling was to oppose any alliance and that she was not going to be stampeded into making a premature decision. She expressed the opinion that '[t]his is a fantastic Internet property. It doesn't deserve everybody trying to pick it and pull it apart', but her cause was not helped unduly by Yahoo!'s 2008Q4 loss of $303 million, the first quarterly loss since 2002 (Vascellaro 2009b). For its part, Microsoft appeared to have withdrawn to mull over other issues such as the possibility of job cuts and the development of Windows 7 (Swisher 2009; Vascellaro 2009a).

Despite the above, the two parties were alleged to have resumed non-specific talks in April 2009 (Waters 2009a). In June, Microsoft launched a revamped version of its struggling search engine branded as Bing, which immediately served to increase its market share by 2 per cent, but hardly posed a threat to Google (Morrison 2009; Nuttall 2009), especially since the majority of Bing users were adjudged to be samplers as against switchers (Kafka 2009). But Yahoo! hardly posed a major threat either, and its expectations of poor trading through to the end of 2009, announced in late July (Lex 2009a), suggested to some that Carol Bartz would do well to reconsider a link with Microsoft.

Several days later she did precisely that, announcing a deal which ceded responsibility to Microsoft for mechanically driven search results both to Bing and to Yahoo! pages. This will save Yahoo! hundreds of millions of dollars over time, while allowing it to retain 88 per cent of the revenue from advertisements placed on its pages over an initial five-year period. During the subsequent five-year period, the Yahoo! share of revenues will lie between 83 per cent and 93 per cent. In addition, Yahoo! will be entitled to revoke the deal should either the combined share of search queries directed to Microsoft and Yahoo! fall below an unspecified level or Yahoo!'s revenue per search fall below an unspecified fixed percentage of that of Google (Menn 2009a). Other clauses in the deal provided further benefits to Yahoo! (Letzing 2009), but the financial markets reacted unfavourably, arguing that Yahoo! should have struck a better deal and that it was turning itself into something uncomfortably reminiscent of AOL (Menn, Waters and Bradshaw 2009).

In contrast, Lex argued that striking $4 billion off the value of Yahoo! in the immediate aftermath of the deal was arguably an overreaction (Lex 2009b), while Carl Icahn appeared, for once, to be happy enough with developments. For its part, Yahoo! was keen to emphasize that its deal with Microsoft was limited in its scope and that it would continue to compete fiercely with Microsoft in other areas (Menn 2009b).

Conclusions

One can only presume that this episode will be viewed retrospectively as a classic example of capitalism at work, involving winners and losers on a massive scale, in what is increasingly looking to be the point in history at which the virtues of capitalism in the raw are coming under severe fire.

However, there is little point in interpreting precisely what it is all going to mean for now because regulatory intervention is inevitable and the outcome of that is uncertain (Waters 2009b). Furthermore, the benefits or otherwise to Yahoo! will take a good many years to become clear (Lex 2009b). In the meantime, Yahoo! needed to project a positive attitude which it singularly failed to do at the end of October 2009, choosing rather to emphasize the potential for further cost cuts (Menn 2009c). Lex was unimpressed, arguing that growth was essential and that Yahoo! was failing to exploit its 600 million users outside the USA who accounted for only 27 per cent of revenues (Lex 2009c). However, Lex went on to admit that Yahoo! needed to collect better information about its users for advertisers and that the users might not be keen on this.

In conclusion, it appears that Yahoo!'s fate remains firmly in its own hands, and its heavily fluctuating share price indicates that there is considerable uncertainty as to what that fate will turn out to be.

References

Angwin, J. 2008. Yahoo's search for salvation. 5 November. Available at http://www.totaltelecom (accessed 6 November 2008).

Bass, D and Thomson, A. 2008. Microsoft to spend $2.5 billion a year to keep pace with Google. 25 July. Available at http://www.washingtonpost.com (accessed 26 July 2008).

BWCS. 2008. If Yahoo can't beat 'em … . 8 January. Available at http://www.bwcs.com (accessed 9 January 2008).

Chaffin, J. 2008. Yahoo developments open up AOL options. 5 May. Available at http://www.ft.com (accessed 6 May 2008).

Chaffin, J. and Nuttall, C. 2008. News Corp joins Yahoo battle. 14 February. Available at http://www.ft.com (accessed 15 February 2008).

CommunicationsDirect. 2008. T-Mobile gets 3G Rolling. 19 September. Available at http://www.communicationsdirectnews.com (accessed 20 September 2008).

Curwen, P. and Whalley, J. 2004. *Telecommunications Strategy: Cases, Theory and Applications*. London: Routledge.

Delaney, K. 2007. Google aims for slice of social networking. 30 September. Available at http://www.totaltele.com (accessed 1 November 2007).

Delaney, K. and Karnitschnig, M. 2008. Google offers to help Yahoo fight off Microsoft. 4 February. Available at http://www.totaltele.com (accessed 5 February 2008).

Delaney, K. and Sharma, A. 2007. Google unveils mobile alliance. 5 November. Available at http://www.totaltele.com (accessed 6 November 2007).

Dow Jones. 2007. Google 'highly likely' to seek carrier revenue-sharing deals. 6 November. Available at http://www.cellular-news.com (accessed 7 November 2007).

----- 2008a. Microsoft threatens Yahoo! board coup. 3 March. Available at http://www.totaltele.com (accessed 4 March 2008).

----- 2008b. T-Mobile USA first to offer Android phone. 15 August. Available at http://www.totaltele.com (accessed 16 August 2008).

Edgecliffe-Johnson, A. and Waters, R. 2008. Google weighs new move on Yahoo. 19 May. Available at http://www.ft.com (accessed 20 May 2008).

Financial Times. 2008a. Yahoo rejects $42bn Microsoft approach. 11 February. Available at http://www.ft.com (accessed 12 February 2008).

----- 2008b. Yahoo says Microsoft has ruled out merger. 28 May. Available at http://www.ft.com (accessed 29 May 2008).

----- 2008c. Microsoft's Balmer rules out new Yahoo bid. 7 November. Available at http://www.ft.com (accessed 8 November 2008).

Forelle, C. 2008. Google closes DoubleClick deal. 12 March. Available at http://www.totaltele.com (accessed 13 March 2008).

Global Insight. 2008. Upheaval in the mobile data market as Microsoft bids for Yahoo! 4 February. Available at http://www.communicationsdirectnews.com (accessed 5 February 2008).

Guth, R. 2008. Microsoft looks to ease concerns on Yahoo bid. 25 February. Available at www.totaltele.com (accessed 26 February 2008).

Harvey, M. 2008. Why can't Yahoo! find the solution to the Microsoft takeover tangle? *The Times* (London), 7 November, 66.

Hodgson, J. 2008a. Microsoft has 'better options' than buying Yahoo. 16 June. Available at http://www.totaltele.com (accessed 17 June 2008).

----- 2008b. Ballmer comments send Yahoo shares skyward. 16 October. Available at http://www.totaltele.com (accessed 17 October 2008).

----- 2009. Ballmer says Yahoo deal 'a thing of the past'. 8 January. Available at http://www.totaltele.com (accessed 9 January 2009).

Kafka, P. 2009. Microsoft's Bing problem is Google. 14 July. Available at http://www.totaltele.com (accessed 15 July 2009).

Kardos, D. 2008. Microsoft willing to discuss deal if Yahoo replaces board. 7 July. Available at http://www.totaltele.com (accessed 8 July 2008).

Karnitschnig, M. and Delaney, K. 2008. Microsoft sticks to guns in Yahoo bid. 20 February. Available at http://www.totaltelecom (accessed 21 February 2008).

Karnitschnig, M. and Guth, R. 2008. Whole of Yahoo offered to Microsoft at $33 a share. 2 July. Available at http://www.totaltele.com (accessed 3 July 2008).

Karnitschnig, M. and Vascellaro, J. 2008. Microsoft walks away, Yahoo enters Google ad pact. 13 June]. Available at http://www.totaltele.com (accessed 14 June 2008).

Kendall, B. 2008. Google, Yahoo agree to delay ad deal. 6 October. Available at http://www.totaltele.com (accessed 7 October 2008).

Letzing, J. 2008. Canada probing Yahoo, Google advertising pact. 26 August. Available at http://www.totaltele.com (accessed 27 August 2008).

----- 2009. Microsoft, Yahoo lay out terms of search ad deal. 5 August. Available at http://www.totaltele.com (accessed 6 August 2009).

Lex. 2007a. Google. 11 July. Available at http://www.ft.com (accessed 12 July 2007).

----- 2007b. Microsoft wins face-off. 25 October. Available at http://www.ft.com (accessed 26 October 2007).

----- 2007c. I, Google. 6 November. Available at http://www.ft.com (accessed 7 November 2007).

----- 2008a. Why Microsoft wants Yahoo. 8 February. Available at http://www.ft.com (accessed 9 February 2008).

----- 2008b. Yahoo's last stand. 10 February. Available at http://www.ft.com (accessed 11 February 2008).

----- 2008c. Yahoo's Asian assets. 3 March. Available at http://www.ft.com (accessed 4 March 2008).

----- 2008d. Microsoft/Yahoo. 6 April. Available at http://www.ft.com (accessed 7 April 2008).

----- 2008e. The fight for Yahoo. 10 April. Available at http://www.ft.com (accessed 11 April 2008).

----- 2008f. Yahoo beats forecasts. 23 April. Available at http://www.ft.com (accessed 24 April 2008).

----- 2008g. Yahoo and Icahn. 13 May. Available at http://www.ft.com (accessed 14 May 2008).

----- 2008h. Icahn and Yahoo. 15 May. Available at http://www.ft.com (accessed 16 May 2008).

----- 2008i. Microsoft and Yahoo. 19 May. Available at http://www.ft.com (accessed 20 May 2008).

----- 2008j. Yahoo's delay. 23 May. Available at http://www.ft.com (accessed 24 May 2008).

----- 2008k. Icahn and Yahoo. 11 June. Available at http://www.ft.com (accessed 12 June 2008).

----- 2008l. Yahoo's deal. 13 June. Available at http://www.ft.com (accessed 14 June 2008).

----- 2008m. Microsoft's vistas. 24 July. Available at http://www.ft.com (accessed 25 July 2008).

----- 2008n. Just browsing. 2 September. Available at http://www.ft.com (accessed 3 September 2008).

----- 2008o. Google Android. 22 September. Available at http://www.ft.com (accessed 23 September 2008).

----- 2008p. Yahoo/Google. 3 November. Available at http://www.ft.com (accessed 4 November 2008).

----- 2008q. Microsoft will return. 9 November. Available at http://www.ft.com (accessed 10 November 2008).

----- 2009a. Yahoo. 21 July. Available at http://www.ft.com (accessed 22 July 2009).

----- 2009b. Yahoo/Microsoft. 2 August. Available at http://www.ft.com (accessed 3 August 2009).

----- 2009c. Yahoo. 1 November. Available at http://www.ft.com (accessed 2 November 2009).

Li, K. 2008. Yahoo shake-up could delay talks with AOL. 19 November. Available at http://www.ft.com (accessed 20 November 2008).

Menn, J. 2009a. Yahoo renews vow to fight Microsoft. 25 August. Available at http://www.ft.com (accessed 26 August 2009).

----- 2009b. Yahoo-Microsoft search deal has escape clauses. 5 August. Available at http://www.ft.com (accessed 6 August 2009).

----- 2009c. Cautious Yahoo outlook disappoints investors. 29 October. Available at http://www.ft.com (accessed 30 October 2009).

Menn, J., Waters, R. and Bradshaw, T. 2009. Microsoft and Yahoo seal deal. 29 July. Available at http://www.ft.com (accessed 30 July 2009).

Minto, R. 2008. Latest Microsoft browser challenges Google. 27 August. Available at http://www.ft.com (accessed 28 August 2008).

Morrison, S. 2008a. MySpace passes Yahoo as top US Internet display ad site. 27 August. Available at http://www.totaltele.com (accessed 28 August 2008).

----- 2008b. Yahoo investor proposes new Microsoft deal. 13 October. Available at http://www.totaltele.com (accessed 14 October 2008).

----- 2008c. Icahn says Yahoo should do search deal with Microsoft. 4 November. Available at http://www.totaltele.com (accessed 5 November 2008).

----- 2009. Microsoft's Bing off to good start – ComScore. 10 June. Available at http://www.totaltele.com (accessed 11 June 2009).

Murti, B. and Kardos, D. 2008. Icahn says Yahoo distorted facts of Microsoft bid. 14 July. Available at http://www.totaltelecom (accessed 15 July 2008).

Nuttall, C. 2008a. Growth fears hit Google shares. 26 February. Available at http://www.ft.com (accessed 27 February 2008).

----- 2008b. Yahoo lifts forecasts after sales top $1.3bn. 22 April. Available at http://www.ft.com (accessed 23 April 2008).

----- 2008c. Calming investors is Yahoo's next task. 4 May. Available at http://www.ft.com (accessed 5 May 2008).

----- 2008d. Yahoo postpones showdown with Icahn. 23 May. Available at http://www.ft.com (accessed 24 May 2008).

----- 2008e. Yahoo put in tight position with Google deal. 13 June. Available at http://www.ft.com (accessed 14 June 2008).

----- 2008f. Yahoo rejects 'opportunistic' Microsoft-Icahn offer. 13 July. Available at http://www.ft.com (accessed 14 July 2008).

----- 2008g. Newcomer lines up to take on Google. 28 July. Available at http://www.ft.com (accessed 29 July 2008).

----- 2008h. Icahn takes his seat at Yahoo table. 21 September. Available at http://www.ft.com (accessed 22 September 2008).

----- 2008i. Yahoo to axe 10 per cent of workforce. 21 October. Available at http://www.ft.com (accessed 22 October 2008).

----- 2009. High hopes resting on Bing. 29 July. Available at http://www.ft.com (accessed 30 July 2009).

Nuttall, C. and Waters, R. 2008a. Calm descends as Yahoo proxy storm ends. 30 July. Available at http://www.ft.com (accessed 1 August 2008).

Parker, A., Edgecliffe-Johnson, A. and Palmer, M. 2008. Microsoft offers $44.6bn for Yahoo. 1 February. Available at http://www.ft.com (accessed 2 February 2008).

Pimental, B. 2008. Yahoo battle renews Microsoft, Google rivalry. 22 April. Available at http://www.totaltele.com (accessed 23 April 2008).

Swisher, K. 2009. Yahoo-Microsoft search deal could be near. 14 January. Available at http://www.totaltelecom (accessed 15 January 2009).

Taaffe, J. 2007. Master of spin. *Total Telecom Magazine*, November, 18–21.

Talley, K. 2007. Google rises above $700 for the first time; more gains seen. 30 October. Available at http://www.totaltele.com (accessed 1 November 2007).

Taylor, P. and Parker, A. 2008. Android is set to take on smartphone market. 2 September. Available at http://www.ft.com (accessed 3 September 2008).

Taylor, P., Waters, R. and Palmer, M. 2007. Will Android be Google's transformer? 6 November. Available at http://www.ft.com (accessed 7 November 2007).

Vascellaro, J. 2008a. Odds dim for Google–Yahoo deal. 30 October. Available at http://www.totaltele.com (accessed 1 November 2008).

Vascellaro, J. 2008b. Yahoo, Google recast ad alliance. 4 November. Available at http://www.totaltele.com (accessed 5 November 2008).

----- 2009a. Bartz eases some hurdles to Microsoft deal. 15 January. Available at http://www.totaltele.com (accessed 16 January 2009).

----- 2009b. Yahoo posts loss, new chief plots strategy. 28 January. Available at http://www.totaltele.com (accessed 29 January 2009).

Vascellaro, J., DiColo, J. and Tibken, S. 2008. Yahoo names Icahn to board to end proxy battle. 21 July. Available at http://www.totaltele.com (accessed 22 July 2008).

Vascellaro, J. and Marr, M. 2008. Yahoo pursues talks with potential partners. 3 July. Available at http://www.totaltele.com (accessed 4 July 2008).

Waters, R. 2008a. Yahoo shares rise above offer value. 7 February. Available at http://www.ft.com (accessed 8 February 2008).

----- 2008b. Bid for Yahoo withers into 'stalemate'. 1 April. Available at http://www.ft.com (accessed 2 April 2008).

----- 2008c. Microsoft steps up pressure on Yahoo. 5 April. Available at http://www.ft.com (accessed 6 April 2008).

----- 2008d. Yahoo board huddles over Microsoft bid. 5 April. Available at http://www.ft.com (accessed 6 April 2008).

----- 2008e. Microsoft faces critical decision over Yahoo. 27 April. Available at http://www.ft.com (accessed 28 April 2008).

----- 2008f. Microsoft ponders next step in bid for Yahoo. 1 May. Available at http://www.ft.com (accessed 2 May 2008).

----- 2008g. Microsoft and Yahoo talks intensify. 2 May. Available at http://www.ft.com (accessed 3 May 2008).

----- 2008h. Microsoft walks away from Yahoo deal. 4 May. Available at http://www.ft.com (accessed 5 May 2008).

----- 2008i. Price gap proved too big to bridge. 4 May. Available at http://www.ft.com (accessed 5 May 2008).

----- 2008j. Yahoo under pressure after deal collapses. 4 May. Available at http://www.ft.com (accessed 5 May 2008).

----- 2008k. Google warms to ad deal with Yahoo. 9 May. Available at http://www.ft.com (accessed 10 May 2008).

----- 2008l. Microsoft rules out post-Yahoo buying spree. 21 June. Available at http://www.ft.com (accessed 21 June 2008).

----- 2008m. Microsoft braced for post-Gates era. 20 June. Available at http://www.ft.com (accessed 21 June 2008).

----- 2008n. Microsoft steps up battle over Yahoo board. 7 July. Available at http://www.ft.com (accessed 8 July 2008).

----- 2008o. Error distorts Yahoo vote on Yang. 5 August. Available at http://www.ft.com (accessed 6 August 2008).

----- 2008p. Google launches Internet browser. 1 September. Available at http://www.ft.com (accessed 2 September 2008).

----- 2008q. Ad trade body opposes Google-Yahoo alliance. 7 September. Available at http://www.ft.com (accessed 8 September 2008).

----- 2008r. Critics leave Google's Android phone on hold. 9 September. Available at http://www.ft.com (accessed 10 September 2008).

----- 2008s. Google earnings resist advertising downturn. 16 October. Available at http://www.ft.com (accessed 17 October 2008).

----- 2008t. Yahoo finds itself in a strategic dead-end. 5 November. Available at http://www.ft.com (accessed 6 November 2008).

----- 2009a. Microsoft and Yahoo renew partnership talks. 10 April. Available at http://www.ft.com (accessed 11 April 2009).

----- 2009b. Venture to face regulatory scrutiny. 30 July. Available at http://www.ft.com (accessed 1 August 2009).

Waters, R. and Allison, K. 2007. MySpace joins Google in Facebook fight. 1 November. Available at http://www.ft.com (accessed 2 November 2007).

----- 2008a. Microsoft nears hostile bid for Yahoo. 21 February. Available at http://www.ft.com (accessed 21 February 2008).

----- 2008b. Microsoft agrees alliance with Facebook. 24 July. Available at http://www.ft.com (accessed 25 July 2008).

Waters, R. and Authers, J. 2008. Yahoo warms to Microsoft's advances. 12 February. Available at http://www.ft.com (accessed 13 February 2008).

Waters, R. and Chaffin, J. 2008a. Microsoft hits back at Google. 4 February. Available at http://www.ft.com (accessed 5 February 2008).

----- 2008b. Microsoft's rivals circle but Yahoo lacks options. 10 April. Available at http://www.ft.com (accessed 11 April 2008).

Waters, R. and Edgecliffe-Johnson, A. 2008a. Google weighs in against Microsoft. 3 February. Available at http://www.ft.com (accessed 4 February 2008).

----- 2008b. Microsoft would stop short of full buy-out. 19 May. Available at http://www.ft.com (accessed 20 May 2008).

Waters, R. and Nuttall, C. 2008a. Web giants take sides in battle for Yahoo. 9 April. Available at http://www.ft.com (accessed 10 April 2008).

----- 2008b. Microsoft renews approach to Yahoo. 18 May. Available at http://www.ft.com (accessed 19 May 2008).

----- 2008c. Yahoo in search deal with Google. 12 June. Available at http://www.ft.com (accessed 13 June 2008).

----- 2008d. Yahoo chief Yang to step down. 18 November. Available at http://www.ft.com (accessed 19 November 2008).

Waters, R. and Tait, N. 2008. Google-Yahoo pact could face European probe. 15 September. Available at http://www.ft.com (accessed 16 September 2008).

Waters, R. and Taylor, P. 2007. Google android steals march on mobile rivals. 5 November. Available at http://www.ft.com (accessed 6 November 2007).

Wingfield, N. and Vascellaro, J. 2008. Ballmer kills hopes for Yahoo bid. 20 November. Available at http://www.totaltele.com (accessed 21 November 2008).

CHAPTER 9

The Role of 3G New Entrants in Promoting Competition

Introduction

It is now time to ask how new entrants – that is, those without pre-existing 2G licences and networks – have fared because, in principle, the issuing of more 3G licences than there are 2G incumbents is a potentially beneficial method of introducing competition. In this respect it should be borne in mind that although all new entrants are small by definition, starting with zero subscribers, some incumbents are also small either in terms of absolute subscriber numbers or relative to other incumbents or both.

Much of what follows is concerned with events in Europe, broadly defined. This is essentially because, as noted previously, a considerably larger number of 3G licences were issued in Europe between 2000 and 2005 than in any other region. Although the experience in Europe, where the outcome is known, may not be replicated in other regions where it remains too early to form a judgement, the odds must remain firmly in favour of that being the case. A number of country case studies will accordingly be used to test this hypothesis.

First-Mover Advantage

Before turning to examine the database, it may be useful to address the issue of whether, in principle, new entrants can expect to prosper in a sector such as high-speed mobile. In other words, is there an early-mover advantage?

It is not necessarily a good idea in principle to be first into a high-tech market.[1] The most obvious reason is that if (as is normal) the technology works imperfectly when first launched, those launching it may find this has a negative impact upon their reputations. In addition, the costs associated with bringing the technology to the attention of the public produce both specific (to the company) benefits as well as generic (to all companies using the technology) benefits. Perhaps, therefore, it may be a wise strategy to wait until

1 It is also possible to analyse new entry in relation to vendors. However, this involves a large amount of statistical manipulation of data on patents and their citation and is accordingly at odds with the approach used elsewhere in the book. An extensive analysis of the period to 2005 can anyway be found in He, Lim and Wong (2006) which is essentially concerned with how Ericsson, Nokia, and Samsung caught up with Motorola. The case studies are too brief to be of value, and (somewhat surprisingly) the paper does not specifically refer to 3G or, indeed, the sequential generations of technology, so there remains an opportunity to conduct further research in this area.

the technology has become reasonably glitch-free and has been widely advertised, before entering the market.

However, there is one particular virtue of being an early mover in the mobile market, which is that you can get customers to sign a contract lasting a year or more, and offer inducements such as subsidized handsets to get them to renew the contract. The new entrant, in contrast, must find (probably expensive) methods of getting existing subscribers on other networks to churn (Klemperer 1987a, 1987b), or of attracting potential subscribers to sign up with an untried network.

A review of the theoretical and empirical studies covering what they call the early-mover advantage is to be found in Bijwaard, Janssen and Maasland (2008). They point out that the vast majority of the empirical studies find that 'market pioneers have substantially higher market shares than later entrants', although there are dissenting voices that emphasize the failure of these studies to treat entry as an exogenous rather than an endogenous variable. The authors claim that their own study is pioneering in that it uses data from one particular sector where purchases are infrequent and that, by utilizing data from the 1990s, it overcomes the problem of whether entry is exogenous on the grounds that this is a consequence of operators only being able to enter markets when licences are issued.

The authors admit that because data (which anyway are restricted to Europe) are only available for market penetration levels, the number and names of active operators and their market shares, it is not possible to discriminate empirically between the various explanations for first-mover advantages. The authors nevertheless conclude that 'the impact of the penetration rate is quite strong, significant and highly non-linear and that it is best to enter as early as possible'. Expressed rather crudely, if the penetration rate is reasonably high and there are three sizeable incumbents, a new entrant into the 2G market in Europe has a negligible chance of making profits.

The authors do not incorporate 3G into their analysis because it was too recent a phenomenon at the time of writing, but there is no very obvious reason why a new entrant should find it easier to build a market presence via 3G than via 2G – indeed, quite the opposite. It is accordingly time to examine some of the relevant facts.

3G Licensing in Europe

The starting point for our analysis is Table 9.1. This table depicts the end-2009 state of 3G licensing across Europe, which is defined here as encompassing the 27 Member States of the European Union (EU), the European Economic Free Trade Area, prospective accession countries to the EU and all other countries or territories having some form of independent government within the post-Communist understanding of Europe. In total, there are 47 independent entries – referred to as countries for convenience in what follows – in the table.

Drawing on Table 9.1, the first observation that can be made is that almost every country within Europe has issued 3G licences. As of 31 December 2009, 43 countries had awarded 3G licences. The first 3G licence in Europe was awarded in March 1999 by Finland, which was joined at the forefront of 3G in April 1999 by the Isle of Man. A significant proportion of the listed countries awarded their 3G licenses during 2000 and 2001, in good part because of the timetable for the launch of 3G laid down by the European Union. Since 2001, 3G licensing has steadily permeated the rest of Europe.

A second observation that can be made is that 23 countries opted to increase the number of companies in the mobile market by issuing more 3G licences than there were 2G incumbents at the time. Most countries increased the number of 2G mobile licences by one when choosing how many 3G licences to issue, although Austria, Germany, Luxembourg and Norway opted to issue two additional 3G licences.

Table 9.1 3G licensing across Europe. 31 December 2009

Country/ Territory	2G licences[1]	3G licences available	3G licences awarded	Method[1] (BC = beauty contest)	Date	3G licence winners[1]
Albania[2]	3	–	–	–	–	–
Andorra[2]	1	1	1	Assigned	Jan 2005	STA
Austria	4	4-6	6	BC + auction	Nov 2000	Hutchison 3G, max.mobil, Mobilkom Austria, ONE, tele. ring,[12] 3G Mobile[13]
Belgium	3	4	3	Auction	Feb 2001	KPN Mobile 3G, Mobistar, Proximus
Bosnia-Herz.[2]	3	3	3	Assigned	Mar 2009	Eronet, GSM BiH, Mobilna Srpske
Bulgaria	3	3	3	Auction (1) Assigned (2)	Mar 2005[5]	MobilTel, Viva Ventures, GloBul
Croatia[2]	2	3	2	Tender – assigned[4]	Oct 2004	T-Mobile, VIPnet
Croatia[2]	3	1	1	Assigned	Dec 2004	Treca Sreca
Cyprus (S)	2	2	2	Auction – assigned[4]	Dec 2003	Investcom, CyTA
Czech Repub.	3	3	2	Auction – assigned[4]	Dec 2001[6]	EuroTel Praha, RadioMobil
Czech Repub.	3	1	1	Assigned	Feb 2005	Oskar
Denmark	4	4	4	Auction	Sep 2001	Hi3G Denmark, Orange,[14] TDC, Telia Denmark
Denmark	3	1	1	Auction – assigned[4]	Dec 2005	Sonofon[14]
Estonia	3	3	3	BC – assigned[4]	Jul 2003[7]	Eesti Telecom, Radiolinja, Tele2
Estonia	3	1	1	Tender	Dec 2006	Grosson/Renberg/RealGroup/ ProGroup[15]
Faroe Isles[2]	2	–	–	–	–	–
Finland	3	4	4	BC	Mar 1999	Radiolinja, Sonera, Suomen 3G,[16] Telia Finland
France	3	4	2	BC – assigned[4]	Jul 2001	Orange, SFR
France	3	2	1	BC – assigned[4]	Sep 2002	Bouygues Télécom
France	3	1	0	Tender	Oct 2007	–
France	3	3	1	BC – assigned[4]	Dec 2009	Free Mobile
Germany	4	4–6	6	Auction	Jul 2000	E-Plus Hutchison, Group 3G, Mannesmann, MobilCom Multimedia[17], T-Mobile, Viag Interkom
Gibraltar[2]	1	–	–	–	–	–
Greece	3	4	3	Auction – assigned[4]	Jul 2001	CosmOTE, Panafon, Stet Hellas
Guernsey[2]	1	2	1	BC – assigned[4]	Mar 2003	Wave Telecom
Guernsey[2]	2	1	1	BC	Sep 2006	Guernsey TeleNet
Hungary	3	4	3	Tender	Dec 2004	Pannon, T-Mobile, Vodafone
Hungary	3	1	0	Tender	Mar 2009	–[18]
Iceland[2]	3	4	3	BC – assigned	Mar 2007	Og fjarskipti, Novator, Síminn[19]

Table 9.1 ***Continued***

Ireland[2]	3	4	3	BC – assigned	Jun 2002[8]	Hutchison 3G Ireland, mmO_2, Vodafone
Ireland	3	1	1	BC	Nov 2005	Smart Telecom[20]
Ireland	3	1	1	Assigned	Mar 2007	eircom (Meteor)[20]
Isle of Man[2]	1	1	1	Assigned	Apr 1999	Manx Telecom
Isle of Man[2]	1	2	2	Assigned	May 2006	Cable & Wireless, Wire9 Telecom
Italy	4	5	5	BC + auction	Nov 2000	H3G, IPSE 2000, TIM, Wind, Omnitel
Italy	4	1–3	3	TE	May 2009	TIM, Wind, Vodafone[21]
Jersey[2]	1	4	3	Assigned[4]	Dec 2005	Jersey Telecom, Cable & Wireless, Jersey AirTel,
Latvia	2	3	2	Auction – assigned[4]	Sep 2002	LMT, Tele2
Latvia	3	1	1	Auction	May 2005	Bité
Liechtenstein[2]	4	4	3	Assigned	Jul 2001[9]	mobilkom, Tele2/Tango, Viag Europlattform
Liechtenstein[2]	4	1	1	Assigned	Oct 2003[10]	Liechtenstein TeleNet
Lithuania	3	3	3	BC – assigned	Feb 2006	Bité, Omnitel, Tele2
Luxembourg	2	4	3	BC – assigned[4]	May 2002	EPT, Orange,[22] Tele2
Luxembourg	2	1	1	Tender – assigned	Jul 2003	LuXcommunications
Macedonia[2]	3	up to 4	1	Tender	Jan 2008	OTE
Macedonia	3	3	1	Tender – assigned[4]	Dec 2008	T-Mobile
Macedonia	3	2	0	Tender	Jul 2009	–
Malta	2	3	3	BC – assigned[4]	Aug 2005	Go Mobile, Vodafone, 3G Telecoms
Monaco[2]	1	1	1	Assigned	Jun 2000	Monaco Telecom
Montenegro[2]	2	1	1	Tender	Mar 2007	m:tel
Montenegro[2]	3	2	2	Assigned	May 2007	ProMonte, T-Mobile
Netherlands	5	5	5	Auction	Jul 2000	3G-Blue, Dutchtone, KPN Mobile, Libertel-Vodafone, Telfort[23]
Norway[2]	2	4	4	BC	Dec 2000	Broadband Mobile, NetCom GSM, Telenor, Tele2
Norway[2]	2	2[3]	1	Auction – assigned[4]	Sep 2003	Hi3G Access
Norway[2]	3	1[3]	1	Auction – assigned	Dec 2007	Mobile Norway
Poland	3	4	3	Tender – assigned[4]	Dec 2000	PKT Centertel, Polkomtel, Polska Telefonica Cyfrowa
Poland	3	1	1	Tender	May 2005	Netia (P4)
Portugal	3	4	4	BC	Dec 2000	ONI-Way,[24] Optimus, Telecel, TMN
Romania	4	4	2	BC	Nov 2004	MobiFon, Orange
Romania	4	2	2	BC	Oct 2006	RCS&RDS, TeleMobil[25]
San Marino	1	–	–	–	–	–

Table 9.1 *Concluded*

Serbia[2]	2	3	3	Assigned	Aug 2006	Telekom Srbija, Telenor, mobilkom Austria
Slovakia	2	3	3	Auction – assigned	Jul 2002	EuroTel, Orange, Profinet.sk[26]
Slovakia	2	1	1	Tender	Aug 2006	Telefónica
Slovenia	3	3	1	Auction – assigned[4]	Nov 2001[11]	Mobitel
Slovenia	3	3	2	Auction	Sep 2006	Si.mobil, T-2
Slovenia	3	1	1	Assigned	Apr 2008	Tusmobil
Spain	3	4	4	BC	Mar 2000	AirTel, Amena, Telefónica, Xfera
Sweden	4	4	4	BC	Dec 2000	Europolitan, Hi3G Access, Orange Sverige,[27] Tele2
Switzerland[2]	3	4	4	Auction	Dec 2000	Dspeed, Orange, Swisscom, Team 3G[28]
Turkey[2]	3	4	3	Auction – assigned[4]	Dec 2008	Avea, Turkcell, Vodafone
UK	4	5	5	Auction	May 2000	BT3G, Hutchison 3G, One-2-One, Orange, Vodafone
Ukraine[2]	5	1	1	Assigned	Nov 2005	Ukrtelecom

Sources: Annual reports, company websites, regulators' websites, other websites, media reports

Notes:

1. The number of 2G licences is applicable to the time of the event. The determination of method is complicated in a number of cases. Some of these are commented on in the notes, but space does not allow for a full exposition. Other sources may accordingly give a different version. Licensees are cited under the names used when the licence was first issued.
2. Not an EU Member State.
3. Two of the 2003 licences had been returned by the original licensees – Broadband Mobile in August 2001 and Tele2 in November 2002 (which became the licence acquired by Hi3G Access) with the other returned licence being bought by Mobile Norway (50 per cent owned by Tele2) which had recently acquired a GSM licence.
4. The initial intended licensing method was abandoned in favour of an assignment given the shortage of applicants.
5. The three licences were not awarded at the same time or through the same method. MobilTel was awarded its licence in March 2005 after a tendering process was completed, while Viva Ventures and GloBul were assigned their licences in April 2005.
6. The award of two licences in December 2001 was actually the third occasion on which the Czech Republic had attempted to award 3G licences. The previous two attempts at a tender, in September and October 2001, both failed to attract bidders.
7. The three licences were not awarded at the same time. Eesti Telecom and Radiolinja received their licences in July 2003, and Tele2 in August 2003.
8. The three licences were not awarded at the same time. Hutchison 3G Ireland received its licence in June 2002, mmO_2 in August 2002 and Vodafone Ireland in September 2002.
9. The three licences were not awarded at the same time. Viag Europlattform finally accepted its licence in March 2001, while Tele2/Tango and mobilkom received their licences in July 2001
10. Telecom FL initially refused the offer of a licence. Its owner, Swisscom, then sold the company to the government in July 2003 and when the transfer was completed in October, the now renamed Liechtenstein TeleNet accepted the licence.
11. At the second attempt. An auction planned for May 2001 attracted no bidders.
12. As a condition for its acquisition of tele.ring in April 2006, T-Mobile was obliged to dispose of tele. ring's two sets of 5 MHz paired 3G spectrum, of which at least one had to go to existing licensee Hutchison 3G Austria.

13. 3G Mobile sold its Austrian licence to mobilkom Austria in December 2003. Half the spectrum (5 MHz paired) was sold on to T-Mobile.
14. When TeliaSonera acquired Orange Denmark it was obliged to return one of its two 3G licences which was re-auctioned.
15. Originally awarded to Grosson Capital in November which failed to pay, so it was subsequently offered to Renberg Investments (which declined), RealGroup (which failed to pay) and ProGroup.
16. The licence held by Tele2 was revoked in July 2005.
17. Mob*il*Com returned its licence in December 2003.
18. The tender was cancelled due to adverse market conditions.
19. One licence was awarded in April.
20. The licence acquired by Smart was finally revoked in November 2006 and offered to eircom which had bought 2G incumbent Meteor.
21. The licence returned by IPSE 2000 was split into three blocks and offered to incumbents (one or two blocks) and new entrants (up to three blocks). No new entrants applied and incumbents were awarded one block apiece.
22. Orange (which had yet to roll out its 2G network) returned its licence in December 2004.
23. The licence, acquired by KPN when it bought Telfort, was returned in July 2009.
24. The licence was revoked in January 2003 and the spectrum divided up among the other licensees.
25. The first instalment was not paid until January 2007 when the award became official. TeleMobil (Zapp Mobile) was sold with its licence to OTE in June 2009.
26. Although it technically won the licence, Profinet.sk did not make the required down payment and the licence was revoked in August 2002.
27. Orange Sverige sold its licence to Svenska UMTS-nät in December 2003, but this was not sanctioned by the regulator. In November 2004, the regulator recalled the licence.
28. The licence was revoked in April 2006 and handed back in April 2007.

It is possible to identify 37 3G new entrants across Europe. A new entrant is defined here either as a 3G licensee that does not have already a 2G licence or as a bidding consortium not majority-owned by a 2G licensee in the market where it is bidding for a 3G licence. On this basis, listed by country and using the names in current usage with licensee names in brackets, the 3G new entrants are:

- Austria: Hutchison 3G, 3G Mobile
- Croatia: Tele 2 (Treca Sreca)
- Cyprus (south): MTN (Investcom)
- Denmark: Hi3G Denmark
- Estonia: Bravocom
- Finland: Telia Finland
- France: Free Mobile
- Germany: Group 3G, MobilCom Multimedia
- Guernsey: Wave Telecom, AirTel (Guernsey TeleNet)
- Iceland: Nova
- Ireland: Hutchison 3G, Smart Telecom
- Italy: Hutchison 3G, IPSE 2000
- Jersey: Jersey AirTel
- Latvia: Bité
- Luxembourg: LuXcommunications, Orange
- Malta: 3G Telecoms

- Montenegro: m:tel
- Norway: Broadband Mobile, Hi3G Access
- Poland: Netia P4
- Portugal: ONI-Way
- Romania: RCS&RDS
- Serbia: mobilkom Austria
- Slovakia: Profinet.sk
- Slovenia: T-2, Tusmobil
- Spain: Yoigo (Xfera)
- Sweden: Hi3G Access, Orange Sverige
- Switzerland: 3G Mobile Suiza
- UK: Hutchison 3G

In addition, it should be borne in mind that one 2G/3G licensee in Liechtenstein, Telekom FL, was sold to the government and relaunched as Liechtenstein TeleNet, took over the 3G licence. Further, that when Smart Telecom in Ireland forfeited its 3G licence, it was acquired by eircom which had recently acquired a 2G licensee Meteor. Neither are accordingly treated here as 3G new entrants. A final exception is Zapp Mobile in Romania which, at the time it acquired a 3G licence, was already operating CDMA (code division multiple access) in the 450 MHz band – it (and its licence) is currently owned by OTE.

A closer inspection of the 37 new entrants allows a clear distinction to be made between those that have been able to launch their 3G services and those that have, for whatever reason, failed (so far) to do so. As shown in Table 9.2, just 21 of the 37 new entrants had launched their 3G services by the end of 2009 and, as the notes to Table 9.1 demonstrate, a significant proportion of the sample no longer have licences and hence never will launch. However, this table also highlights the fact that four companies – Sonera (now part of TeliaSonera), Telefónica, France Télécom/Orange and Hutchison Whampoa – originally set out to use the 3G licensing process to enter new markets.

The tribulations of Sonera, Telefónica and France Télécom/Orange are set out in detail in Curwen and Whalley (2006). It is evident from Table 9.2 that only Hutchison Whampoa remains as a major active 3G new entrant. Hutchison Whampoa is present as a new entrant in seven European markets and has launched networks in all cases bar Norway.

Hutchison Whampoa is clearly an oddity within the European mobile industry. First of all, it is axiomatic that, since Hutchison Whampoa has no installed 2G customer base to fund its expansion into 3G, it must rely on other sources for funds. The foray into 3G has been financed by a very deep-pocketed parent company that seems willing to suffer huge short-term losses and to persevere against the odds that have led incumbents like Sonera and Telefónica to abandon most or all of their 3G new entrant investments. Secondly, the roll-out strategy adopted by Hutchison Whampoa is almost diametrically opposite to that of every incumbent. By and large, incumbents decided early on that they were earning massive revenues from their 2G networks, the 3G technology was immature and that handsets were either unavailable and/or clunky and that if one incumbent was holding back there was every incentive for the others to follow suit. In contrast, Hutchison Whampoa needed to obtain a revenue flow as early as possible and hence chose to be the first to launch in every market if humanly possible.

Table 9.2 European 3G new entrants. 31 December 2009

Country	Company	Date service launched[1]	Number of subscribers 31/12/08	Main shareholders[2]
Austria	Hutchison 3G	April 2003	657,000	Hutchison Whampoa
Austria	3G Mobile	Never[3]	–	Telefónica
Croatia	Tele2	–	–	Tele2
Cyprus (S)	MTN	Dec 2004	16,000	MTN
Denmark	Hi3G Denmark	Nov 2003	410,000	Hutchison Whampoa, Investor
Estonia	Bravocom	–	–	Bravocom Mobil
Finland	Telia Finland	Oct 2004	906,000	TeliaSonera
France	Free Mobile	–	–	Iliad
Germany	Group 3G	Never[4]	–	Sonera, Telefónica
Germany	MobilCom Multimedia	Never[5]	–	France Télécom, MobilCom
Guernsey	Wave Telecom	Jul 2004	5,000	Jersey Telecom Group
Guernsey	AirTel	Mar 2008	1,000	Bharti Group
Iceland	Nova	Dec 2007	26,000	Novator
Ireland	H3G Ireland	July 2005	479,000	Hutchison Whampoa
Ireland	Smart Telecom	Never[6]	–	Private investors
Italy	H3G Italy	Mar 2003	8,517,000	Hutchison Whampoa
Italy	IPSE 2000	Never[7]	–	Sonera, Telefónica
Jersey	Jersey AirTel	June 2007	Negligible	Bharti Group
Latvia	Bité	June 2006	54,000	TDC
Luxembourg	LuXcommunications	Mar 2004	24,000	Mobistar (Orange)
Luxembourg	Orange	Never[8]	–	Orange
Malta	3G Telecoms	Feb 2009	–	Melita
Montenegro	m:tel	July 2007	39,000	Telekom Srbija, Ogalar
Norway	Broadband Mobile	Never[9]	–	Sonera, Enitel
Norway	Hi3G Access	–	–	Hutchison Whampoa, Investor
Poland	Netia P4	Mar 2007	2,010,000	Netia
Portugal	ONI-Way	Never[10]	–	ONI, Telenor
Romania	RCS&RDS	Dec 2007	1,373,000	RCS&RDS
Serbia	mobilkom Austria	July 2007	Negligible	Telekom Austria
Slovakia	Profinet.sk	Never[11]	–	Profinet
Slovenia	T-2	–	–	Zvon Ena Holding
Slovenia	Tusmobil	Jun 2008	Not known	Tus
Spain	Yoigo	Dec 2006	970,000	Vivendi Universal, Sonera, ACS
Sweden	Hi3G Access	Apr 2003	822,000	Hutchison Whampoa, Investor
Sweden	Orange Sverige	Never[12]	–	Orange
Switzerland	3G Mobile (Suiza)	Never[13]	–	Telefónica
UK	Hutchison 3G	Mar 2003	4,820,000	Hutchison Whampoa, NTT DoCoMo

Source: Calculated by authors using a wide range of sources

Notes:

1. 'Launch' is taken here to be the date when the service is first made available, usually to corporate customers, via data cards in laptops.
2. When licensed.
3. The licence was sold to mobilkom Austria in December 2003.
4. Shut down in 2002.
5. Licence returned to regulator in 2003.
6. The licence was revoked and offered to eircom in 2007.
7. Licence revoked in 2006.
8. Licence returned in 2004.
9. Licence returned in 2001.
10. Licence revoked in 2003.
11. Licence never formally issued due to licence condition breach.
12. Licence revoked in 2004.
13. Licence revoked in 2006.

Such a strategy was, and remains, very risky, not least because it assumes that the initial outlays can be recouped once a subscriber base has been accumulated. But Hutchison's subscriber base remains relatively small in global terms at roughly 20 million, of which the majority are accounted for by just two networks. As a consequence, it cannot spread its costs over as large a subscriber base as can its 2G rivals, nor for that matter can it use a large installed subscriber base to achieve significant scale economies.

Hutchison Whampoa is also disadvantaged by the inherent tension of growing its subscriber base through handset subsidies and competing largely on price. Handset subsidies are costly and only pay for themselves after several years of subscriber revenue. However, this payback period is extended when prices are lower than those of competitors.

Given the huge cost of rolling out networks, it was Hutchinson's intention to float minority stakes in the more promising networks, but this plan was effectively abandoned in early 2006 when it was finally acknowledged that with little product differentiation, investors would not find the strategy of buying market share particularly attractive. True, a 10 per cent stake in the Italian business was sold to an investment bank for €420 million in February 2006, but only after a €7 billion ($8.3 billion) float was cancelled – the original valuation was €12 billion (Guerrera and Lau 2006; Michaels 2006). By the year end the outlook appeared to be so poor that commentators were touting the prospect that Hutchison might either exit the European market or merge with incumbents (Cellular-news 2006; TelecomDirectNews 2006), but despite the periodic re-emergence of these rumours, no significant progress has so far been made.

Prospects for Smaller Operators in European Countries

At this point it is worth considering the extent to which the potential for new entry is governed by structural features of the market. Expressed rather crudely, the issue is whether there is room in a given market for more than a specified number of profitable operators. It is helpful to begin by examining mobile market shares as determined by the number of subscribers. These can be divided into a number of categories as shown in Table 9.3. The first row defines the range in terms of the total number of subscribers in a country, expressed in millions (but excluding those on CDMA450 networks); the second is the number of countries within that range; and the third the average number of subscribers in millions.

Table 9.3 Aspects of market concentration in Europe. June 2009

Range (M)	0.0–0.1	0.1–1	1–5	5–10	10–20	20–50	50–80	80–110
Number	7	5	9	8	9	2	5	2
Average (M)	0.05	0.34	2.43	6.81	13.67	35.48	60.83	98.36
Monopoly	4	0	0	0	0	0	0	0
Duopoly	1	2	1	0	0	0	0	0
Duopoly+	1	2	4	2	0	0	0	0
> 3 networks	1	0	0	2	2	2	3	2

Source: Compiled by authors

Roughly 80 per cent of the entries fall within the 1 to 20 million range, but there are also a significant number of tiny markets with total subscriber numbers below 100,000. Even in relatively developed Europe, markets in excess of 20 million remain uncommon, at least for now.

In principle, one would expect the smallest market to be the most highly-concentrated as there does not appear to be room for economically viable competition, and, indeed, all four monopolies are to be found in this category. However, the duopolies are more widespread. When defined technically as a two-network market, there are only four, all in small markets, but if a more informative definition is used whereby it includes any market with two networks accounting for over 90 per cent of the subscribers – referred to as duopoly+ in Table 9.3 – then the picture changes somewhat in that there are nine cases of duopoly+ spread over the four smallest ranges.

Table 9.3 also indicates where there are more than three networks in operation – 12 markets in total, almost all (tiny Liechtenstein is an exception because of bordering networks) spread evenly across the markets with more than five million subscribers. The reasonable assumption is that the market share of the largest three networks will be in excess of 90 per cent in every case, but there are two main exceptions other than Liechtenstein, namely Germany and the UK. Realistically, only the UK is a true exception in that there are two networks that were originally licensed only for the 900 MHz band and which failed to become sufficiently established to fend off the entry of the two networks in the 1800 MHz band. Indeed, each would have very close to a quarter of the market apiece (as measured by subscribers) were it not for 3G licensee Hutchison.

So the overall picture presented in Table 9.3 is on the face of it fairly emphatic: unless the size of the market is below roughly half a million, the optimum number of networks is either three or four, with four replacing three as the total number of subscribers crosses the 20 million threshold. However, this leaves one major unanswered question, namely whether this is a natural construct that has arisen as an extrapolation of early history – that is, there were, say, four networks at the beginning and there still are today – or the result of a restructuring of the market due either to new entry or, more likely, the disappearance of former incumbents.

The new entry issue has been discussed in detail above and the conclusion was straightforward: only Hutchison has made real inroads as a new entrant and Hutchison only holds 3G licences with negotiated 2G roaming rights. Furthermore, its subscribers are significant only in Italy and the UK where the overall markets are very large, so even there it is not significant in terms of market share. Nor, for that matter, is the other relatively successful new entrant, Netia P4 in Poland, which was also awarded E-GSM (enhanced global system for mobile) spectrum in January 2009.

Hutchison may or may not prove to be successful in selling its European assets, but for now it remains a modest fourth-place operator in Austria, Denmark, Ireland, Italy and Sweden while occupying only fifth place in the UK. Setting aside Hutchison and ignoring Liechtenstein, we have two fourth-place networks with more than 10 per cent of the market in the form of Telefónica in Germany (13.4 per cent) and Orange in the UK (21 per cent). The fourth-place networks with less than 10 per cent of the market are as follows:

- Poland where Netia P4 is a 3G new entrant.
- Romania where RCS&RDS is a 3G new entrant.
- Spain where Yoigo is a 3G new entrant.
- Ukraine where Beeline has no 3G licence.

Yoigo is owned by TeliaSonera, while the others are not directly associated with incumbents although Beeline is owned by VimpelCom which has a minority stake in incumbent Kyivstar. Insofar as the first three have 3G subscribers that would be useful to the 2G incumbents, they could reasonably be expected to be acquired by one or other of them. However, given the current depressed values of networks, and the relatively small numbers of subscribers involved, it is unlikely that any incumbent will be willing to pay the sort of premium to the ruling market price that would lead to a sale.

In fact, it may reasonably be argued that if networks are to disappear, it will not be the small fry like the above where the cost of acquisition is relatively high and the potential for scale economies is low, but rather the networks that have market shares in excess of 10 per cent, but are not sufficiently profitable for existing owners.

It is logical to start by looking at the UK because there are five networks and competition is intense (partly due to the activities of Hutchison, as noted above). As noted, Orange is (only just) the smallest of the four main networks, as defined by number of subscribers, but that does not appear to be troubling France Télécom. In fact, it is Deutsche Telekom that is troubled, having written down its UK network by €1.8 billion – virtually all of the goodwill – in May 2009 (Wiesmann 2009; Wiesmann and Parker 2009). The solutions proposed included outright sale, muddling along in some sort of (rather better focused) relationship with Hutchison or a merger with another network (Cellular-news 2009a). What transpired was a proposed joint venture with Orange, labelled as ever as a 'merger of equals' (Lennighan 2009).

The UK may be fiercely competitive, but it is still a little alarming to read that a network with 17 million subscribers is in serious trouble – after all, nearly 40 European markets are smaller in total than this one network. And it is not as though things are about to improve because the European Commission has been waging a largely successful war to drive down mobile termination rates (MTRs) and roaming rates (Global Insight 2009b) with the result that mobile call prices fell by an average of 23 per cent in 2008 (Global Insight 2009a). Furthermore, T-Mobile parent Deutsche Telekom is one of the largest telcos in the world and T-Mobile is one of the world's largest mobile groups, so scale economies must be more readily available to both than to most other companies.

Experience Outside Europe

It may be argued that Europe provides the bulk of the evidence concerning the prospects for 3G new entrants, but that this evidence does not apply to circumstances in less-developed countries with penetration levels still comfortably below 100 per cent. Equally, there are developed countries with somewhat unique circumstances. For these reason, some additional case study material is considered below, although it is clearly not possible to be truly representative.

INDIA

It is useful to begin by contrasting the above with the experience of India where the country is divided into 22 (previously 23) service areas (circles) with varying numbers of competitors in each one, the government is constantly at loggerheads with the regulator (the TRAI), the rules on foreign ownership and their interpretation are constantly in dispute, the state-owned operators are given priority on a random basis, spectrum is too scarce to meet the needs of 3G licensees and everyone is bemused by the constantly changing regulations.

Consider, first, the issue of ownership. It is hard to believe that, not so long ago, operators had reservations about investing in India. For example, France Télécom sold its stake in BPL Mobile in December 2004 and the then AT&T Wireless sold a 32.9 per cent stake in Idea Cellular in September 2005. However, *inter alia*, foreign investors were restricted at the time to a minority 49 per cent stake in an Indian operator, huge infrastructure-based investments were required despite a prospective very low ARPU (average revenue per user) and the regulatory environment was volatile. Nevertheless, the sheer size of the potential market and the expectation of the authorization of 74 per cent foreign control, which duly took place in October 2005, stimulated a surge of foreign interest in Indian operators during 2005 and 2006. For example:

- Maxis Communications of Malaysia bought 74 per cent of Aircel during 2006.
- Telekom Malaysia bought a 49 per cent stake in Spice Telecom in June 2006, having failed to buy Aircel in December 2005.
- Vodafone acquired 10 per cent of Bharti AirTel from Warburg Pincus in November 2005.

However, the seismic event was the acquisition of a majority stake in Hutchison Essar by Vodafone at the end of 2007. Vodafone's offer won the day by placing an effective valuation on the entire company of $18.8 billion. As a consequence, Vodafone was obliged to sell part of its existing stake in Bharti AirTel – in which Singapore Telecom also held a roughly 30 per cent stake – at a price which valued the whole of Bharti at close to $30 billion.

Valuations at this level caused other foreign operators to sit up and take notice, but the situation was far from straightforward. In the first place, by no means all existing operators held licences in all the then 23 circles. Secondly, some used GSM (global system for mobile) – the largest being Bharti AirTel, state-owned BSNL, Idea Cellular and the now Vodafone Essar – while others used CDMA, in particular Reliance Communications. Thirdly, the sector was restructuring rapidly. For example, Spice set out to merge with Idea Cellular, giving Telekom Malaysia a 20 per cent stake in the merged entity.

Most significantly of all, the unified access services (UAS) licences regime, introduced in December 2003, had opened up the possibility of new entrants. Since there were already multiple competitors in every circle, this had not led to a rush for new licences, especially since there was a marked preference for GSM even among most existing CDMA licensees. But when the prospect of obtaining GSM spectrum in at least some circles was held out to any company that applied for a unified licence before 25 September 2007, the queue of successful applicants contained, *inter alia*, UniTech, Swan Communications, S. Tel, Datacom Solutions, Agrani Telecom and ByCell (Haslam 2009).

In September 2008, Etisalat had agreed to pay $900 million for a 45 per cent stake in Swan Telecom which possessed licences in 13 circles and had been awarded spectrum in 10. In November, Telenor agreed to pay $1.1 billion for a 60 per cent stake in national operator UniTech Wireless. What this meant in effect was that any start-up operation, provided it possessed some licences and spectrum, could expect to be valued at $2 billion. Not being deterred, DoCoMo promptly moved to pay $2.7 billion for a 26 per cent stake in national operator Tata Teleservices.

The news was met with astonishment. True, Vodafone had also apparently overpaid, but in practice it had transformed the now Vodafone Essar into an efficient operation with 55 million subscribers, and its head-to-head struggle with its even larger rivals, Bharti AirTel and Reliance, was rapidly driving down tariffs. Nevertheless, Russia-based Sistema set out to join in after buying roughly 72 per cent of national operator Shyam TeleLink, while Turkcell was negotiating a 51 per cent stake in Datacom Solutions (valuing it at $2 billion overall).

Although the rapid growth of 2G subscriptions was arguably sufficient to attract new entrants, a further pull was exerted by the potential for acquiring 3G licences. It is not possible to explain fully the Byzantine processes that have led to the latest utterances on which operators are to get 3G licences, bearing in mind that it was originally intended to issue licences in 2003. The first point to bear in mind is that most GSM operators have been issued with, on average, 7 MHz of 2G spectrum – far less than in, for example, Europe and clearly not enough to roll out 3G services. Secondly, as ever, the military is sitting on a lot of highly desirable spectrum which it does not want to release. Thirdly, the government has prioritized the needs of state-owned BSNL and MTNL.

The TRAI published its recommendations for 2G/3G in May 2005 which, *inter alia*, included provisions that there should be an annual licence fee for 3G spectrum and that new entrants should only be allowed to apply for spectrum not already assigned to 2G operators. In March 2006, the Department of Telecommunications (DoT) recommended that the amount of 3G spectrum assigned should be proportional to the number of 2G subscribers. The TRAI responded by arguing that 2G (for voice) should be viewed entirely separately from 3G (for data). In April 2007, the DoT overrode the TRAI by recommending that only four licences should be issued in the 2100 MHz band with two reserved for BSNL and MTNL. By 2008, the DoT and TRAI seemed intent on little other than immediate contradiction of each other's statements.

The position at the end of 2008 was roughly as follows. According to Government of India (2008) the (unspecified) spectrum was to be auctioned in January 2009, but this was postponed until December. The initial auction would probably take place either in 14 circles (said the DoT) or all 22 (said the Finance Ministry which appeared to be at constant loggerheads with the DoT). Foreign companies with a stake of 74 per cent or less in a Bidding Entity, which had to acquire a UAS licence (which would not be automatic) prior to its launch, would be allowed to bid for 3G licences provided they paid a supplementary fee and acquired a local partner with a 10 per cent stake in the Entity. However, every single fee remained a matter of fierce dispute.

During 2009, a fundamental dispute arose over the number of potential licensees. Between 13 and 19 applications for licences were pending in every circle, and as a consequence the DoT asked the TRAI to restrict the number of licences (TeleGeography 2009b). The matter was placed in the hands of the Empowered Group of Ministers (Cellular-news 2009b). The eventual agreed plan was that of the initial three to five licences in each circle (other than Rajasthan (0), North East (1) and West Bengal (2)), each consisting of 5 MHz paired (Cellular-news 2009c), MTNL (Mumbai and Delhi) and BSNL (other circles) would be awarded theirs immediately – they both launched in early 2009 in limited areas with little success (TeleGeography 2009c).

The minimum price per 5 MHz paired was set at a compromise $714 million in August 2009. Successful applicants (other than MTNL and BSNL) would be required to pay 25 per cent of their bid immediately, but would only be awarded their spectrum as and when it became available. The armed forces were becoming increasingly intransigent about releasing spectrum so this process appeared likely to take until 2011 to be completed (TeleGeography 2009d) – in October 2009 they told the DoT that they would be postponing the release of the agreed 20 MHz of spectrum from December 2009 to June 2010. Existing operators would have to pay an additional 1 per cent of their annual revenues to the government (but would benefit from tax exemptions) whereas new 3G-only entrants would probably have to pay 3 per cent – this was confirmed in September 2009 (Dow Jones 2009). In addition to the above, one slot of 1.25 MHz paired was made available in each circle for cdma2000 1xEV-DO operations at a minimum price of $178.5 million.

It became evident that the auction would not take place in December because a sequence of 10 steps, including a mock auction, had first to be completed and it had yet to commence (Gabriel 2009). It is officially now scheduled for January 2010, probably with three licences on offer, but with possibly one of these complete with its spectrum. The other two (or all three) will receive spectrum in September 2010 (Wood 2009). Hence, in summary, a situation existed where huge valuations were placed upon new entrants that intended to apply for 3G licences which they might or might not win and, even if they did win them, might not be assigned any spectrum for at least a year.

THE USA

Historically, spectrum had not been seen as a valuable resource. As a consequence, a band of spectrum that could have been used for 3G in the USA – the 700 MHz band

– was occupied by broadcasters, while 1710–1860 MHz was heavily occupied by the military, 2110–2150 MHz was used by schools and health care centres and 2500–2690 MHz was reserved for the Multipoint Distribution Service (MDS) and the Instructional Television Fixed Service (ITFS). This meant that when the FCC (Federal Communications Commission) came to auction spectrum specifically designated for 3G services, it would be obliged to offer it 'encumbered' in most areas. The winners – assuming anyone wanted it in the first place – would then have to persuade (that is, bribe) the owners to release it, but if the owners proved unwilling to do so because, for example, in the case of broadcasters this would cause them to lose their cable channel and hence, in effect, go out of business, the spectrum would remain useless for 3G for years to come.

Not surprisingly, therefore, there was no prospect of conducting a single 3G auction in the manner of European Union countries. Furthermore, given the paucity of national operators, there was no prospect of holding an auction solely for nationwide licences since there could be no new entry and hence there would be no competitive bidding. The situation proved almost impossible to untangle until recently and incumbents, whether national or regional, simply rolled out high-speed services using their 2G spectrum despite the capacity limitations this imposed (Curwen 2005).

Recently, there has been progress in fostering new entry, albeit largely at a sub-national level, but it must be borne in mind that there must ultimately be a combination of bandwidths, including 2G, if high-speed data networks are to be introduced into the USA. Auction 66 (Federal Communications Commission 2006) comprised so-called AWS (advanced wireless service)-1 licences, consisting of 734 cellular market area (CMA), 352 economic area (EA) and 36 regional economic area grouping (REAG) licences. Six blocks, each of 5 MHz or 10 MHz paired, were offered in the 1710–1755 MHz and 2110–2115 MHz bands. There were 168 companies qualified to bid. After 161 rounds of bidding, the auction concluded in mid-September with 104 bidders winning some spectrum. While the most successful bidder was a 2G incumbent, T-Mobile USA, which bought 120 licences for $4.2 billion, a cable company consortium, SpectrumCo, also did very well and the other 2G incumbents made little effort to bid competitively as they did not need the spectrum. Furthermore, the licences obtained by regional operators MetroPCS and Leap Wireless were located such that the big three (AT&T, Sprint Nextel and Verizon Wireless) now faced one or other of these low-cost operators in every major market. Overall, therefore, some semblance of new entry was achieved.

In 2007, the FCC announced an auction (Auction 73) in the 700 MHz band (Federal Communications Commission 2007) comprising five blocks as follows: A = 6 MHz paired (EA); B = 6 MHz paired (CMA); C = 11 MHz paired (REAG); D = 5 MHz paired (national); E = 6 MHz unpaired (EA). Any unsold spectrum would be reoffered in Auction 76. Block C spectrum could be reconfigured into a national licence with a minimum bid of $4.64 billion, but 'open access' conditions would apply that prevented the winner(s) from restricting the type of handsets or software used on the resultant network. Altogether, 261 rounds of bidding raised a total of $19.6 billion, and the Block C spectrum was sold in pieces as these were worth more than the highest bid for the entire block. Although Verizon Wireless was the biggest winner at a cost of $9.6 billion, EchoStar, bidding via Frontier Wireless, won almost enough licences in the A and E blocks to provide national coverage. Once again, therefore, some semblance of new entry was achieved. However, the reality is that these auctions have not seriously threatened the position of the four national operators.

CANADA

The situation in Canada, where foreign ownership rules are relatively stringent, was not historically all that different from that in the USA given the small number of national operators combined with relatively large regional operators. Equally, Bell Canada, Rogers Wireless and Telus Mobility were initially able to launch high-speed services using 2G spectrum.

In November 2007, the government set the rules for the auction of 105 MHz of advanced wireless service (AWS) spectrum to take place on 27 May 2008. Of this, 40 MHz (Blocks B, C and D) would be set aside for new entrants which would need to demonstrate that they were entities 'including affiliates and associated entities' that 'held less than 10 per cent of the national wireless market based on revenue'. Blocks B and C would be sold in the form of 14 regional licences (eight provincial and six large service areas) with the other blocks divided into 59 regional service areas.

The auction concluded in July 2008 after 331 bidding rounds, raising a total of C$4.25 billion for the 282 licences sold. The most successful new entrant was Globalive Communications, an ISP (Internet service provider) backed by the Sawiris family of Egypt, with Data & Audio-Visual Enterprises (DAVE) Wireless, Shaw Communications and Quebecor also spending substantial sums. Nevertheless, the three heaviest spenders by a wide margin were the three national incumbents which seem set to use the new spectrum for LTE (long-term evolution) while rolling out HSPA (high-speed packet access) using their 2G spectrum (TeleGeography 2008). Hence, once again, the reality is that these and other auctions have not seriously threatened the position of the national operators.

AUSTRALIA

In mid-March 2001, the Australian government announced that it would be auctioning 15-year 3G licences, operational from October 2002. A simultaneous multi-round auction would be utilized. The available spectrum in the 2 GHz band, which covered 95 per cent of the population, was divided up such that 45 MHz paired plus 20 MHz unpaired was on offer in Canberra; 60 MHz paired plus 20 MHz unpaired was on offer in other state and territory capitals (in both cases with a metropolitan bidder's cap of 15 MHz paired and 5 MHz unpaired); and 20 MHz paired was on offer elsewhere (with a regional bidder's cap of 10 MHz paired). The 58 lots were arranged such that two national licences were available with a further two made possible by aggregating up the requisite lots. Registered bidders comprised: (1) established 2G incumbents bidding as Telstra 3G Spectrum, C&W Optus Mobile and Vodafone Pacific; (2) fixed-wire operators which were rolling out 2G mobile networks at the time, bidding as Hutchison Telecommunications (Australia) and AAPT Spectrum (controlled by Telecom New Zealand); and (3) new entrants CKW Wireless (a subsidiary of US-based ArrayComm) and 3G Investments (Australia) (controlled by Qualcomm). The established incumbents plus 3G Investments made the maximum down payment, indicating a wish to bid for a national licence, whereas Hutchison and AAPT bid slightly less and CKW a very small amount. However, AAPT withdrew in mid-March.

The auction ended after six days and 19 rounds with 48 of the lots under offer. The national licences went to C&W Optus – taken over by SingTel in August 2001 – which also bought other spectrum and paid $123 million, and Vodafone Pacific which acquired slightly more spectrum than Optus and paid $126 million. However, it was Telstra that paid the most – roughly $150 million – to buy more than enough spectrum for full national coverage. In addition, Hutchison paid $97 million for spectrum covering the five largest cities, 3G Investments acquired 10 MHz paired in the capital cities for $79 million and CKW bought 5 MHz unpaired in the same cities for under $5 million (which it intended to use exclusively for data services).

In May 2001, Telecom New Zealand announced strategic alliances to develop 3G services with both Hutchison Whampoa – via its subsidiary Hutchison Telecommunications (Australia) – and Microsoft. In the former case, the joint venture would be called Hutchison 3G Australia (H3GA) in Australia and Telecom 3G in New Zealand. In both cases, the Hutchison brand would be used. In January 2002, H3GA announced that in April it would be acquiring most of the reclaimed 2G assets built for the now bankrupt One.Tel by Lucent. This was alleged to provide H3GA with 60 per cent of the base stations it would need to roll out 3G, thereby permitting services to be launched in Sydney, Melbourne and Brisbane, possibly as early as the beginning of 2003. To expedite an extended roll-out Hutchison signed a roaming agreement with Vodafone covering both 2G and 3G services outside the metropolitan areas.

SingTel Optus, Telstra and Vodafone launched in 2005. In June 2009, Vodafone and the now renamed Hutchison 3G Australia merged to form a 50/50 joint venture called Vodafone Hutchison Australia (VHA), but this still comprised the smallest of the three incumbents which was essentially why its formation presented no major issues for the competition authorities (TeleGeography 2009a). In summary, therefore, the 3G auction had a negligible effect upon the industry structure, with the wholly new entrant acquiring spectrum only in the cities, one of the recently licensed 2G operators dropping out and the other ending up as half of VHA.

TURKEY

In Turkey, the incumbents Turkcell and Telsim had the 2G market to themselves until new PCN (personal communications network) licences were awarded in April 2001 to Aria (part-owned by Telecom Italia) and fixed-wire incumbent Turk Telecom (trading as Aycell) to commence services. The 3G licences were supposedly to be offered at the end of 2002, with the number of licences determined by the number of bidders, but no rules were published until 2007 when the government announced in February that four 20-year licences would be issued with minimum prices set respectively at €252 million ($344 million) for 45 MHz, €224 million for 45 MHz, €194 million for 35 MHz and €140 million for 25 MHz. A sealed bid tender was pencilled in for 24 May after which a second round of verbal bidding would take place. At this point the incumbents were Turkcell, Vodafone (formerly Telsim) and Avea (formerly Aria, now merged with Aycell), with bidding documents also taken up by Orange. However, the only bidder when the documents were opened in September proved to be Turkcell, which offered €321 million ($440 million) plus VAT after the government pressured it to raise its bid from an initial €311 million. It had placed bids for all four licences. The other operators made the point that mobile number portability (MNP) should have been introduced beforehand and the licence was cancelled on 19 September.

With MNP pencilled in for 13 November 2008, a possible new auction shortly thereafter was mooted. Meanwhile, Avea deployed a 3G network in anticipation of MNP. In September 2008, it was announced that a new tender would take place within six months with minimum prices set at €285 million, €250 million, €214 million and €178 million. In December, Turkcell won the 'A' licence with a bid of €358 ($462) million, Vodafone won the 'B' licence with a bid of €250 ($323) million and Avea won the 'C' licence with a bid of €214 ($276) million. No new entrant bid for the fourth licence.

SAUDI ARABIA

Saudi Arabia licensed a second GSM network in August 2004, raising $3.26 billion in the process (the then highest sum ever paid for a GSM licence), with applicants required to include 51 per cent local ownership. The licence was won by the Mobily consortium headed by Etisalat which simultaneously paid only $200 million for a 3G licence. It launched its 3G service in May 2006. The incumbent, Saudi Telecom, was allowed to apply for a licence at the same price provided it matched the technical plans proposed by Etisalat, and was duly awarded its licence in July 2005. It launched with a W-CDMA (wideband code division multiple access) service in June 2006. In June 2007, a third 2G+3G licence was awarded to Saudi Mobile (headed by Zain) with an August 2008 launch which relied partly on a roaming agreement with Mobily. Clearly, this is one case where new entry has truly transformed the market (albeit in the unrepresentative context of an initial monopoly), but it can reasonably be argued that it was the new 2G licences rather than the new 3G licences that were the more significant from a competitive viewpoint.

It may be noted that a similar situation prevails in Oman where Nawras Telecom entered the market controlled exclusively by Oman Mobile via a 2G+3G licence issued in June 2004.

Conclusion

Although large numbers of new entrants have been formed to take advantage of the licensing of 3G in Europe, Hutchison Whampoa has emerged as the only significant new entrant that remains active in the European mobile communications market. Telefónica, TeliaSonera and France Télécom/Orange have all withdrawn either wholly or in part for a variety of reasons. In contrast, Hutchison Whampoa has so far launched services in six markets. However, while a substantial subscriber-base has indeed been accumulated, it has been achieved at considerable financial cost, raising doubts as to the long-term viability of the company.

It is evident that the biggest difficulty facing new entrants in Europe is that the optimum number of licensees is four and even then the smallest of the four may be struggling. Hence, to enter as a fifth licensee is unlikely to be profitable. Yet, as we know, this lesson has not been learned in many less-developed countries, especially those in Africa where, perhaps fortunately, high-speed licences have yet to be awarded at all in many cases, let alone to new entrants. Also, fortunately, where provision has been made for new entrants, they have generally not been forthcoming

Admittedly, the Indian market is vastly bigger than any in Europe, but 2G incumbents are well established and the system of circles has parallels with individual European countries. That being so, it is very hard to believe that new 3G licensees have much of a future if the total number of operators in any given circle exceeds five or six. But even if, excluding the state-owned operators, only three or four 3G licences are available in most circles, such new entrants as can obtain spectrum that are valued in billions of dollars are going to be very hard-pressed to devise a strategy that will justify those valuations.

The other case studies tend to reiterate the point that there is no single model for inducing new entry. Indeed, it is evident that it is only really in situations – now almost non-existent – where there is initially a single incumbent that competition can be significantly effected by new entry via high-speed licences, and that since they are invariably 2G + 3G in such circumstances, it is the 2G licence that is the major influence.

References

Bijwaard, G., Janssen, M. and Maasland, E. 2008. Early mover advantages: An empirical analysis of European mobile phone markets. *Telecommunications Policy*, 32(3–4), 246–61.

Cellular-news. 2006. Is a Hutchison 3G withdrawal from Europe imminent? 16 October. Available at http://www.cellular-news.com (accessed 17 October 2006).

----- 2009a. Orange tried to buy T-Mobile UK. 19 August. Available at http://www.cellular-news.com (accessed 20 August 2009).

----- 2009b. India may limit numbers of 3G licenses. 31 May. Available at http://www.cellular-news.com (accessed 1 June 2009).

----- 2009c. India finally sets 3G licence fees. 28 August. Available at http://www.cellular-news.com (accessed 29 August 2009).

Curwen, P. 2002. *The Future of Mobile Communications: Awaiting the Third Generation*. Basingstoke: Palgrave.

----- 2005. The Prospects for 3G in the USA: The view from Europe. *info*, 7(3), 16–32.

Curwen, P. and Whalley, J. 2006. Third generation new entrants in the European mobile telecommunications industry. *Telecommunications Policy*, 30(10–11), 622–32.

Dow Jones. 2009. India to take 3% of revenues from new telcos' 3G services. 9 September. Available at http://www.totaltele.com (accessed 10 September 2009).

Federal Communications Commission. 2006. *Auction of Advanced Wireless Services Licences Scheduled for June 29, 2006: Comment sought on reserve prices or minimum opening bids and other procedures. AU Docket No. 06-30 of 31 January 2006*. Washington DC: FCC.

----- 2007. *Auction of 700 MHz band Licences. FCC Public Notice DA07-4514 of 2 November 2007*. Washington DC: FCC.

Gabriel, C. 2009. Indian 3G auction delayed yet again, as 2G operators get cut-throat. 7 October. Available at http://www.rethink-wirelesscom (accessed 8 October).

Global Insight. 2009a. European mobile call prices fell by an average of 23% in 2008. 3 June. Available at http://www.communicationsdirectnews.com (accessed 4 June 2009).

----- 2009b. EU Council gives final approval to roaming charge cuts. 9 June. Available at http://www.communicationsdirectnews.com (accessed 10 June 2009).

Government of India. (2008). *Auction of 3G and BWA Spectrum. Information Memorandum*. Delhi: Department of Telecommunications. 12 December.

Gruber, H. 2005. *The Economics of Mobile Telecommunications*. Cambridge, England: Cambridge University Press.

Guerrera, F and Lau, J. 2006. Rare disappointment for Hutchison Whampoa. 15 February. Available at http://www.ft.com (accessed 16 February 2006).

Haslam, C. 2009. A guide to India's telecom market. 19 April. Available at http://www.lightreading.com (accessed 20 April 2009).

He, Z-L., Lim, K. and Wong, P-K. 2006. Entry and competitive dynamics in the mobile telecommunications market. *Research Policy*, 35(8), 1147–65.

Klemperer, P. 1987a. Markets with consumer switching costs. *Quarterly Journal of Economics*, 102(2), 375–94.

----- 1987b. Entry deterrence in markets with consumer switching costs. *Economic Journal*, 97(Conference), 99–117.

Lennighan, M. 2009. Friday review: Franco-German bonding. 11 September. Available at http://www.totaltele.com (accessed 12 September 2009).

Michaels, A. 2006. Market woes derail 3 Italia IPO plans. 12 June. Available at http://www.ft.com (accessed 13 June 2006).

TelecomDirectNews. 2006. Speculation deepens on Hutchison Whampoa's possible sale of the 3 Group. 22 November. Available at http://www.telecomdirectnews.com (accessed 23 November 2006).

TeleGeography. 2008. AWS auction ends with 15 winners, CAD4.25 billion raised. 22 July. Available at http://www.telegeography.com (accessed 23 July 2008).

----- 2009a. ACCC approves Vodafone, H3G merger proposals. 29 May. Available at http://www.telegeography.com (accessed 1 June 2009).

----- 2009b. DoT calls for limits on number of operators; new entrant stakes locked in for three years. 27 July. Available at http://www.telegeography.com (accessed 28 July 2009).

----- 2009c. Virgin and Spice both bid for MTNL franchise. 4 September. Available at http://www.telegeography.com (accessed 5 September 2009).

----- 2009d. Committee created to oversee defence forces' spectrum release. 10 September. Available at http://www.telegeography.com (accessed 11 September 2009).

Wiesmann, G. 2009. D Telekom resists calls to sell T-Mobile UK. 7 May. Available at http://www.ft.com (accessed 8 May 2009).

Wiesmann, G. and Parker, A. 2009. Pressure rises on T-Mobile UK. 30 April. Available at http://www.ft.com (accessed 1 May 2009).

Wood, N. 2009. Indian government insists 3G auction on schedule. 11 December. Available at http://www.totaltele.com (accessed 12 December 2009).

CHAPTER 10 Issues in Mobile Industry Structure – Operators

Introduction

We have already determined in Chapter 9 that the use of high-speed data licences to effect a positive impact upon competition has borne negligible fruit. Overall, with few exceptions, and even fewer of note, these licences have been taken up by 2G incumbents. The fact that not all incumbents have either wanted, or been able, to acquire licences is of some interest, as are the reasons why, again in a modest number of cases, licensees have chosen (so far) not to launch their networks.

However, the critical issue is whether the structure of the mobile industry is optimal in the context of high-speed connectivity. There are two main aspects to this, relating to equipment vendors and to operators. The former are the subject of Chapter 6, while this chapter analyses the operators.

The crucial lesson arising from the discussion of new entry is that there appears to be room for a finite number of operators that can survive profitably within a given market. The establishment of 3G does not appear to have provided many opportunities – at least so far – for that number to be increased. Indeed, the question now appears to be whether the need to recoup the cost of licensing and rolling out high-speed networks has meant that the optimal number of networks in a given country has decreased. Optimality in this context can be analysed theoretically, but this does not seem to be a fruitful way to proceed in a sector that is constantly evolving in response to an environment that contains major sources of uncertainty. As is the custom elsewhere in this book, the issue will accordingly be addressed by reference to databases.

Market Concentration

EUROPE

It is best to begin by examining Europe, widely defined, given that both 2G and 3G licensing occurred there at a relatively early date, and there has accordingly been more time for the industry structure to move towards equilibrium of sorts. Table 10.1 contains subscriber data relating to December 2008 – those for December 2009 are as yet unavailable and quarterly data are less reliable. The annual data should themselves not be treated as gospel given that the numbers provided by regulators, operators and consultancies often differ by surprisingly large amounts. In addition, with penetration levels almost always in excess of 100 per cent, it is evident that a good many people carry either multiple handsets or multiple SIMs (subscriber identity modules). However, given the difficulty in establishing that this phenomenon is associated with one network rather than another, it is reasonable to assume – in the case of Europe, but not necessarily the whole world – that the distribution of subscribers between networks is not significantly affected.

Table 10.1 Mobile market[1] concentration in Europe. 31 December 2008

Country[2]	Total subscribers (million)	% of market: Top Operator	% market: Top 2 operators	% market: Top 3 Operators
Albania	2.78	50.2	90.7	100
Andorra	0.06	100	n/a	n/a
Austria	10.63	42.3	74.2	93.8
Belgium	11.42	43.0	76.5	100
Bosnia/Herzeg.	3.18	44.9	79.8	100
Bulgaria	10.63	50.7	89.2	100
Croatia	5.86	45.5	88.0	100
Cyprus (South)	1.02	85.4	100	n/a
Czech Republic	13.57	40.0	78.8	100
Denmark	6.88	46.0	72.4	94.1
Estonia	1.62	48.0	79.0	100
Faroe Isles	0.05	69.0	100	n/a
Finland	6.90	38.8	75.6	100
France	56.58	46.3	83.0	100
Germany	105.47	37.1	71.4	86.6
Gibraltar	0.03	100	n/a	n/a
Greece	18.71	42.2	73.0	100
Guernsey	0.08	66.3	95.1	100
Hungary	12.24	43.9	79.0	100
Iceland	0.38	54.7	92.7	100
Ireland	5.53	41.6	72.7	91.4
Isle of Man	0.76	97.4	100	n/a
Italy	91.26	38.1	70.7	90.5
Jersey	0.20	80.7	95.2	100
Latvia	2.29	48.4	94.6	100
Liechtenstein	0.04	40.0	67.1	83.9
Lithuania	5.03	40.0	78.3	100
Luxembourg	0.70	53.1	88.5	100
Macedonia	2.37	58.3	89.8	100
Malta	0.38	50.9	100	n/a
Monaco	0.02	100	n/a	n/a
Montenegro	1.23	41.1	76.1	100
Netherlands	19.17	48.4	76.1	100
Norway	4.98	60.6	92.3	100
Poland[3]	43.69	33.0	55.3	95.4
Portugal	15.05	46.2	83.3	100
Romania	27.28	38.0	73.4	94.9
San Marino	0.04	100	n/a	n/a
Serbia	9.75	61.5	90.7	100
Slovakia	5.73	51.1	91.2	100
Slovenia	2.05	66.7	94.5	100
Spain	52.49	45.0	76.5	98.2
Sweden	11.55	46.2	75.3	93.0
Switzerland[4]	8.68	61.8	82.2	100
Turkey	66.14	56.3	81.6	100
UK	76.25	25.6	50.7	72.7
Ukraine	52.70	44.7	78.1	93.5

Source: Calculated by authors from a variety of sources

Notes:

1. The table excludes CDMA450/800 networks.
2. Russia is excluded due to inconsistent methods of counting used by operators.
3. The regulator, UKE, estimates that the inactivity rate is as high as 17.7 per cent.
4. The second- and third-place operators, Orange and TDC, agreed a joint venture (subject to regulatory permissions) in November 2009.

As can be seen, the subscriber numbers vary from in excess of 100 million to less than 100,000, so the range encompasses almost all other countries in the world – China is the only real exception – and can accordingly be regarded as representative of developed (as well as many emerging) countries. Table 10.1 can be summarized as follows, where any market with two networks accounting for over 90 per cent of the subscribers is referred to as duopoly+.

Table 10.2 Aspects of market concentration in Europe. 31 December 2008

Range (M)	0.0–0.1	0.1–1	1–5	5–10	10–20	20–50	50–80	80–110
Number	7	5	9	8	9	2	5	2
Monopoly	4	0	0	0	0	0	0	0
Duopoly	1	2	1	0	0	0	0	0
Duopoly+	1	2	4	2	0	0	0	0
> 3 networks	1	0	0	2	2	2	3	2

There are 47 countries in total, and of these 12 have more than three networks. In six cases, there are three 2G/3G operators plus one 3G operator – essentially the Hutchison effect discussed in Chapter 9 – and another was added in Romania in 2009 when 2G-only operator CosmOTE bought 3G-only operator TeleMobil. The same pattern is also found in Sweden although TeliaSonera is technically a 3G MVNO (mobile virtual network operator). Only in the UK are there four 2G/3G, plus one 3G-only, operators. The others comprise Germany with four 2G/3G, Liechtenstein with three 2G/3G plus one 2G-only and the Ukraine where there are more 2G operators than anywhere else but the only identifiable W-CDMA (wideband code division multiple access) licensee is Ukrtelecom.

The most common construct is where the market comprises precisely three 2G/3G operators, and the most common distribution where the largest two operators account for 75–80 per cent of total subscribers. But that still leaves 10 cases where the third-largest operator accounts for less than 20 per cent of the market, which does not translate into a strong financial position if competition is fierce. In other cases the situation is messier because even if there are three 2G/3G operators, they have not all launched networks (in some cases because the licences were issued only in 2009).

In summary, it is fair to say that historically there was considerable support for a four-operator structure in a European market with more than a few million subscribers, and that this partly underpinned the tendency to try to use 3G licensing as a means to maintain operator numbers at, or raise operator numbers to, that level. But that strategy did not succeed, indicating that there was little belief in the prospects of a fourth operator with a greenfield operation, albeit one using W-CDMA. In more recent times, and particularly in the wake of the credit crunch, support for a three-operator market (in conjunction with a vibrant set of MVNOs) appears to be growing. In Austria, for example, where between four and six 3G licences were on offer in 2000 (but only four were awarded), Telekom Austria has recently stated (Morris 2009) that whereas it would be inappropriate for market leader mobilkom to be involved in consolidation, consolidation was favoured by investors, and mergers between third- and fourth-place operators – Orange and Hutchison in the case of Austria – should be supported.

At this point it is reasonable to ask two questions. First, is the European model likely to come to pass elsewhere in the world? Second, how is this related to the behaviour of the largest international operators? Let us now turn to address the first question.

REST OF WORLD

It has been convenient to roll up into Europe parts of what was treated separately as Eastern Europe and CIS in previous chapters, especially given that Russia is difficult to analyse due to its regional structure for licensing. Let us now turn to examine certain other regions.

North America

North America presents its own problems for much the same reason as Russia – the national operators compete with much smaller regional operators. However, the three national operators in Canada have always dominated the market and they currently account for roughly 21 million of the 22 million subscribers. In contrast to Europe, one uses W-CDMA while the other two predominantly use cdma2000 1xEV-DO (although they have very recently also launched HSDPA (high-speed downlink packet access)), but the fierce competition increasingly revolves around the launch of touch-screen smartphones – Rogers has clearly gained a temporary advantage on the back of its introduction of the 3G iPhone in July 2008.

However, whereas Canada's market structure is largely set in stone, this is certainly not the case in the USA. It is, in practice, not that long ago that there was great antipathy towards the consolidation of the market where large numbers of regional/local operators thrived or survived as the case might be. But despite the relatively large market size compared to individual European countries, the economic rationale for M&A (merger and acquisition) activity was too strong to resist (Curwen and Whalley 2008: Chapter 8). Although the commencement of this activity dates back 10 years, it is continuing currently. What has emerged so far is a market structure with four large national operators which together account for roughly 90 per cent of all subscribers. With market penetration fast approaching 100 per cent, a vicious price war broke out in 2009 in consequence, as typified by the $50 flat-rate unlimited use price plan (Cheng 2009; Gabriel 2009; TeleGeography 2009a).

The new Obama administration is keen to flex its anti-trust muscles in order to differentiate itself from its predecessor, but it is a mystery to most commentators why it thinks the wireless sector needs more regulation (Bartash 2009). Indeed, the prevailing view appears to be that three national operators are better than four – which the FCC (Federal Communications Commission) will almost certainly cavil at should the issue arise – as evidenced by the rumour that struggling T-Mobile USA is casting its eye over struggling Sprint Nextel even though they use incompatible technologies for 2G and 3G. Interestingly, T-Mobile tried to quash the rumour in October 2009, using the argument that there were only four operators in a large market and did not therefore appear to be too many, and that it had been suffering because it had not (unlike its rivals) launched 3G until recently (TeleGeography 2009c). Given Sprint's value-destroying takeover of

Nextel, it is difficult in the extreme to make a case for a T-Mobile merger with Sprint. But Verizon Wireless has the best network and AT&T has gained subscribers (but not necessarily profits) via its exclusive deal with Apple, so the prospects are hardly rosy for the two laggards which face intense competition in the pre-paid market from regional heavyweights Leap Wireless and Metro PCS.

Latin America

The contrast with Latin America (excluding the majority of the Caribbean where most markets are small and Digicel is the dominant operator) is of interest. The position from January 2004 to end-2007 is discussed in Curwen and Whalley (2008: Chapter 6). Prior to 2004, BellSouth, the then regional US operator, had held an extensive portfolio of Latin American assets, but it sold these to Telefónica in 2004. The other crucial factor was the decision by Telecom Italia to withdraw from all countries bar Brazil – it had eight holdings (including Cuba) at the time – which it has not quite achieved as yet. The figures in Table 10.3 are approximate given the usual disagreements about the precise number of subscribers, and it should be borne in mind that whereas América Móvil controls virtually all of its subscribers, that is the case neither for Telefónica which, for example, shares its Brazilian network with Portugal Telecom, nor Telecom Italia. The much smaller Millicom International controls the majority of its subscribers in every country in which it operates.

It is very clear from the table that América Móvil is a dominant force in the region, and that only Telefónica offers any real resistance to its growing dominance. The absence of other European operators is largely a reflection of the use of Spanish and Portuguese – the English, French and Germans had significant parts of their empires in Africa and the trade route countries of Asia and largely ignored Latin America. Historically, Latin America was the fairly happy 'backyard' of USA operators, but the departure of BellSouth effectively signalled the death knell for US interest in the region (and for the time being, any other region although AT&T has retained the roughly 8 per cent[1] stake in América Móvil inherited from its days as SBC).

Not surprisingly, the regional giants have led the move into 3G and beyond. In Table 10.3, the symbol [Y] indicates the launch of 3G and/or 3.5G and there are very few omissions. Furthermore, even where CDMA (code division multiple access) was initially used for 2G – as was often the case – there has been a move to overlay this with GSM (global system for mobile) leading on to W-CDMA/HSPA (high-speed packet access)/LTE (long-term evolution). In effect, therefore, the regional giants have shut the door firmly in the face of new entrants. Accordingly, the only real issue is whether regulators will become sufficiently alarmed at the prospects of (near-)duopoly structures so as to take remedial action. For now, despite the occasional threat of nationalization as in Bolivia, this does not appear to be the case and, in any event, it is hard to see who (apart, maybe, for the state) would be interested in taking stakes in relatively small operators in the countries where the regional giants are firmly entrenched.

1 As so often, the media get this wrong – see PriMetrica (2009) which refers to a 23.4 per cent stake. But that relates to the voting shares, equivalent to 24.5 per cent of the AA shares which comprise only 30.2 per cent of the total capital. In any event, the figure is too low since AT&T owns a minimum of 7.9 per cent of América Móvil (because AT&T inherited 26.3 per cent of the AA shares from SBC) and, according to some sources, exactly 8.2 per cent.

Table 10.3 Selected Latin America. Percentage of subscribers in direct subsidiaries. Spring 2009

	América Móvil[1]	Telecom Italia[3]	Telefónica[4]	Millicom
Argentina	35 [Y]	28 [Y]	34[5] [Y]	–
Bolivia	–	–[4]	–[5]	31 [Y]
Brazil	26 [Y]	24 [Y]	30[5][Y]	–
Chile	19 [Y]	–	43 [Y]	–
Colombia	66 [Y]	–	25 [Y]	9 [Y]
Costa Rica	–	–	–	–
Dominican Rep.	48 [Y]	–	–	–
Ecuador				
El Salvador	72 [Y]	–	27 [Y]	–
Guatemala	27 [Y]	–	18	37 [Y]
Honduras	37 [Y]	–	21 [Y]	42 [Y]
Mexico	24 [Y]	–	–	75 [Y]
Nicaragua	72 [Y]	–	20 [Y]	–
Panama	71 [Y]	–	29	–
Paraguay	-[2]	–	42 [Y]	–
Peru	10 [Y]	32 [Y]	1[5]	49 [Y]
Puerto Rico	35 [Y]	–	62 [Y]	–
Uruguay	32 [Y]	–	–	–
Venezuela	22 [Y]	–	39 [Y]	–
	–	–	38[Y]	–

Notes:

1. América Móvil controls virtually all of its subscribers.
2. América Móvil launched in March 2009.
3. Telecom Italia shares its Brazilian network with Portugal Telecom. It is also present in Cuba.
4. The situation in Bolivia is unclear. The network was threatened with nationalization in 2007 and the issue was sent for arbitration in 2009. The outcome is not yet known.
5. Telefónica also has an indirect stake in Telecom Italia so its direct plus indirect stakes in these countries are larger than those stated in the column.

However, it is of no small interest that in Brazil, much the largest market in Latin America with over 150 million subscribers (only Argentina, Chile, Colombia, Mexico, Peru and Venezuela can claim even one-tenth of that number), there was a clear desire on the part of the authorities to cobble together a fourth operator capable of taking on the 'big three'. Realistically, the only possibility was for Telemar (Oi) to take over the much smaller Brasil Telecom even though they had competing fixed-wire networks in a number of regions where, by law, they could only have one as a merged entity. However, given that the government supported the takeover, the problems were promptly made to disappear (Stewart 2008) although regulatory approvals were delayed until December 2008 and the takeover was initiated only in January 2009 and completed in May (Teleco 2009).

The networks in Costa Rica are a state-owned monopoly (but not for much longer), and there are a number of modest iDEN (integrated digital enhanced networks) networks elsewhere in Latin America, but if these are disregarded it is evident, in conclusion, that the optimum number of networks is primarily determined by the number of 'big four' operators that are present.

Africa

Given that most of the 55 African markets are fairly modest in size – only Algeria, Egypt, Morocco, Nigeria and South Africa had more than 20 million subscribers in June 2009 while the entire continent accounted for only 415 million – it is entirely reasonable to suppose that anything in excess of three operators, let alone four, is probably excessive. In four of the five largest markets there are indeed three 2G incumbents but, oddly, Nigeria has elected to spread licences like confetti (the consequences of which are as yet unknown).

The position in a number of countries is somewhat hazy since there are licensees that have yet to launch, others that may never launch and several with licences that have (allegedly) been, or are in the process of being, revoked (possibly subject to appeal). All told, there are roughly between 170 and 180 licensees which indicates that the average network has only roughly 2.2 million subscribers, but the average is not very meaningful given the immense disparities in market size – the 'big five' markets account for over 50 per cent of the total subscribers, so the average network in other countries has barely one million. It is true that growth is relatively rapid, but even so it is virtually impossible to justify the existence of more than 10 licensees in Uganda, especially since the regulator ran out of GSM spectrum to allocate two years ago (Barigaba 2009). It is also hard to justify, for example, six networks in the DR Congo, Somalia and Tanzania or five networks in Benin, Ghana, Guinea Republic, Ivory Coast, Sierra Leone and Sudan – and some of these countries are continuing to licence as are countries that have fewer licensees but also fewer than one million subscribers per existing licensee (as in, for example, Gabon, The Gambia and Madagascar). The six monopolies that still remain are arguably just as bad in their own way, but there are surprisingly few duopolies – when allowance is made for the issue of a third licence without, as yet, a corresponding launch – so there appears to be a commonly held view that three or more operators are needed to secure a suitable degree of competition. This view gains plausibility if ongoing rapid market growth is factored in, but at the end of the day there are relatively few citizens comfortably above the breadline in many African countries, so new licensees are going to be forced to scrabble for low-income, low-expenditure customers.

Given that only roughly 35 operators have so far launched high-speed data services based on W-CDMA – Africa is largely the preserve of GSM although there are also a handful of 1xEV-DO networks – and these are essentially 2G incumbents, such networks appear to be seen for now as a method to gain competitive advantage over rivals. Although some 3G licences have been issued to new entrants, these have accompanied 2G licences and the additional cost of the 3G element of the licence has tended to be modest. A factor here, as discussed below, is that Africa is undergoing a significant restructuring of the telecommunications sector with an as yet unclear result, and whereas a number of dominant operators are likely to appear in due course, as in South America, the effect is necessarily going to be somewhat diluted because there are so many countries. Given that economies arise by operating within a cluster of (near-)neighbouring countries, the bigger operations can, unlike in South America, avoid competing with each other to some extent although they clearly cannot ignore the largest five markets altogether.

Middle East

The Middle East needs to be addressed in terms of its core countries. Turkey has been assigned to Europe for convenience and Iraq (because of the current conflict), Lebanon (because there are no licensees as such) and Palestine (because of licensing issues with Israel, which is itself atypical) are best set aside. Iran is also problematic because there are regional networks. A notable feature of the 10 or so core countries is that recent growth has been so rapid – in many cases there are large numbers of non-citizens anxious to call home – that most exhibit penetration levels well in excess of 100 per cent. This might suggest that there is neither the need, nor an economic case, for new licences, but they are nevertheless being issued.

However, the overall position remains conservative compared to other regions because monopolies and duopolies were the norm until fairly recently. Only the Yemen has four operators (of which one uses CDMA) and Unitel, the most recent, is arguably redundant. In other cases, the second (as in Qatar) or third (as in Bahrain) licence is of recent vintage and the effects of an additional licensee are as yet unclear. High-speed licences are predictably common, and appear largely to perform a defensive role in relation to strong competitors (as in Bahrain, Israel, Kuwait, Oman, Saudi Arabia and the UAE). They are proving very successful, so there is little prospect of a new entrant disturbing the status quo by pressing ahead with high-speed technology.

Asia Pacific

The Asia Pacific region is the most varied. Its constituent countries vary in terms of subscribers from China, with over 600 million, to Nieu and the Cocos Islands with fewer than 500. It is desirable to ignore China because its mobile sector structure is an artificial construct determined by the state. There are also more than 10 islands with small populations and a monopoly supplier, as well as several islands with duopolies where Digicel has entered, which can contribute little of value to the analysis – in practice, only Guam and the Northern Marianas have multiple tiny networks (for no very good reason other than the desire of Japanese tourists to phone home using compatible handsets).

This leaves a core sample set of 26 countries, the largest of which is India with roughly 350 million subscribers and the smallest is Bhutan with roughly one-tenth that number. Aside from India, only Indonesia and Japan have more than 100 million with Pakistan just below that figure. As Japan is a developed country, there is an expectation based on European experience of four operators whereas in practice DoCoMo has been so dominant that there has historically only been room for two competitors – although a third (eAccess) has not long ago been licensed as a 3G new entrant. The position in Malaysia is also of interest with the three substantial incumbents brushing off the threat posed by 3G new entrant Ufone – initially minority-owned by DoCoMo and KTF although the latter has recently withdrawn because Ufone is making very little progress.

But are the other three large emerging markets more akin to North American/European experience or to that of Nigeria? The answer, in practice, is the latter (at least for now). The size of the Indian market might indicate space for five or six operators with more or less national coverage despite the contradictory trend in the somewhat smaller USA, but there are in fact already 14 (although most operate for now in a limited number of circles) and that is before 3G licences are awarded. Furthermore, Indonesia boasts eight and Pakistan has seven. This suggests that emerging markets – especially where there are no overarching bodies such as the European Union – spend a period in something akin to a free-for-all. The issue is accordingly whether this is a temporary phenomenon and, if so, how long it will last.

Because of the atypical circumstances of many countries including Vietnam (state ownership, but multiple networks), Thailand (the CAT/TOT concession system) and Hong Kong (small, hyper-competitive and oversupplied with networks), it is difficult to determine what sort of equilibrium will become commonplace. However, something can be gleaned by looking at the countries that are more highly developed than all bar Hong Kong. Thus, Malaysia (as noted) has three 2G incumbents plus one 3G-only, (the much smaller) Singapore has three, South Korea has three and Taiwan has four plus one 3G-only. A prediction might accordingly be that markets with less than 100 million subscribers will stabilize with three or four operators while larger markets will have five or six.

In Indonesia, the largest three operators account for more than 90 per cent of subscribers and although the figure in Pakistan is much lower at roughly 75 per cent, adding in the fourth operator causes this figure to rise comfortably above 90 per cent. In fact, whether the emerging or developed countries are examined, the above prediction broadly holds true, and from it can be derived a secondary prediction that there will be a shuffling of assets between operators as economies of scale increasingly come to bear in a high-speed world.

The Role of Operators

We have already established that high-speed licences are largely the preserve of 2G incumbents in advanced countries, and that new entrants in emerging countries have not made much more progress so far. Furthermore, we have established that it is uncommon for a 2G incumbent to overlook the opportunity to acquire a high-speed licence. In Europe, for example, TeliaSonera did apply for a licence, but surprisingly failed to win one in the Swedish beauty contest although it appears to be doing reasonably well as a MVNO. Macedonia, where Vip does not have a licence, is the only significant exception and it seems that new licences there are treated with disdain – the attempt at a tender in August 2009 attracted no interest whatsoever. Does this mean that European operators can no longer see any point in having a licence? This is not altogether clear since, first, whereas the three licences on offer in Bosnia-Herzegovina were awarded in March 2009, only the three 2G incumbents applied and none of them was a subsidiary of the major European operators; secondly, although there were four declared bidders for the previously unallocated fourth licence on offer in Hungary, the tender was cancelled in March (CommunicationsDirect 2009); and thirdly, France offered up a fourth licence towards the end of 2009, but there was no great rush to pre-qualify and only one company eventually agreed to bid. As for Georgia to the east, a request by a potential bidder for an auction to take place in October was granted, only for bidders to be notable by their absence at the appointed time (TeleGeography 2009b)

The other element that now needs to be factored in is the existence of multi-country and, in particular, multi-regional operators. This is because they are likely to have a major say in resolving inappropriate market structures such as over-provision of networks. Although operators are bound to claim that they acquire overseas assets in accordance with some master plan, assets cannot be bought unless they are available at the right (albeit often somewhat inflated) price, and if that does not apply the temptation may be to settle for second best. Furthermore, a strategy for expansion overseas strictly requires a sequence of purchases in the right place and at the right price.

The sequence of purchase does not tend to happen and, in any event, factors such as changes at the top of companies invariably lead to reassessments of previous strategies and a desire to dispose of poorly performing assets. The financial health of an operator is also highly relevant in respect of its trading of overseas assets. Table 10.4 illustrates the situation in relation to the operators with the largest number of countries served as of 31 December 2009. These countries are allocated to regions in accordance with the method used in earlier chapters. The table also shows how the number of countries served has varied since 2003.

Some reservations are in order. First, to be included in the country tally the investment must involve an operational network. As a consequence, potential entry arrangements such as international roaming or brand franchising are excluded as they do not provide a sufficient degree of investment to merit inclusion. Secondly, where the operator has multiple investments in the same country, only a single presence is recorded. Subject to this proviso, each equity stake, regardless of its size, is recorded as a single presence in a country. Furthermore, a presence is recorded only when a network has been rolled out and is attracting subscribers; the mere acquisition of a licence is not counted. However, a mobile virtual network operation is counted because it brings with it subscribers and thus income.

Table 10.4 Presence in countries. End 2009 and changes 2009/2008, 2008/2007, 2007/2006, 2006/2005, 2005/2004, 2004/2003

Company	Number of countries	Number of countries by region						
		Western Europe[1]	Eastern Europe[2]	Middle East[3]	Asia Pacific[4]	Latin America[5]	North America[6]	Africa[7]
France Télécom	40 [+2, -1, +3, 0, 0, -2]	12	1	2	1	6	1	17
Vodafone[8]	37 [+1, +1, 0, -1, +3, 0]	14	1	4	7	–	1	10
Telefónica[9]	34 [0, +1, +3, +5, +3, +5]	9	–	–	4	16	–	5
MTN[11]	22 [0, 0, 0, +11, +4, 0]	1	–	3	1	–	–	17
TeliaSonera[12]	22 [0, +3, +2, +1, +1, -2]	8	4	4	6	–	–	–
Zain[10]	22 [+1, +2, 0, +1, +14, +2]	–	–	5	–	–	–	17
Telenor	21 [+3, 0, 0, +3, +2, -2]	4	5	2	10	–	–	–
Deutsche Telekom	20 [+5, +2, -1, 0, -3, 0]	11	6	–	–	–	3	–
Etisalat	19 [+2, +2, +3, 0, +7, +1]	–	–	4	5	–	–	10
América Móvil	18 [+1, 0, +2, +1, +3, +2]	–	–	–	–	16	2	–
Millicom	14 [-2, 0, -1, +1, +1, 0]	–		–	1	6	–	7
Hutchison	13 [-1, -1, +1, 0, 0, 0]	6	–	–	7	–	–	–
Qatar Telecom[13]	13 [+1, 0, +10, 0, +1, 0]	–	–	6	5	–	–	2
Singapore Telecom	12 [+1, +1, +3, 0, +1, -1]	2	–	–	9	–	–	1
NTT	11 [+1, +1, 0, +1, -1, -1]	–	–	–	11	–	–	–
Orascom[14]	11 [+3, -6, 0, 0, +4, +1]	–	–	–	3	–	1	7
Portugal Telecom	9 [0, 0, 0, 0, +1, 0]	1	–	–	2	1	–	5
VimpelCom	9 [+2, 0, +1, +2, +2, 0]	–	2	2	5	–	–	–
Tele2	8 [-1, -4, -1, +1, +1, 0]	6	2	–	–	–	–	–
Axiata[15]	8 [0, -1, -1, +1, +2 -6]	–	–	1	7	–	–	–
Telekom Austria	8 [0, 0, +3, 0, +1, 0]	3	5	–	–	–	–	–
Turkcell	8 [0, +1, 0, 0, +1, 0]	–	3	4	1	–	–	–
Vivendi Universal	8 [+1, 0, +2, 0, +1, -2]	1	–	–	–	–	–	7
OTE	7 [-1, 0, +2, -1, 0, 0]	3	4	–	–	–	–	–
Saudi Telecom	7 [-1, +8, 0, 0, 0, 0]	–	–	3	3	–	–	1
Mobile TeleSystems	6 [0, 0, +1, 0, +1, +1]	–	3	1	2	–	–	–
Telecom Italia	6 [0, 0, -1, -1, -4, -1]	1	–	–	–	5	–	–
SK Telecom	5 [-1, -1, 0, +3, 0, 0]	–	–	–	4	–	1	–
Vodacom	5 [0, 0, 0, 0, 0, 0]	–	–	–	–	–	–	5
Total	23 [+17, +8, +31, +28, +46, -5]	82	36	41	94	50	9	111

Notes:

1. Western Europe includes the 27 EU current Member States plus residual Western Europe.
2. Eastern Europe includes the Balkans, Belarus, Moldova, Russia and the Ukraine.
3. The Middle East includes the Arabian peninsula, Armenia, Azerbaijan, Cyprus (N), Georgia and Turkey.
4. Asia includes Kazakhstan, Australia and New Zealand.
5. Central and South America plus most of the Caribbean.
6. North America mainly constitutes Canada, Puerto Rico and the USA.

7. Defined geographically to include Egypt.
8. Network Partnership Agreements are not included within the Vodafone total.
9. Telefónica took over the networks of O_2 during 2006. The comparison is with the sum of the separate networks in 2005, 2004 and 2003.
10. Zain, formerly The Mobile Telecommunications Co., is based in Kuwait. In May 2005, it acquired 85 per cent of Celtel, a pan-African operator with 14 networks.
11. MTN, which is based in South Africa, took over the networks of Investcom during 2006. The comparison is with the sum of the separate networks in 2005, 2004 and 2003.
12. Telia Sonera's home market is defined as Finland plus Sweden.
13. Qatar Telecom owns 51 per cent of al Wataniya, or simply Wataniya, which is based in Kuwait.
14. Orascom Telecom Holding is based in Egypt. It is 50 per cent + 1 share owned by Weather Investments which also owns two stakes in Western Europe, the first of which, in Italy, it bought in 2006 and the second of which, in Greece, it acquired in 2007.
15. Prior to 1 April 2009, the company was known as Telekom Malaysia International, which had been split off from Telekom Malaysia.

The use of holding companies complicates the process of determining exactly how many countries in which an operator is present. If an operator has invested in a holding company, should this be counted as a single investment or should each of the holding company's subsidiary investments be counted? The approach adopted here is to count all of the subsidiary investments. Given that, broadly speaking, it is not possible to determine the control that each operator exercises over its investments, it is assumed that each investment is an active one unless the mobile operator has expressly stated that the holding is purely for investment purposes. The same is also true for the holding of stakes in other operators. Although it is possible to argue that a minority subsidiary of a minority-owned operator should not be counted as a country presence due to the indirect nature of the stake, it may reasonably be counter-argued that the strategy being pursued is precisely to gain a presence in a spread of countries via this method rather than to attempt to buy direct stakes in tightly held companies at exorbitant prices.

All told, 29 operators are listed. AT&T is excluded because despite an 8.0 per cent stake in América Móvil, which in principle provides it with wide coverage in Latin America, this is best regarded as an investment. Equally, the likes of Russia's Altimo or Singapore's Temasek Holdings are excluded on the grounds that they do not actually operate any networks. Predictably, four of the top five in the table are European – they are among the longest established and have colonial associations – while USA-based operators are non-existent, having largely withdrawn in recent times to their home market (Curwen and Whalley 2008: Chapter 8). However a particular matter of interest is the appearance of Etisalat, MTN and Zain among the top 10, of which more below. It may also be noted that Asia-based operators tend to have limited empires – amongst other matters, large differences in culture, language and religion between adjoining countries in Asia militate against the creation of a tight grouping of assets. Furthermore, although DoCoMo, which has relatively large Asian holdings, aspired to expand beyond Asia, it was notably unable to make a success of these and all of them have now been sold.

Before moving on to examine the table in more detail, it is useful to remind ourselves that overseas expansion requires takeovers which tend to go in cycles. There are so many of these, with the majority involving 'only' hundreds of millions of dollars, that it is not possible to provide a full list. Hence, Table 10.5 is restricted to the period post-2005 and to the largest bids. Despite this, it shows very clearly how the cycle works, with a massive upturn in M&A activity post-2004 when the worst effects of the 2001–2002 meltdown had been purged from balance sheets, followed by a collapse of activity as the credit crunch began to bite. It is also worth noting that whereas activity has clearly picked up towards the end of 2009, this is not the same thing as consummated activity since a significant number of bids do not reach fruition even in the absence of competing offers. A final point is that outright takeover bids are the tip of the iceberg. Most activity comprises the buying of additional stakes to add to those already held, in the process possibly transferring control to the buyer. These purchases generally involve smaller amounts of money.

Table 10.5 Illustrative major[1] takeovers by/of operators. 2005–2009[2]

Bidder	Target	Date	Value $billion
Telecom Italia	Telecom Italia Mobile	January 2005	19.0[3]
Alltel	Western Wireless	January 2005	4.4 + 1.5 debt
SBC Communications	AT&T	January 2005	16.0[4]
TeliaSonera	Turkcell	March 2005	3.1[5]
Alfa Telecom	Turkcell	March 2005	3.2[5]
Blackstone Group et al.	Wind	March 2005	n/a[6]
Weather Investments	Wind	March 2005	n/a[6]
MTN	Celtel	March 2005	2.7
MTC	Celtel	March 2005	3.4[7]
Telefónica	Ceský Telecom	March 2005	3.6[8]
Swisscom	Ceský Telecom	March 2005	3.4[8]
Kohlberg Kravis Roberts	Auna	April 2005	13.6[9]
Telekom Austria	MobilTel	June 2005	2.0
France Télécom	Amena	July 2005	12.1[10]
Telefónica	O_2	October 2005	31.3
Apax Partners et al.	TDC	November 2005	11.5 + 3.5 debt[11]
Sprint Nextel	Alamosa Holdings	November 2005	3.4 + 0.9 debt
Vodafone	Telsim	December 2005	4.6
Sprint Nextel	Nextel Partners	December 2005	6.5[12]
Temasek Holdings	Shin Corp.	January 2006	3.8[13]
Sonae.com	Portugal Telecom	February 2006	12.8
VimpelCom	Kyivstar	February 2006	5.0 + 0.5 debt
AT&T (SBC)	BellSouth	March 2006	67.0 + 17.0 debt
SoftBank	Vodafone Japan	March 2006	14.8 + 0.8 debt[14]
Cerberus/Providence	Vodafone Japan	March 2006	16.4 + 0.8 debt[15]
Telmex	Verizon Latin America	April 2006	3.7
Babcock & Brown	eircom	April 2006	3.1 + 2.3 debt[16]
América Móvil	Verizon Dominicana	April 2006	2.0
China Mobile	Millicom International	April 2006	5.3[17]
Investcom	Millicom International	April 2006	n/a[18]
MTN	Investcom	May 2006	5.5[19]
Verizon Communications	Verizon Wireless	May 2006	48.0 + 6.0 debt[20]
Kohlberg Kravis Roberts	Vivendi Universal	October 2006	51.5[21]
Naguib Sawiris	Weather Investments	December 2006	2.5[22]
Etisalat	TIM Hellas	December 2006	4.9[23]
Turkcell	TIM Hellas	December 2006	4.6[23]
Telmex	TIM Brasil	December 2006	8.0[24]

Table 10.5 ***Concluded***

Bidder	Target	Date	Value $billion
Vodafone	Hutchison Essar	December 2006	17.0[25]
Maxis/Texas Pacific	Hutchison Essar	December 2006	13.5[26]
Essar Group	Hutchison Essar	December 2006	11.0[26]
Swisscom	Swisscom Mobile	December 2006	3.5[27]
Weather Investments	TIM Hellas/Q Telecom	February 2007	4.4[28]
Qatar Telecom	al Wataniya	March 2007	3.7[29]
Providence Equity Ptnrs	BTC	March 2007	2.4[30]
Telco consortium	Olimpia	April 2007	4.5
Binariang	Maxis Communications	May 2007	4.6[31]
Goldman Sachs/TPG	Alltel	May 2007	27.5
AT&T	Dobson Communications	July 2007	2.8 + 2.3 debt
OTPP et al.	BCE	July 2007	32.6 + 15.9 debt[32]
MetroPCS	Leap Wireless	September 2007	5.5 + 2.0 debt[33]
T-Mobile USA	SunCom Wireless	September 2007	1.6 + 0.8 debt
OTE	CosmOTE	November 2007	2.8[34]
Telemar	Brasil Telecom	March 2008	3.5[35]
France Télécom	TeliaSonera	June 2008	47.0[36]
Verizon Wireless	Alltel	June 2008	5.9 + 22.2 debt
Vivendi Universal	Zain Africa	June 2009	10.5[37]
Consortium	Zain	September 2009	13.7[38]
Telefónica	T-Mobile UK	September 2009	5.6[39]
Vodafone	T-Mobile UK	September 2009	5.6[39]
Bharti AirTel	MTN	September 2009	14.0[40]
Eternad-e-Mobin	TCI (Iran)	September 2009	7.8[41]

Notes:

1. Valued at over $2 billion including debt taken over. The figures quoted are mostly those that applied at the end of the first day of the offer and hence are often significantly different from the value of the offer when it was finally agreed, let alone when the transaction was completed.
2. Not all bids have yet received regulatory approval.
3. Maximum assuming the purchase of two-thirds of the outstanding ordinary shares and all of the outstanding savings shares.
4. The agreed bid was unlikely to close for over one year from acceptance.
5. For a 27 per cent stake that would give TeliaSonera 64 per cent overall and control.
6. In May 2005, Weather was awarded 62.75 per cent of Wind for €3.1 billion in cash + 26 per cent of Weather which would be given a 50 per cent + one share stake in Orascom Telecom valued at €1.96 billion. In January 2006, the rest of Wind was transferred for a further €328 million in cash and newly issued Weather shares. The total value of the transaction is given as €12.14 billion.
7. An initial $2.84 billion for an 85 per cent stake followed by a further $520 million for the other 15 per cent within two years.
8. For a 51.1 per cent stake. A bid was also made by a consortium including France Télécom.
9. They withdrew when France Télécom entered the bidding.
10. France Télécom won the bidding for Amena, initially buying 80 per cent for $4.1 billion in cash + $3.6 billion in shares + $2 billion debt taken on – $2 billion in tax credits. The fixed-wire assets were then sold to Ono for $2.2 billion.
11. Apax Partners and others combined as the Nordic Telephone Company allegedly to make an unspecified bid in October which was rejected, but returned formally in November.
12. The agreement reached in December was that Sprint Nextel would purchase the outstanding 68 per cent of the shares in Nextel Partners during the first half of 2006.

13. The initial $1.9 billion offer was for the 49.6 per cent stake held by the Thai Prime Minister's family, and in March a further 46.7 per cent was acquired via an open offer at the same price per share.
14. The name was subsequently changed to SoftBank Mobile.
15. Vodafone ignored the tabled intention to bid.
16. Which would leave Babcock & Brown with 65 per cent and the Employee Share Ownership Trust with 35 per cent. Debt taken on was later quoted as $4.7 billion.
17. Rejected in July.
18. Wthdrawn after takeover bid by MTN. Other offers for Millicom have been lodged but no details are available.
19. With an irrevocable acceptance of 70.6 per cent.
20. Alleged. For the 45 per cent outstanding stake. Rejected in principle.
21. The price was made public, but the bid was never made formal.
22. Paid to Enel for the outstanding 26.1 per cent of the company.
23. It was subsequently alleged that the highest bid was €3.2 billion. All were rejected by Apax Partners and Texas Pacific Group.
24. Alleged. Allegedly raised to roughly $10 billion.
25. Vodafone was bidding for the whole of Hutchison Essar, but by law could not end up with more than 52 per cent of the company. It eventually proved victorious with an offer of cash worth $11.1 billion + $2 billion of debt for a 67 per cent stake, but 15 per cent of this had to be held by Indian nationals.
26. The stake under offer elsewhere was 67 per cent of Hutchison Essar. Essar already held the other 33 per cent.
27. For the 25 per cent outstanding stake.
28. The bid was €3.4 billion.
29. For a 51 per cent stake.
30. For a 90 per cent stake.
31. Ananda Krishnan, via Binariang, offered to buy the outstanding 40 per cent of Maxis not owned by his company Usaha Tegas.
32. The technical buyer was Serefina Holdings, owned 71 per cent by BC Partners. The deal collapsed in December 2008.
33. The unsolicited bid was rejected as too low.
34. For the outstanding 32.17 per cent of the shares by way of a share swap for OTE shares.
35. For an initial 60.5 per cent of the voting shares. Altogether, this and related share purchases are expected to cost $7 billion.
36. Indicative offer that was immediately rejected. The offer was said to be worth $40.5bn, $41.3bn and $47bn – in every case by the *Financial Times* on 5 June.
37. For a 65 per cent stake. The offer was rejected in July.
38. For a 46 per cent stake which comprised a majority of the voting shares and hence control.
39. Both rejected.
40. For a 49 per cent stake. Rejected.
41. For a 50 per cent + 1 share stake.

As of now, there are seemingly two main forces at work. In the first place, the weaker operations in a variety of countries are being merged or taken over. Illustrative of this are the unofficial discussions between T-Mobile USA and Sprint Mobile. Secondly, operators are consolidating their holdings, generally on a regional basis. Illustrative of this is the behaviour of Millicom and, until recently, of Zain in Africa (see below). The bottom line is that when empires are created in a somewhat hit-and-miss manner because licences and assets become available on a random basis, and when this occurs in the context of ever increasing competition and pressure on margins, it sooner or later becomes evident to operators that profitability is largely derived from a limited number of networks.[2] At this point, they may begin to wonder why they continue to own the others.

Turning once again to Table 10.4, a number of observations are in order. The first is that, even allowing for some necessary ambiguities about how to define regions, there is clearly no such thing as a 'global' operator, despite claims to the contrary. Although France Télécom does appear in every column, the North American network operates on a small island off the coast of Canada and its presence in four regions is modest at best. In practice, the company is more accurately described as a tri-regional operator, or even as bi-regional given that its Latin American assets essentially consist of networks on a number of small Caribbean islands.

Being bi-regional or tri-regional is, as Table 10.4 reveals, quite commonplace, although the table does not indicate the relative importance of the home market. Of the 29 operators, eight generate less than 10 per cent of their proportionate (adjusted in accordance with the share of equity held) subscribers domestically whereas six generate more than 70 per cent. Nor does it indicate the relative importance of each region in terms of generating subscribers. For example, Telefónica derives over 95 per cent of its proportion of subscribers (adjusted for the size of its equity holding) from Western Europe and Latin America while Vodafone derives more than half of all such subscribers from Western Europe (Curwen and Whalley 2008: 229–30) as do seven others including France Télécom. The two Russian operators are heavily dependent upon Eastern Europe; and for Turkcell and Saudi Telecom, they are dependent on the Middle East.

Latin America is very important for five operators, comprising the four discussed previously and Portugal Telecom, whereas Africa dominates the operations only of MTN, Vodacom and (for now) Zain. However, Africa is the only region outside Western Europe where Vivendi International is to be found, and the same goes for Orascom outside the Asia-Pacific region.

If, in summary, we take at least 85 per cent of proportionate subscribers obtained from two regions as a reasonable yardstick for a bi-regional operator, then a surprising 26 of the 29 most internationally minded among them fall into this category. Only Etisalat, TeliaSonera and Vodafone can truly be said to have a reasonably wide spread of holdings, and this provides support for the proposition that international operators are increasingly trying to rationalize their assets on the basis of geography – although this, in turn, tends to bring with it common languages, cultures and so forth.

2 Many operators now identify core markets – as in the case of Telecom Italia (Italy and Brazil) or TDC (anything inside Nordic region) – and advertise their willingness to sell non-core assets 'at the right price' (which may be quite low).

It is worth investigating this proposition in relation to Africa and Asia-Pacific, two regions that have not previously been scrutinized in detail in this Chapter.[3]

Table 10.6 Mobile operator presence in Africa: Number of countries 2003–2009

	2003	2004	2005	2006	2007	2008	2009[1]	Subs (mn)[2]	African subs as % total[2]
MTN	7	7	11	17[6]	17	17	17	56.4	79.8
Zain	0	0	14[4]	15	15	16	17	35.0	74.6
Vodacom	5	5	5	5	5	5	5	33.6	100
Vodafone	9	9	9	9	9	10	10	32.0[8]	11.4
Orascom	8	7	5	5	5	6	7	23.4	37.6
France Télécom	11	11	11	11	14	15	16	18.1	18.4
Millicom	5	5	7	7	7	7	7	8.9	34.4
Vivendi Universal	4	3	4	4	6	6	7	8.7	45.2
Etisalat	2	2	8[5]	8	9	11	11	7.3	24.8
Portugal Telecom	5	5	5	5	5	5[7]	5	3.8	15.0
Qatar Telecom	0	0	0	0	2	2	2	3.5	14.0
Telefónica	5	5	5	5	5	5	5	2.9[9]	1.9
Comium	0	1	2	2	4	4	4	2.2	100
Axiata[3]	8	2	2	2	1	0	0	0	0
Atlantique Télécom	6	6	6[5]	–	–	–	–	–	–
Celtel International	13	13	14[4]	–	–	–	–	–	–
Investcom	5	4	5	6[6]	–	–	–	–	–

Source: Compiled by authors from a wide variety of (mainly operators') websites

Notes:

1. Numbers at the year end. This column takes account of all agreed transfers of assets up to the specified date. These had not necessarily taken place in practice by that point due to such matters as approvals by the relevant company boards and regulators.
2. As of 31 December 2008. By subscribers we refer here not to gross subscriber numbers but to 'proportionate' or 'equity' subscribers: that is, the gross figure is weighted according to the proportion of the equity held by the company in question.
3. Prior to 1 April 2009, the company was known as Telekom Malaysia International, which was in turn spun off from Telekom Malaysia.
4. In March 2005, the then MTC made an agreed offer for 85 per cent of Celtel, with the other 15 per cent to follow within two years. The subscribers attributed to MTC were therefore 85 per cent of the total for Celtel. In May 2007, the stake was raised to 100 per cent.
5. The six additional networks were the result of taking a 50 per cent stake in Atlantique Télécom in April 2005. The stake was raised to 70 per cent in 2007. There is doubt about a further network in the Central African Republic.
6. MTN acquired 100 per cent of Investcom in August 2006 at which point it operated six networks.

3 But see Curwen and Whalley (2008: Chapter 3, 4).

7. The situation is a bit confusing. Portugal Telecom has a management contract for Mascom in Botswana and appears to have a 3.1 per cent stake in a licence in DR Congo after selling most of its stake in October 2007. There is also considerable doubt as to the legitimacy of its stake in Guinea-Bissau's Guinée Telecom which it lists as an asset, but gives no subscriber data in its annual accounts. None of these are included.
8. Including 50 per cent of the total for Vodacom.
9. Only Morocco is not via its 9.9 per cent of the total for Portugal Telecom.

Africa is interesting because it consists of large numbers of countries in varying stages of development and it is a region where structural equilibrium is still some way into the future. Additional data are presented in Table 10.6. The first obvious point is that only six operators derive significant numbers of proportionate subscribers from the region, and that four of these are based in Africa. Vodafone now owns a majority stake in Vodacom so its presence in this group is no great surprise, but it also has substantial interests in Egypt, Ghana and Kenya. In contrast, France Télécom has lots of mostly quite small interests and its view of these is slightly odd. Although they contribute relatively little compared to its Western European holdings, France Télécom has claimed (implausibly) that they are comparable to Telefónica's stakes in Latin America where Telefónica has nearly as many proportionate subscribers as France Télécom has in total. Furthermore, it has denied any interest in the huge markets of Nigeria or South Africa.

One can readily understand why these two operators are keen to have stakes outside their saturated European markets, although France Télécom is for now generating a rather small number of proportionate subscribers per network, both absolutely and compared to Vodafone. On the other hand, neither MTN nor Zain have much higher averages. However, there is one significant difference, which is that they both make attractive targets for any non-African operator wishing to build up an African portfolio in one fell swoop.

In the case of Zain, this is a little curious because until mid-2009 the company was very much regarded as predator rather than target and, in any case, no single private stakeholder could legally hold more than 5 per cent of its shares prior to September 2009. However, despite its wealth, its standard policy of paying over the odds to acquire stakes and licences ran into difficulties during the credit crunch of 2008–09, and in July 2009 it was revealed that it had opened negotiations to sell off a majority of its African holdings to France's Vivendi Universal. In the event, talks quickly broke down (Parker 2009a) with India's Reliance Communications stepping into the breach (Wood 2009), only for a completely different deal to be announced in September, this time involving a 46 per cent (controlling) stake in Zain itself.

This offer was curious not only because it involved a Malaysian billionaire's alBukhary Group and India's Vivasi Group – with possibly the limited support of India's BSNL and MTNL – but also because the stake was being offered by minority shareholders led by the Kharafi Group of Kuwait (Haslam 2009). Given a sharp drop in the Zain share price and legal action to prevent the takeover, it remains highly uncertain whether or not it will succeed. Either way, analysts remain puzzled as to why a consortium containing no network operators is acting in this way, and BSNL and MTNL have now made it clear that they intend to launch a joint bid should the Vivasi consortium fail to agree terms (TeleGeography 2009e).

It is of no small interest that Reliance Communications had attempted, albeit unsuccessfully, to acquire a stake in MTN in May 2008 (Leahy and Parker 2008). Subsequently, MTN's comments about consolidation in Africa indicated that it saw itself in the role of consolidator (Burgis 2009), but in so doing it had underestimated the determination of India's Bharti AirTel to acquire a controlling stake. Rebuffed during 2008 when merger discussions broke down over the issue of ultimate control, Bharti reappeared with a second proposal in May 2009 which involved an initial cross-shareholding, to be followed at some unspecified time by a full merger (Telecompaper 2009). The proposed deal was extremely complicated (Lex 2009) and, predictably, politics rapidly intruded. In particular, South African politicians were determined not to surrender a strategic company to foreigners and demanded stock exchange listing arrangements that were largely illegal under existing Indian regulations (Leahy 2009a, 2009b).

It can be argued that the deal was predestined to fail, as it inevitably did in late September, which does rather raise the issue as to why it was pursued in the first place. That aside, the issue of further consolidation in Africa, which is widely regarded as a necessity, can best be described as murky. While existing operators may continue to buy up the licences and occasional (mostly privatized) stakes that gradually seep onto the market, these will have only a modest effect in the near future. But it is hard to see where a predator is to be found who has the resources and political acceptability to take control of an existing pan-African operator. Zain's situation is unstable – it wants to dispose of its African assets, yet is itself under offer by a consortium containing no telcos – and MTN, albeit now free (perhaps temporarily) from foreign predators, has failed to increase its holdings for the past four years. Under the circumstances, Etisalat may be the most obvious contender since it badly needs to scale up in Africa, but any kind of deal with Zain is surely most unlikely and politicians in South Africa are unlikely to welcome its attentions other than as very much a minority partner.

It is of no small interest that operators based in India are currently performing the role of potential predator in Africa. Over the past two years, there has been a massive influx of foreign investment into India, with large stakes changing hands as well as the creation of several greenfield operations. Although the Indian market has grown extremely quickly, it is rapidly approaching the point at which only parts of the relatively poor rural population remain to be accessed and, as noted above, this has induced Bharti and Reliance, together with Essar Holdings, to look abroad to places where their experience in India can be applied to good effect.

However, they have so far achieved remarkably little. The failures in Africa discussed above are notable only for their scale since failures overall have greatly outnumbered successes. But where else can they look? The obvious answer might appear to be Asia since they clearly have nothing to offer in developed countries of any size, but as previously noted there is no tradition of cross-border acquisition in this region. Table 10.5 illustrates this point, and it must be borne in mind that the Asia-Pacific region is unusually heterogeneous – there is little in common between Australasia, China, India and Japan, let alone the CIS countries to the west.

NTT's mobile subsidiary DoCoMo, which as noted has been a complete failure outside Asia, has accumulated its holdings within a tightly restricted eastern section of the region. However, its proportional subscribers – roughly 55 million at the end of 2008, of which just over 30 million can therefore be attributed to parent NTT – are largely to be found in Japan because NTT's strategy has long been to take quite small minority stakes elsewhere.

It is accordingly anything but a regional powerhouse. Meanwhile, Telenor is surprisingly dependent upon the region, although its roughly 20 million proportionate subscribers in Pakistan, together with roughly 12 million in both Bangladesh and Thailand, accounted for the great majority of its total at the end of 2008. Its other holdings are largely through VimpelCom to the west.

Table 10.7 Mobile operator presence in Asia-Pacific: Number of countries 2004–2009

	2004	2005	2006	2007	2008	2009[1]	Subs (mn)[2]	Asia-Pacific subs as % total[2]
Singapore Telecom	6	7	7	8	8	9	82.6	99.9
Vodafone	5	6	6	7	7	7	57.6	20.6
Telenor	4	6	7	7	7	11	51.0	53.6
Axiata[3]	4	6	7	7	7	7	49.1	99.9
NTT[4]	7	8	9	9	10	11	42.8	100
Orascom[5]	2	7	7	8	2	3	38.8	62.4
SK Telecom	4	4	6	6	5	4	31.1	97.1
Qatar Telecom	0	0	0	5	5	5	16.4	65.7
Etisalat	0	0	1	3	3	5	12.1	40.8
Telstra	2	2	2	3	3	3	11.7	100
VimpelCom	1	2	3	3	3	6	8.6	14.6
Saudi Telecom	0	0	0	3	4	4	7.8	28.8
TeliaSonera	1	1	2	4	6	6	6.5	10.9
Hutchison	6	6	6	7	7	7	6.4	29.4
Millicom	5	4	4	3	3	0	3.4	13.1

Source: Compiled by authors from a wide variety of (mainly operators') websites

Notes:

1. Numbers at year end. This column takes account of all agreed transfers of assets up to the specified date. These had not necessarily taken place in practice by that point due to such matters as approvals by the relevant company boards and regulators.
2. As of 31 December 2008. We refer here not to gross subscriber numbers but to 'proportionate' or 'equity' subscribers: that is, the gross figure is weighted according to the proportion of the equity held by the company in question.
3. Prior to 1 April 2009, the company was known as Telekom Malaysia International.
4. NTT largely operates internationally via majority-owned subsidiary DoCoMo.
5. The 2005 and 2008 changes reflect the acquisition and sale of a stake in HTIL respectively.

Axiata is also to all intents and purposes an exclusively Asian operator. Its nine million domestic subscribers at the end of 2008 were heavily outweighed by the 22 million proportionate subscribers in Indonesia, and with only three other holdings of any size (in Bangladesh, India and Sri Lanka) its operations are in essence also tightly restricted geographically. For its part, SK Telecom is important only in its home market, in stark contrast to Singapore Telecom whose home market is small. SingTel's other Asian stakes are all minorities, but those in India and Indonesia yield well above 20 million proportionate subscribers, with the Philippines contributing well over 10 million. Furthermore, SingTel is unusual in having a significant stake in Australasia. It is accordingly rather more of a regional powerhouse than Vodafone which sold its interests in Japan in 2006. Furthermore, Vodafone obtained 15 million proportionate subscribers in 2008 from its very small stake in (pre-restructured) China Mobile HK and over 30 million from its majority stake in India's Vodafone Essar, so its importance in the Asia-Pacific region should not be exaggerated.

Conclusion

The purpose of this chapter has been to provide a survey of the structural, and to a lesser extent, strategic issues facing mobile network operators around the world. By its nature this review has had to be highly restrictive, and the approach has accordingly been to examine each of the main regions in one particular context rather than to try to take a worldwide view throughout.[4]

Ideally, it should be possible to use this review to make forecasts about the structure of the worldwide mobile industry beyond 2010. This has been made considerably more difficult by the credit crunch of 2008–09 which has affected different operators in substantially different ways. There is some controversy as to whether there is going to be a 'double dip' in equity markets which have risen sharply during the latter part of 2009, in good part due to the (alleged) fragility of consumer spending. If spending holds up then the mobile industry will recover during 2010 because there is clear evidence that the ability to communicate while on the move is highly valued throughout the world (although it does not, of course, necessitate the constant upgrading of handsets).

However, with so many markets approaching saturation, operators are understandably tending to the view that vicious price competition is no longer the best way forward and that it needs to be replaced, so far as is possible (subject to regulatory authorization), by cooperation and consolidation. The latter has been considered above, and whereas it is never possible to claim that there is such a thing as an optimal one-size-fits-all market structure, there is some evidence that the number of operators within a market that can expect to survive profitably lies within a fairly narrow range. In consequence, those countries that have structures significantly at odds with such a range – the likes of Cambodia, India, Indonesia and Vietnam spring to mind in Asia – can be expected to restructure in the short term.

Structural change can also take place through the portfolio behaviour of international operators. As noted, it is simply not sensible to expect to accumulate an entirely coherent portfolio of worldwide assets. Licences and stakes appear on the market entirely randomly, and there are generally so many interested bidders for the most attractive among them that failures typically heavily outweigh successes for individual operators. Furthermore,

4 A much fuller, but slightly more dated, review can be found in Curwen and Whalley (2008).

although the odds can be weighted in favour of a particular operator by virtue of access to large cash reserves, a strategy of overpaying can prove to be unsustainable, as exemplified by the situation of Zain during 2009. Not surprisingly, therefore, many operators prefer to concentrate upon enlarging existing stakes while divesting non-core assets to finance this.

The emergence of Middle Eastern operators as the main predators during the latter part of the decade has been something of a surprise, even given their mostly small home markets and strong cash reserves. This may continue for some time even if not all will remain willing participants. The most predatory operators at the end of 2009 are based in India, but it is one thing to make massive bids, quite another to bring them to fruition – ask China Mobile, among others – and so far there has been rather little enthusiasm expressed by target country governments. Not that this is exclusively an African phenomenon. One has only to listen to politicians in Italy railing on about the evil intentions of Telefónica to realize that nationalism is alive and well in the telecommunications sector (Zampano 2009).

Cooperation accordingly appears to offer a better way forward, although virtually every proposal is likely to attract regulatory attention. So far as high-speed networks are concerned, this is likely to take the form of infrastructure sharing,[5] as exemplified by the creation of Cornerstone to manage the shared network of Telefónica and Vodafone in the UK (TeleGeography 2009d). But the economics of network sharing are such that it can – and in many cases will – lead logically to consolidation. Hence, for example, Vodafone and Hutchison agreed in June 2009 to merge their businesses in Australia because neither felt able to invest enough to build out a full network. A similar proposal was tabled by Orange and T-Mobile in the UK (Parker 2009b). It is important to bear in mind, however, that these arrangements are born out of weakness and are more likely to involve third or fourth place operators than their more successful rivals. Nevertheless, the credit crunch has made it difficult for virtually all operators to pay for the upgrades of their networks to HSPA+ that are needed to remain competitive and, in particular, to create the cash reserves for the rolling out of LTE during the period 2010–12.

It is useful, finally, to link up the above comments to the iPhone case study in Chapter 7. As noted, smartphones, especially those with touch-screens, are increasing their share of the device market in leaps and bounds as are dongles in computers. But these devices have their rationale in their ability to download ever larger amounts of data, and such is the pressure this is creating that 3G networks, such as that of AT&T in the USA, are barely able to cope. This, in turn, necessitates the network upgrades to HSPA, HSPA+ and/or LTE that promote network sharing and consolidation. Regulators may not like to admit it, but they will ultimately be obliged to sanction virtually all such proposals and the restructuring of the industry will continue apace

References

Barigaba, J. 2009. Uganda cannot take more GSM operators, says UCC. 4 November. Available at http://www.theeastafrican.co.ke (accessed 5 November 2009).

5 There is a discussion of tower sharing at Telecoms.com (2009).

Bartash, J. 2009. Big Phone companies are unlikely US antitrust targets. 20 July. Available at http://www.totaltelecom (accessed 22 July 2009).

Burgis, T. 2009. MTN expects buying opportunities in Africa. 12 March. Available at http://www.ft.com (accessed 13 March 2009).

Cheng, R. 2009. New low mobile rate reignites fear over US price war. 3 July. Available at http://www.totaltele.com (accessed 4 July 2009).

Communications Direct. 2009. Regulator fails to pick winner in Hungarian mobile tender, cites economic crisis. 17 March. Available at http://communicationsdirectnews.com (accessed 18 March 2009).

Curwen, P. and Whalley, J. 2008. *The Internationalisation of Mobile Telecommunications: Strategic Challenges in a Global Market.* Cheltenham: Edward Elgar.

Gabriel, C. 2009. Prepaid price war and barrage of handsets hit US holiday season. 5 August. Available at http://www.rethink-wireless.com (accessed 6 August 2009).

Haslam, C. 2009. Zain's $13.7B mystery investors. 9 September. Available at http://www.lightreading.com (accessed 10 September 2009).

Leahy, J. 2009a. Doubts raised over Bharti-MTN deal. 13 September. Available at http://www.ft.com (accessed 14 September 2009).

----- 2009b. Bharti merger talks with MTN hit opposition. 16 September. Available at http://www.ft.com (accessed 17 September 2009).

Leahy, J. and Parker, A. 2008. Ambani rivalry could scupper MTN deal. 16 June. Available at http://www.ft.com (accessed 17 June 2009).

Lex. 2009. Bharti/MTN valuation. 27 May. Available at http://www.ft.com (accessed 28 May 2009).

Morris, A. 2009. Further M&As off the agenda for Telekom Austria. 13 October. Available at http://www.totaltele.com (accessed 14 October).

Parker, A. 2009a. Vivendi halts Zain talks on Africa assets deal. 20 July. Available at http://www.ft.com (accessed 21 July 2009).

----- 2009b. Closer connections. 27 October. Available at http://www.ft.com (accessed 28 October 2009).

PriMetrica. 2009. AT&T's Centennial purchase nears completion. 14 October. Available at http://www.telegography.com (accessed 15 October 2009).

Stewart, A. 2008. Brazil's Anatel approves rule changes for Oi-Brasil Telecom deal. 13 June. Available at http://www.totaltele.com (accessed 14 June 2008).

Teleco. 2009. Chronology of the acquisition of BrT by Oi. 21 March. Available at http://www.teleco.com (accessed 22 March 2009).

Telecompaper. 2009 Bharti Airtel, MTN restart talks on merger. 25 May. Available at http://www.telecompaper.com (accessed 26 May 2009).

Telecoms.com. 2009. Tower sharing offers opportunities for emerging markets. 10 November. Available at http://www.telecoms.com (accessed 11 November 2009).

TeleGeography. 2009a. T-Mobile launches USD50 service trial. 20 February. Available at http://www.telegeography.com (accessed 21 February 2009).

----- 2009b. GNCC to hold wireless spectrum auction after DataWave request. 22 September. Available at http://www.telegeography.com (accessed 23 September 2009).

----- 2009c. DT not looking for US consolidation. 13 October. Available at http://www.telegeography.com (accessed 14 October 2009).

----- 2009d. Company managing O2 UK and Vodafone UK shared network sites is up and running. 30 October. Available at http://www.telegeography.com (accessed 31 October 2009).

----- 2009e. BSNL waiting on Vivasi before calling on Zain. 9 November. Available at http://www.telegeography.com (accessed 10 November 2009).

Wood, N. 2009. Reliance eyes Zain's Africa operations. 18 August. Available at http://www.totaltele.com (accessed 19 August 2009).

Zampano, G. 2009. Italian govt says Telefonia stake in Telco is a problem. 29 September. Available at http://www.totaltele.com (accessed 30 September 2009).

CHAPTER 11 *Socio-Economic Impact*

Introduction

The growth of mobile handsets has prompted research into their use and impact. It is probably not an exaggeration to state that this research is very extensive in its scope, and has changed over time as mobile handsets have become more numerous and their capabilities have evolved. This change is aptly illustrated by successive reviews of telecommunications published by *The Economist* that initially discussed the growth of the number of mobile subscribers before shifting onto the opportunities that emerged with the advent of faster networks and the use of mobile handsets in emerging markets (Cukier 2007; Standage 2001, 2003, 2009).

One aspect of the research that has been undertaken has been to explore how mobile handsets have made an impact on a particular country or region. Cohen, Lemish and Schejter (2008), Ito, Okabe and Matsuda (2005) and Horst and Miller (2006), who investigated the use and impact of mobile handsets in Israel, Japan and Jamaica respectively, illustrate the country-focused research that has been undertaken. Mobile handsets are also one of the telecommunication technologies that Qiu (2009) covered as part of his investigation into the working-class use of ICT (information, computing and technology) in China. A wider geographical focus can be found in Southwood (2008), whose focus is on Africa. In contrast, Yoo, Lee and Rowley (2008) and Hjorth (2009) investigated developments within the Asia-Pacific region. While Yoo, Lee and Rowley (2008) examined different issues within specific Asia-Pacific countries, Hjorth (2009) also discussed crosscutting issues such as what is meant by mobility and the relationship between gender and mobile media.

The uses to which mobile handsets have been put have been extensively explored in the literature. A plethora of issues are examined, from seeking to understand how handsets create, maintain and reinforce social relationships (Goggin 2006; Ling 2004, 2008) to analysing the language used in voice and SMS (short message service) conversations (Caron and Caronia 2007; Ling and Pedersen 2005). The use of mobile handsets by children and teenagers is one of the issues discussed in Goggin and Hjorth (2009) and Castells et al. (2007), while Katz (2008) and Goggin (2008) draw attention to how mobile handsets have been used in political campaigns and by religious groups. Less frequently examined in the literature are the role of mobile handsets as fashion items (Katz 2006), and the health implications of using mobile handsets (Burgess 2004).

Quite clearly, therefore, there are many areas where mobile handsets have had a socio-economic impact. As mobile subscribers now outnumber fixed-wire subscribers, it has been suggested that they provide the 'missing link' identified by Maitland (ITU 2002: 13). In addition, the wide-ranging socio-economic impact of mobile handsets enables them to contribute, directly and indirectly, to the achievement of the Millennium Development Goals adopted by the United Nations at the turn of the millennium. As a consequence, the remainder of this chapter will explore the economic impact of mobile handsets as well as some of the new circumstances in which mobile handsets are being applied.

Digital Divide

The 'digital divide' can best be expressed in terms of those who have access to a particular technology and those who do not, with this dichotomy often being labelled as the 'haves' and 'have-nots'. The establishment of the term after the Internet emerged as a major force in the mid-1990s has led to a strong association in the minds of many between the Internet and the digital divide (*The Economist* 2005: Vehovar et al. 2006). However, this is misleading in two respects since not only can the basic distinction between haves and have-nots be applied to technologies other than the Internet, but there are also more aspects to the digital divide than just that of access.

The inherent complexity of the term is evident from a 2001 definition put forward by the OECD (Organisation for Economic Co-operation and Development). It defined the digital divide as 'the gap between individuals, households, businesses and geographic areas at different socio-economic levels with regard to their opportunities to access information and communication technologies and to their use of the Internet for a wide variety of activities' (OECD 2001: 5). This definition, along with that of DiMaggio, Harigttai, Neuman and Robinson (2001), drew attention to two related aspects of the digital divide, namely access and use. In contrast, Fink and Kenny (2003) identified four different interpretations of the digital divide – access to ICT, ability to use ICT, actual use of ICT and impact of use. Given the different aspects of the term, it is not altogether surprising that Rowe (2003) stated that 'the term "digital divide" means different things to different people'.

There is also a tendency to view the digital divide in relation to countries. This may reflect reports such as ITU (2009) or OECD (2001, 2009) which highlighted the differences that exist between countries, but while this is informative and useful in its own right it inevitably obscures similarities and differences that occur both within and between countries. Having noted this, we can see from ITU (2009) that on a global basis there are considerably more mobile than fixed-wire subscribers and that these subscribers are increasingly to be found in developing countries. Mobile devices are just one of the topics that the ITU (International Telecommunications Union) combines together into an ICT development index to help ascertain whether the digital divide has changed over time. Comparing 2007 with 2002, the ITU concluded that the digital divide had persisted, notwithstanding the small decreases that they observed between some of the different categories of countries that they used in their analysis.

It is also worth noting that even when a technology is available, the would-be users must possess a range of skills and experiences that enable them to utilize that particular technology (van Dijk 2005; Tookey, Whalley and Howick 2006; Vehovar et al. 2006). This means not only that the initial access to technology can be held back by a lack of relevant skills and experience, but also that the latter may persist when the former does not. Moreover, as the digital divide is not restricted to a single technology, a wider range of skills is, by definition, required, thus ensuring that its reduction is more challenging and complex than would otherwise be the case. For example, in their study of USA Internet and mobile usage, Rice and Katz (2003) identified three different digital divides

– user/non-user, veteran/recent and continuing/dropout – and found that the factors influencing each of the divides were different.

Economic Development, Mobility and Markets

The relationship between economic development and telecommunications has been investigated in a range of studies. Jipp (1963), in a much-cited early article, observed that tele-density increased with wealth. More recently, Kathuria, Uppal and Mamta (2009) found a similar relationship in their study of mobile penetration rates in Indian states. A broader analysis of the relationship between telecommunications and economic development can be found in Röller and Waverman (2001). They used data from 21 OECD countries over a 20-year period (1950 to 1970) to examine this issue, finding evidence to support the existence of a causal relationship. Significantly, they suggested that the impact that telecommunications has on output may not be linear due to the presence of network effects, with the consequence that the causal relationship has a greater effect in those countries with an already existing high level of telecommunication infrastructure development.

Madden and Savage (1998) also found a positive relationship between telecommunications investment and economic growth. They suggested that changes in economy-wide growth precede changes in economy-wide and telecommunications-specific investment. The relationship between telecommunications and economic development is further complicated when it is remembered that it is bi-directional, or 'mutual causal' in the language of Madden and Savage (2000: 894). Based on an analysis of data for the USA between 1958 and 1988, Cronin et al. (1991) found 'a feedback process in which telecommunications investment enhances economic activity and growth while economic activity and growth stimulate demands for telecommunications infrastructure investment'.

However, as Waverman, Meschi and Fuss (2005) noted, telecommunications effectively meant fixed-wire connectivity at the time when Röller and Waverman conducted their analysis whereas this does not reflect the situation today. Not only are there, as noted, more mobile than fixed-wire subscribers in aggregate, but it is also true of many emerging markets. Waverman, Meschi and Fuss used data for 92 countries during the period 1980 to 2003 to determine whether the introduction and subsequent diffusion of mobile handsets contributed to economic growth. They found that mobile handsets had a positive impact on economic growth, and that this impact was more significant for developing than for developed countries.

Given the relationship between telecommunications and economic development on the one hand, and the contribution to economic growth that mobile devices make on the other, an interesting question to ask is how, in practice, mobile devices have contributed to economic activity. However, before beginning to analyse this, it is first necessary briefly to describe the nature and structure of markets. Markets align buyers with sellers. Since buyers are in need of something that a seller wishes to sell, when a transaction is made resources flow between the seller and buyer. These resources include products, services and information and can either be purchased with money or bartered. If the latter, labour, goods or information are exchanged between the buyer and seller.

Naturally, there are more participants present in a market than just a single buyer and a single seller. Markets also contain intermediaries who perform a diverse array of functions. Intermediaries may provide credit or market a particular good or service. Intermediaries can also provide market-related information such as the price of a particular good in the marketplace, the cost of raw materials or weather forecasts.

Individual transactions between a buyer and seller occur within the wider context of a value chain that links together the different stages in the production of a good or service. For example, a farmer buys fertilizer and sells his produce to a company for processing. In turn, this company sells the processed produce to another party, which may be another food-processing company or a supermarket. The presence of an individual transaction within a value chain complicates the structure of the market as well as adding to the number of participants. Markets are further complicated by the fact that they can occur at different geographical levels. Some markets are local while others are regional, national or international in scope. In the above example, the farmer may buy his fertilizer from a locally-based company but sell his produce to one located in the regional or provincial capital.

Jagun, Heeks and Whalley (2008) identified three characteristics of trading: in the first place, trading tends to be slow, in part due to the journeys required to collect information; secondly, trading is costly since not only are journeys time-consuming but they can also be expensive; and thirdly, trading tends to be risky due to one party being better informed than another. Their analysis of the impact of mobile handsets on aso oke cloth weaving in Nigeria found that their use replaced the need to undertake journeys. In other words, mobile handsets saved time and money for their users. In addition, mobile handsets reduced some of the risks inherent in aso oke cloth weaving, but they did not entirely replace face-to-face meetings due to such matters as the need physically to inspect the cloth and check the design.

Perhaps surprisingly, Jagun, Heeks and Whalley (2008: 58) stated that mobile handsets appeared to have entrenched the position of the intermediaries considered in their case study. The use of mobile handsets did not replace the pivotal role played by intermediaries in the industry's value chain, nor did it affect their ability to provide capital or credit to weavers. However, mobile handsets did encourage the emergence of a new type of intermediary – the coordinator-weaver – who coordinates orders from different intermediaries for several weavers. Although the case study provided only limited evidence that the use of mobile handsets changes the geography of supply chains, it nevertheless suggested that their use affects competitiveness. Quite simply, those weavers who do not have a mobile handset will lose out on orders.

Donner (2007) investigated the use of mobile handsets in Rwanda. Drawing on a survey of users, he found that an inverse relationship existed between the level of education and the proportion of business-related calls made via a mobile handset. In other words, the better-educated made fewer business calls on a mobile handset than the less-educated users. He also found an inverse relationship between the age of the handset user and whether the called party was new to the user's network of contacts – older users were less likely to call new contacts. In his discussion, Donner drew attention to how the call profiles differed depending on when the user acquired a mobile handset. Relatively recent purchasers of a handset made fewer business-related calls than those who purchased a handset earlier on. In addition, the different call profiles of users, which combine business and social calls, were used by Donner to argue that even though mobile calls are relatively expensive, 'the mobile is already a *personal* device' (Donner 2007: 14).

Veeraraghavan, Yasodhar and Toyama (2009) examined how mobile handsets replaced PCs in an Indian sugar cooperative. The cooperative, which was jointly owned by the farmers, centralized the collection, processing and selling of sugar in a single processing plant. PCs, located in kiosks, were used to transmit information between the processing plant and the individual farmers. Many of the original goals of the PC kiosk project were unfulfilled, although the kiosks did enable farmers to save time by not having to visit the central processing plant of the cooperative. According to the authors' analysis, the switch to a SMS-based system was advantageous in several respects: in the first place, mobile handsets were less prone to power supply problems than the PC kiosks; secondly, the use of mobile handsets enabled farmers to receive details of their accounts when being visited in the field by kiosk operators – under the old PC kiosk system, farmers could only receive market-related information during regular office hours; thirdly, farmers could receive information at any time and, moreover, the response time using the SMS-based system was considerably faster than that of the PC kiosks; and finally, the availability of cheap handsets widened participation in the scheme. Nevertheless, the authors went on to identify problems with respect to how text was inputted as well as the reliance of the system on a single centralized server.

Although Molony (2008) identified the advantages that accrued from the use of mobile handsets in agricultural trading, he also identified some limitations. On the positive side, mobile handsets helped farmers to understand the demand that was present in the market and facilitated the faster and more reliable distribution of information. The interviewees cited by Molony also revealed how the use of mobile handsets reduced the costs of trading by removing the need to accompany produce to the market. On the negative side, he argued that the mobile handset cannot always be relied upon as the sole source of information when relationships go beyond swapping details of supply and demand. Face-to-face contact occurs when more detailed information is swapped, and is integral to the development of those relationships involving credit. Interestingly, while Molony wrote that the relationship between the farmer and his wholesale broker is based on trust, he then went on to state that the research did not provide any evidence to suggest that the use of mobile handsets facilitated the development of trust in this relationship.

The impact of mobile handsets on the fishing industry in Kerala, India, was examined by Jensen (2007). He found that the use of such handsets extended the geographical scope of the market, with around one-third of the fishermen studied selling their produce outside of their local market (Jensen 2007: 896). As the introduction of mobile handsets in Kerala was staggered, with some areas receiving them before others, Jensen was able to ascertain whether the introduction in one area had an impact on the other areas. The introduction of mobile handsets into the first region resulted in the significant reduction in the spread of prices compared to when no handsets were available. Price variation also declined in the other two regions identified in the study when mobile handsets were introduced. The staggered introduction of handsets also reduced the amount of waste (unsold fish) in a sequential fashion – that is, waste was reduced in each region as handsets were introduced.

As mobile handsets allow for the improved coordination of markets and information flows, their introduction into Kerala could have resulted in increased price collusion among fishermen. However, Jensen found no evidence of either this or price-fixing (Jensen 2007: 909). Nor did the introduction of mobile handsets encourage large numbers of people to become fishermen – while some entry into the industry could be observed over the period of the study, Jensen suggested that other factors such as capital and social structures limited entry into the industry. In other words, while mobile handsets may create opportunities, it does not necessarily follow that these will be exploited due to the presence of other factors.

Mobile handsets are also integral to the Grameen Village Handset programme in Bangladesh. In essence, Village Handsets are 'public access GSM [global system for mobile] handsets that are owned and operated by local women entrepreneurs' (Kaul et al. 2008: 217). Through the provision of financial assistance by Grameen Bank, the female entrepreneur purchases a mobile handset and a Grameenphone connection. The entrepreneur then operates the mobile handset as a pay handset, thereby enabling her fellow villages to gain access to it (Knight-John 2008: 116). One consequence of the entrepreneur being female is that social structures in Bangladesh have changed, while another is that the profits generated by the handset can be used to fund other economic activities (Sullivan 2007).

Although it has been suggested that the Village Handset programme emerged in response to the specific circumstances to be found within Bangladesh (Knight-John 2008), it is possible to identify several key features that have contributed to its success. One of these is the mutually supportive relationships that exist between what Knight-Wood calls the 'Grameen Family' – that is, between Grameenphone, Grameen Telecom, Grameen Bank and the Village Handset operators. For example, Grameen Bank provides microfinancing to the operators, while Grameenphone provides heavily discounted airtime to the operators. A second factor is that the programme concentrates demand within a village for the mobile handset. As a consequence, even though an individual may not be able to afford access to a mobile handset, the village collectively can do so.

By virtue of the collaboration between the Grameen Foundation and MTN Rwanda, the Village Handset programme was subsequently introduced into Rwanda. Futch and McIntosh (2009) examined how this introduction had fared and found that it did have an impact on the access to telecommunication services reported by entrepreneurs. However, they found little evidence to suggest that this resulted in better prices for farmers or additional revenues for the Village Handset operators. Indeed, they found that operator revenues were such that the resulting profits were insufficient to cover the whole of the debt servicing costs (Futch and McIntosh 2009: 63). It is worth noting that when the Village Handset scheme was introduced into Rwanda, mobile handsets were already available. Futch and McIntosh (2009: 66) suggested that this would have impacted on usage, with those having access to mobile networks seeing little secondary usage due to the presence of other mobile handsets. Although they were able to show that heavier use of the Village Handset operators' handsets occurred at an intermediate distance from other handsets, it was not statistically significant.

One of the opportunities that have emerged from the widespread diffusion of mobile handsets across the Caribbean, namely teleworking, was the focus of Dunn (2009). He showed that teleworking, which is the use of telecommunications to enable employees to work somewhere other than their offices, was not a new phenomenon and had been extensively discussed in the literature. Although relatively few examples of teleworking in operation were recounted, and the definition of teleworking was relaxed, Dunn did suggest two ways through which mobile handsets could contribute to economic activity: in the first place, a mobile handset encourages further economic activity by its users; and, secondly, it provides users with 'address anonymity'. As some in society are marginalized because of where they live, the use of a mobile handset removes the possibility of a particular location being associated with a specific telephone number.

Miller (2007) also explored the use of mobile handsets in the Caribbean. He found that, within Jamaica, mobile handsets were integral to the survival strategies of those with the lowest incomes as they enabled the quick and efficient redistribution of income. Interestingly, Miller (2007: 323) found that the majority of low-income Jamaicans did not make entrepreneurial use of handsets unless such activities already existed prior to their diffusion.

Banking

Technology is used throughout the financial services industry. Indeed the industry is such a heavy user of information technology that it was expected to spend more on technology than any other part of the global economy in 2009 (*The Economist* 2009: 83). The history of information technology use within the industry is also longstanding, with perhaps the two most recent technologies that have been embraced being the Internet and mobile communications. Before focusing on mobile banking, we will briefly review the motives for IT investment and use by banks as this will provide contextual information to help us understand mobile banking from the perspective of both users and banks.

INTERNET BANKING

From the literature, it is possible to identify a variety of reasons why banks have adopted technology. Most obviously, banks have been able to reduce their costs (Calisir and Gumussoy 2008). One way through which costs can be reduced is by technology replacing staff, such that the number of bank employees falls. Costs also fall because technology reduces the cost of each transaction by replacing the manual processing of financial transactions. In support of their argument that technology reduces costs, Calisir and Gumussoy (2008) drew on both Nevens (1999) and Yakhlef (2001). The former suggested that the cost of a financial transaction drops by 80 per cent or more when undertaken electronically, while the latter highlighted the significant cost advantage enjoyed by Internet banks vis-à-vis full service banks. The declining cost of financial transactions was also illustrated by Margherio et al. (1999), who showed that the cost of a representative transaction declined from $1.07 if undertaken in a branch to $0.52 on the telephone, to $0.27 if an automated teller machine was used and to $0.01 via the Internet.

Another reason why the financial services industry has adopted information technology is that it can be used to develop new ways for banks to interact with their customers and for customers to access their accounts. In other words, the use of information technology creates new channels that link together banks and customers. Initially, automated teller machines (ATMs or cash machines as they are more colloquially known) and, more recently, telephone and Internet banking have all provided new ways for customers to access their bank accounts. Through these technologies, customers can access their banks accounts at times that are convenient for them (Durkin et al. 2008;

Mäenpää et al. 2008). The greater convenience enjoyed by customers contributes to their increased satisfaction, in the process making them less likely to move their accounts to a competitor. Interestingly, Ho and Wu (2009) stated that most banks operate a 'click and mortar' that combines electronic and physical ways of interacting with their customers. In other words, banks can use these new channels to complement their existing physical means of interacting with customers through branches in order to push financial products towards their customers as they seek to sell them a wider range of services.

While the benefits of the adoption of information technology by banks can be readily identified, this does not necessarily mean that customers will make use of the various electronic ways of banking that are available to them. Al-Somali, Gholami and Clegg (2009) are not alone in identifying a range of factors that contribute to the acceptance of Internet banking. Drawing on the literature, they identified seven different factors that contribute to the adoption of Internet banking: the quality of the Internet connection; awareness of the services and their associated benefits; trust; social influence; whether customers are resistant to change; IT literacy; and demographic characteristics. An extended version of the technological acceptance model was used to explore how these factors shaped the adoption of Internet banking in Saudi Arabia. The technological acceptance model was originally suggested by Davis (1989) and argues that attitudes towards a new technology are shaped by the interplay between perceived usefulness on the one hand and perceived ease of use on the other. The analysis of Al-Somali, Gholami and Clegg suggested that not all of these factors play a significant role in influencing the adoption of Internet banking. Their analysis suggested that four of these factors – the quality of the Internet connection, awareness of the services and their associated benefits, social influence and IT literacy – have a significant impact on the adoption of Internet banking in Saudi Arabia.

Agarwal, Rastogi and Mehrotra (2009) also explored the adoption of Internet banking, although their focus was on India. While their literature review raised similar issues to that of Al-Somali, Gholami and Clegg (2009), their findings were broader in scope. They found that the most important factor that raises awareness and encourages the adoption of Internet banking by a customer is whether friends and family are already users. Advertisements also influence the adoption of Internet banking, and were found to be particularly important for those adopters with a savings account. They also suggested that banks could improve the attractiveness of Internet banking by making the services that they offer more effective, in part through using this medium to resolve complaints with alacrity. Interestingly, they found that, among those already using Internet banking, security and trust played a larger role in determining customer satisfaction than the combination of services that was available. It is suggested that this may be because Indians are, by nature, conservative, thereby favouring security over the range of services that are available.

Alsajjan and Dennis (2010) confirm that trust encourages the use of Internet banking. More broadly, they find that intentions towards Internet banking are attitudinal in character, with differences occurring between the British and Saudi Arabian students surveyed. The influence of risk – of which security/privacy is just one example – on the adoption of Internet banking was explored by Lee (2009). In addition to security-related risk, Lee identified four other types of risk: financial risk, which is the loss when a transaction goes wrong; social risk, which is the disapproval of friends and family; time/convenience risk, which is the loss of time and the inconvenience incurred from delayed payments or difficult-to-navigate websites; and performance risk, which is that the website malfunctions. Lee found that all five different types of risk were negatively associated with the intention to adopt Internet banking, while perceived usefulness had a significant impact.

The influence of product complexity on the adoption of Internet banking was investigated by Durkin et al. (2008). After noting that one aspect of product complexity is the number of steps required to undertake a transaction and that users are familiar with transactions that they make regularly, the authors suggested a three-fold categorization of financial products into transaction, insurance and specialist services. These differ in terms of the number of steps required to complete a transaction and the extent to which the outcome is known, with the first category having fewer steps and more certainty than the third. From their survey of customers of a UK bank, the authors reaffirmed the importance of issues such as security, trust, IT literacy and convenience in encouraging the adoption of Internet banking. Interestingly, they found that those customers requiring financial products with medium complexity did not regard the lack of face-to-face interaction for simple products as problematic, while those with complex financial product requirements demonstrated a preference for face-to-face interactions with the bank.

From this review of Internet banking it can be seen that a range of factors encourage adoption. One such factor is increased convenience, while another is security/trust. IT factors also contribute, in terms of both the quality of Internet access as well as the level of the user's IT literacy. Routine financial transactions lend themselves to Internet banking, whereas more complex ones do not.

MOBILE BANKING

What is mobile banking? According to Kim, Shin and Lee (2009: 268), mobile banking 'enables customers to access their bank accounts through mobile devices to conduct conventional and more advanced financial transactions'. In contrast, a narrower definition was suggested by the focus of Ivantury and Pickens (2006) and Porteous (2007). Both suggested that the device used in mobile banking is the mobile handset. In other words, mobile banking to them is the undertaking of financial transactions using a mobile handset – more sensibly referred to as a mobile device in this context given the use of dongles in laptop/netbook computers. A broad range of transactions can be undertaken using a mobile device. For example, Porteous (2006) distinguishes between m-payments and mobile banking, with the former consisting of financial transactions enacted through mobile devices and the latter involving a broader range of services. In other words, mobile banking includes m-payments.

Sullivan (2007) offered a different categorization of the financial transactions facilitated by mobile banking. For example, money can be stored in a handset and then spent accordingly. Alternatively, the handset can be used to send and receive remittances, or to makes purchases and pay bills. Although Sullivan (2007) and Standage (2009) provided illustrations of each of these different types of financial transactions, it is perhaps worth highlighting remittances if for no other reason than to show their scale. According to Coyle (2007), the inflow of remittances for some countries is larger than official aid and they are, in aggregate, comparable to flows of foreign direct investment. Given that remittances often involve small amounts, the use of mobile banking as a transfer mechanism is attractive because this serves to circumvent traditional transfer methods that are disproportionately costly.

Various aspects of mobile banking have been discussed in the literature. Laukkanen (2007) compared Internet and mobile banking, focusing on customer value perceptions. He found that the use of the standard mobile handset (often now called a feature phone to distinguish it from a smartphone) was problematic, given its unsuitable display and small keyboard. In the light of the role of security in Internet banking, a somewhat surprising conclusion of his study was that security was not a major obstacle to the use of mobile banking. Instead, the interviewees were more concerned about making mistakes while using the service. It is worth noting that, in his conclusion, Laukkanen suggested that technological advances will remove some of the barriers to the adoption of mobile banking, although the input mechanism is likely to remain troublesome.

Kim, Shin and Lee (2009) focused on one of the key issues identified by the research on what encourages the adoption of Internet banking, namely trust. Adopting a similar approach as much of the Internet banking research noted above, they found that those surveyed in Korea had a low degree of trust in mobile banking and were not reassured by contract terms, the offers of guarantees and so forth. Their results suggested that trust in mobile banking could be enhanced through users becoming aware of its convenience and efficiency, and that the trusting predisposition of some individuals makes them more likely to adopt than others. However, they did qualify this second point by stating that perceived benefits play a greater role in encouraging adoption than whether the person was trusting by nature. In addition, the initial level of trust displayed by their survey respondents was not influenced by the reputation of the mobile banking service provider. As this is at odds with online shopping research, they went on to state that it could be the case that the reputation of the mobile banking service providers is important only up to a point, beyond which it does not affect trust levels.

Trust is one of the issues discussed by Donner and Tellez (2008) in their exploratory study that they conducted in urban India. This study examined informal credit mechanisms among small enterprises and the role that could be played by mobile handsets. Although they suggested several ways in which the mobile handset can contribute to the relationship between small enterprises and clients, when it comes to credit its role is somewhat limited. Their interviews led Donner and Tellez to state that face-to-face interaction between a small enterprise and its clients provides the former with the best opportunity to assess the latter's trustworthiness.

Ratten (2008) argued that the adoption of WAP (wireless application protocol)-based mobile banking is driven by behavioural intentions, which can be understood by examining the environment in which individual adopters are located. The difficulty of inputting text is one of the issues touched on by Peevers, Douglas and Jack (2008). Given the slowness of inputting text using a standard mobile keypad, they suggested that the format used in SMS-based banking should be simple and succinct. Through a series of experiments, which required participants to use different text input formats to request financial products, they found that a version of SMS-based banking using numbers or abbreviations was substantially quicker than one with 'free form' text. Moreover, older users found 'free form' less user-friendly than the other two input formats and, more generally, 'free form' was felt to be a slow way of inputting text.

BANKING THE UNBANKED

Standage (2009) and Sullivan (2007) are not alone in drawing attention to the growth and extent of mobile banking in emerging markets. They identified a range of examples of mobile banking in emerging markets, including G-Cash in the Philippines, WIZZIT in South Africa and M-PESA in Kenya. These examples are often discussed in the context of transformational mobile banking – that is, mobile banking which allows the unbanked to gain access to banking services for the first time through the use of mobile handsets (Porteous 2006). In contrast, additive mobile banking is when the mobile handset is another channel through which banking transactions occur. In Figure 11.1, additive mobile banking is located in quadrant 1 and transformational banking in quadrant 2.

The transformational nature of mobile banking in emerging markets hinges on a simple fact, namely that bank branches are not to be found in many locations. As a result, individuals in many emerging markets do not have access even to basic financial services (Coyle 2007). This is particularly true for those living in rural areas where banks have often struggled to establish a presence and for those who lack the necessary documentation to open a bank account (*Appropriate Technology* 2008). Mobile banking promises to change this. Mobile handsets are more common than bank accounts and Internet subscribers, with the consequence that they become the channel through which the unbanked gain access to banking services (*Africa Research Bulletin* 2008, 2009). It has been suggested that mobile banking will have far-reaching implications. *Appropriate Technology* (2008: 22) quotes Elizabeth Littlefield, chief executive of the Consultative Group to Assist the Poor, as stating that 'mobile handset banking and other emerging applications will revolutionize financial markets in rich ands poor countries alike' while Sullivan (2007: 127) describes mobile banking as 'the first big "killer application" for cell handsets'. A slightly more sombre assessment can be found in Standage (2009), where it was suggested that mobile operators do not expect to make much money from mobile banking, but instead use the service to attract new subscribers or to retain existing ones. Having said this, he did illustrate the range of ways in which mobile banking is developing in emerging markets.

	Banked	Unbanked
Have a mobile handset	1	2
Do not have a mobile handset	3	4

Figure 11.1 Banks and mobile handsets

Source: Adapted from Porteous (2007: 20)

The reality, of course, is likely to be found somewhere between these assessments. As mobile banking is still in its infancy, it is difficult to assess how its future may develop other than to say that it is likely to grow considerably in the coming years. It is possible, however, to observe the appearance of mobile banking in many emerging markets. Kramer and Paul (2006) identified mobile banking in a range of African, Middle Eastern and Asian countries. Based on his assessment of mobile banking in Africa, Porteous (2006) was able to identify the emergence of four different models:

1. 'pure' bank driven
2. joint venture
3. non-bank led
4. non-bank driven.

The underlying business model in each case is different. As the name of the first business model implies, the prominent actor in this model is the bank. In this business model, the service is not tied to any specific mobile operator, unlike in the second model where it usually is. In addition, the brand adopted in the second model could be that of the non-bank joint venture partner. This is also the case in the third business model, where the service might or might not be tied to a single mobile operator. In the fourth business model, the service is tied to a specific mobile operator and the brand adopted is that of the non-bank or mobile operator.

According to Porteous, additive mobile banking illustrates the first business model while MTN Mobile Money is an example of the second. M-PESA and WIZZIT are both examples of non-bank led mobile banking, while Globe in the Philippines illustrates the fourth model. The first two business models differ from the latter two in terms of where cash can be accessed. In the first two models, cash can only be deposited or withdrawn via a bank. To this the third model adds a network of agents, while the fourth model dispenses with the bank altogether and allows cash to be deposited and withdrawn through the mobile network as well as other channels. What this means, in practice, is that the channels through which customers can access their accounts varies considerably.

That differences exist between the mobile banking business models that have emerged is also evident from Williams and Torma (2007). They surveyed three mobile banking services – WIZZIT, M-PESA and Globe – and compared and contrasted them in terms of their functionality and a series of underlying characteristics. Their functionality is assessed in terms of information, transaction and payment. All three services are largely the same, with only WIZZIT not providing payment services. Perhaps of greater interest are the differences that occur with respect to their underlying characteristics. According to the authors, WIZZIT and Globe operate open systems that enable transactions to be made to accounts held on other networks. Both are also interoperable – that is, they are not tied to a specific mobile operator. In contrast, M-PESA operates a closed system and is not interoperable. In the case of WIZZIT and Globe, a bank holds customer deposits. In contrast, customer deposits in M-PESA are held in a float although this is, in turn, deposited with a bank. A broader range of cash deposit and withdrawal mechanisms are available for WIZZIT than the other two mobile banking services, which limit such activities to just authorized agents.

Various studies have examined the development and impact of additive mobile banking. Ivantury and Pickens (2006) explored the adoption and use of mobile banking among low-income South Africans. Their survey of users and non-users found that low-income South Africans do use the mobile banking services of WIZZIT and rate the service highly for convenience, cost and security. Users also value the cheapness of the service since it is considerably cheaper than other banking channels. The users of WIZZIT also appear to be untypical since not only do they appear to be better off than most poor South

Africans, but they are also 'more financially and technologically sophisticated' (Ivantury and Pickens 2006: 5). In other words, WIZZIT users are not among the poorest in South African society and seem comfortable using technology. Interestingly, among the non-users surveyed, a majority of respondents felt that the monthly charge was affordable, leading Ivantury and Pickens to state that 'WIZZIT could have powerful appeal if more people knew about it' (Ivantury and Pickens 2006: 7). This lack of awareness is one of the barriers to mobile banking adoption identified by Porteous (2007), with a second being that those with bank accounts simply do not trust mobile banking. Having said this, the drivers of mobile banking that he identified were similar to those noted elsewhere, namely convenience and time-saving.

Sullivan (2007) charted the evolution of mobile banking in the Philippines. According to Sullivan, Smart Communications launched the first of a series of related products in 2000 that enabled subscribers to purchase prepaid SIM (subscriber identity module) cards in low increments. However, this product – Smart Buddy – was not as successful as anticipated, with the consequence that a new product – PureTxt 100 – was introduced. Not only was this a SMS-only product, but it also reduced further the increment level to $1.80. Due to this rate reduction, the cost of SMS was significantly less than that of a handset call. In 2003, Smart Load – a new product which combined voice and SMS – was launched. Once again, this new product further reduced the incremental amount. Significantly this product allowed for airtime to be transferred from resellers to subscribers and soon become an electronic-only system when Smart abandoned the use of cards. As a consequence, Sullivan claimed that the owners of local shops were transformed into the providers of credit.

Globe Telecom, Smart's competitor in the Philippines, has also introduced a mobile banking service. Globe introduced G-Cash in 2004 and positioned the product as a mobile wallet (Williams and Torma 2007). G-Cash supports a wide range of services such as the payment of bills and taxes, the purchase of goods or services from partner organizations, the transfer of money to another subscriber and loan repayments (Sullivan 2007). As G-Cash does not require a subscriber to have a bank account, it removes one of the key barriers to participation and thus widens its potential market.

Perhaps due to the disparate economic environments that can be found in emerging markets, it comes as no surprise that different models of mobile banking have appeared. From their survey of mobile banking, Kramer and Paul (2006) identified a series of lessons that can be drawn. The first lesson is that it is important to know your customers. They stated that the economic potential of those at the base of the economic pyramid has not been appreciated, but that this was beginning to change as research was being undertaken and companies were targeting this market. They went on to suggest that this market required the development of new business models if it was to be successfully targeted and that initiatives should be developed from the ground up.

Naturally, mobile banking is not without its risks. Although it has been suggested that mobile banking is subject to many of the same vulnerabilities as Internet banking (Bezuidenhoudt and Porteous 2008), the use of a mobile handset adds an additional dimension to the risks associated with mobile banking. The risks and vulnerabilities that they identified are broad in scope, which in part reflects the value chain that is necessary to deliver mobile banking services. This value chain combines handsets with mobile operators and banks as well as other intermediaries. Some of the risks inherent to mobile banking are technical in character, emanating from the handset as well as the extent to which the service is dependent on, or independent from, a particular mobile operator. Other risks are operational, while yet others are more generic in nature and include the loss of the handset used for mobile banking.

Bezuidenhoudt and Porteous also drew attention to the risks that emerge from the business model that has been adopted. Interestingly, they distinguished between the risks associated with additive and transformational business models. Given the target market of transformational business models, the prevalence of basic handsets may be a source of risk whereas more advanced handsets are likely to occur in additive business models. Another source of risk is whether mobile banking is one of several channels through which the bank interacts with its customers. A bank with several channels may be cautious in the development of its mobile banking service for fear of reputational damage if problems emerge. In contrast, if mobile banking is the only channel between the bank and its customers, the failure of the mobile network will have far-reaching implications.

Government Services

Mobile handsets can also play a role in the delivery of government services. Drawing on the literature, it is possible to identify a wide range of e-government definitions which according to Helbig, Gil-Garcia and Ferro (2009: 90) 'range from the descriptive to value-laden'. That aside, a useful definition is suggested by Guida and Crow (2009: 283) who wrote that the term 'is generally used to describe the delivery of services via network technologies to citizens, businesses and government agencies such as customs clearance, tax payments or license processing. It encompasses both the "front-office" and "back-office" use of ICTs to support the business processes that a government performs, typically the provision of information and services.'

Such a definition is useful for two reasons. In the first place, it does not restrict e-government to any specific technology such as fixed-wire, mobile or the Internet. Secondly, Guida and Crow's definition draws attention to the extensive nature of e-government in practice, since it affects both the internal processes of the government as well as its interactions with individuals and businesses. A diverse array of services can be offered as part of e-government, ranging from the provision of information to providing documentation and licences or allowing payments to be made or tax to be collected (Guida and Crow 2009).

But why have governments begun to deliver some of their services electronically? Although a variety of factors have encouraged the growth of e-government, Trkman and Turk (2009: 420) suggested that most of the identifiable benefits of e-government fall into one of four broad categories – cost reduction, reduced lead times, increased flexibility and improved quality. These are more or less echoed by Guida and Crow (2009: 287), although they did add two additional factors – on the one hand, e-government allows the delivery of services to areas that were previously underserved such as rural or less densely populated areas while, on the other hand, it also increases participation which, in turn, strengthens democracy. Nevertheless, Misuraca (2007) noted that the use of ICT in Africa by females was less than that by males. As a consequence, if participation is a goal of e-government then this imbalance needs to be addressed. Sinha (2009) explored the relationship between gender and ICT in Bhutan, and placed ICT use within a wider context of societal norms, cultural values and education/literacy levels. If nothing else, this research reiterated an earlier point that, for a variety of reasons, the availability of a technology does not automatically translate into its use.

Andersen (2009) explored the role that e-government can play in reducing corruption. According to Andersen (2009: 201), the mechanism through which e-government reduces corruption is relatively straightforward, namely it 'reduces contact between corrupt officials an increases transparency and accountability'. His econometric analysis of a range of countries between 1996 and 2006 found that for non-OECD countries an increase in e-government reduced corruption.

The adoption of e-government can also be encouraged by the structurally significant position of the government. A government can direct its departments and agencies to interact electronically with both individuals and businesses, thus encouraging them to adopt e-government services (Henriksen and Damsgaard 2007). This is not to suggest that adopting e-government is straightforward. Heeks (2006) stated that many e-government projects failed due to their poor implementation and management, a sentiment echoed by Guida and Crow (2009). It also takes time to implement e-government, with Henriksen and Damsgaard (2007) showing how it had been introduced and implemented in Denmark over a number of years. Misuraca (2007) identified illiteracy as a barrier to the adoption of e-government, and suggested that the creation of local content will help in its adoption, though this is difficult due to the plethora of languages that can be found across Africa.

One way to deliver e-government services is through tele-centres. Tele-centres promote the diffusion of ICT and have an impact on the economic development of communities through improving access to market-related information (Best and Kumar 2008; Falch and Anyimadu 2003; Rajalekshmi 2007). Rajalekshmi (2007) examined a tele-centre project in the Indian state of Kerala, paying particular attention to the relationship between trust and the delivery of e-government services. The project combined the establishment of tele-centres with an education programme to improve ICT literacy and the creation of local content.

Rajalekshmi's analysis found that one of the most popular e-government services was e-pay, the Internet-based payment of government-related payments. In contrast, less trust was placed in the intermediary providing sector-specific information. In other words, the users of the tele-centres seem to place more trust in the government than in the intermediary running the tele-centre. Interestingly, entitlements-based services were not available electronically because the appropriate systems within government had not been developed. Moreover, as some of the services provided by tele-centres are not trusted by those who use them this throws into doubt the assertion that tele-centres need to be multi-purpose in order for them to make an impact on a particular community.

Best and Kumar (2008) investigated the sustainability of tele-centres. They examined how tele-centres established as part of The Sustainable Access in Rural India (SARI) project had fared. Although 36 privately owned and operated tele-centres were established between November 2001 and February 2004, 32 of them had closed by May 2005. Best and Kumar's analysis identified several factors that contributed to the closure of the tele-centres. One factor was the lack of technical support, with those receiving better support staying open for longer. As the quality of technical support declined, tele-centres found it increasingly difficult to offer Internet-based services with the consequence that they received fewer customers. This, in turn, undermined their financial viability. The financial viability of tele-centres was also undermined by their inability to offer voice services

– although such a service was promised it never materialized because the government did not issue the appropriate licence to the project (Best and Kumar 2008: 41).

At the start of the project, the tele-centres did provide some e-government services, but these stopped at the start of 2003 when the government was reorganized (Best and Kumar 2008: 42). This also reduced the amount of revenue generated by the privately owned and operated tele-centres. As some of the tele-centres in the project were operated by the Dhan Foundation, a non-governmental organization, Best and Kumar (2008) were able to compare the two groups. They found that the Dhan Foundation contributed financial support to the tele-centre that enabled them to continue operating even though revenues fell considerable short of the costs incurred (Best and Kumar 2008: 43).

Ntaliani, Costopoulou and Karetsos (2008) suggested how mobile government, which they regarded as a subset of e-government, could be introduced into the agricultural sector. To facilitate this introduction they proposed a framework comprised of two elements – the identification and usage of potential mobile government services and technical solutions to ensure that such services are available. With respect to the services that could be offered, they drew a distinction between, on the one hand, generic agricultural services such as details of new products and policies, weather forecasts and, on the other hand, mobile government services like the renewal of licences, tele-diagnosis and forums.

Through applying the aforementioned framework, they were able to identify not only the services that producers within the Greek organic agricultural sector would require, but also how these could be technically delivered. The producers also stated that any mobile government service should enjoy low costs and be ubiquitous and easy to use. Given these requirements, as well as the range of services that were identified and the widespread availability of mobile handsets, it is perhaps no surprise that SMS was chosen as the preferred technical solution. According to the authors' analysis, SMS would be able to support six different types of services – product information, epidemic alerts, weather alerts, legislation news, administrative information and business news (Ntaliani, Costopoulou and Karetsos 2008). That these services could be delivered via SMS was shown by the use of a pilot project, which also illustrated the cost effectiveness and scalability of the technology.

Healthcare

Kwankam, Pablos-Mendez and Kay (2009) provided an overview of the e-health landscape. Like e-government, there does not appear to be a single agreed definition of e-health, although Kwankam, Pablos-Mendez and Kay (2009) suggested that it is possible to identify this as improvements in health through the use of ICT. As a consequence, the scope of e-health is very broad. On the one hand, it includes the use of ICT to improve the administrative processes underpinning the delivery of healthcare while, on the other hand, it also includes the use of ICT in treatment through tele-medicine.

Given this scope, it is understandable that several advantages arising from the application of ICT to healthcare have been identified. Chandrasekhar and Ghosh (2001) suggested that health improvements occur in three different ways – by the continued education and training of healthcare workers, by the delivery of services to poor and remote locations and by improved governance that, in turn, enhances the delivery and availability of healthcare. In contrast, Thompson and Dean (2009) suggested that the application of ICT to healthcare allows for improvements in efficiency, productivity and quality. O'Hara and Stevens (2006) highlighted the administrative benefits of applying ICT to healthcare.

The application of ICT to healthcare could also help developing countries counteract their lack of resources. Mechael (2009) drew on World Health Organization data to show that developing countries face a shortage of doctors, nurses and midwives. ICT would allow for those resources that developing countries do have to be used more efficiently and effectively. ICT would enable doctors to extend the geographical scope of treatments through, for example, the use of video-conferencing facilities or email. In Bhutan, for example, X-rays have been taken in one hospital and then photographed using a digital camera so that they can then be emailed to another hospital where a specialist was available (Kezang and Whalley 2007: 78).

Although there are many advantages to be gained from using ICT in healthcare, there are probably an equivalent number of obstacles to be overcome. Using the experience of the UK, Avison and Young (2007) identified a range of reasons why UK healthcare ICT projects have failed, including inadequate project management, organizational uncertainty and the lack of fit between the project and the processes it was supposed to support. The problems experienced as new projects are introduced may also be compounded by ineffective communication between the various participants (Juciute 2009).

Avison and Young (2007) also noted that healthcare is not like other sectors due to the range of specialists involved in the treatment of patients and the key role played by face-to-face contact. The fragmented nature of healthcare provision means that coordination is complex, and it may not be possible to use ICT on the scale necessary to make it cost-effective (Avison and Young 2007; Thompson and Dean 2009). Although these problems are themselves highly significant, Chandrasekhar and Ghosh (2001) identified further obstacles that emanate from the developing status of a country. One such obstacle is the lack of ICT, while another is cost. A lack of literacy, both in general and, more specifically, in respect of ICT is a third obstacle that needs to be overcome.

Having said this, the application of ICT to healthcare can be observed in a wide range of countries. Micevska (2005) investigated the relationship between ICT and healthcare. The first part of this investigation was an econometric analysis that used data from the World Bank and International Telecommunications Union. Micevska found that improvements in life expectancy were positively correlated to an increase in the number of fixed-wire lines in a country. In other words, the more fixed-wire lines a country has, the better its life expectancy. The second part of her analysis focused on two countries – Bangladesh and Laos – and used survey and interview data. With respect to Bangladesh, Micevska (2005: 65) found that the share of household income spent on healthcare was positively correlated to the use of the telephone to obtain information and save time. In contrast, the results for Laos shed light on household priorities. In 2000, more households placed greater importance on the provision of a hospital than on the availability of telecommunications infrastructure, but just a year later this was reversed (Micevska 2005: 66). Interestingly, a substantial majority of those households without access to a telephone stated that a hospital was more important (Micevska 2005: 67).

Vodafone (2006) suggested three broad areas where mobile handsets could play a role in the provision of healthcare. In the first place, mobile handsets could be used to tackle inefficiencies by improving communication between doctors and patients. Secondly, healthcare effectiveness could be improved by the monitoring and management of patients with chronic conditions. Thirdly, mobile handsets could be used to lower the barriers to participating in healthcare that might exist in the minds of some groups within society.

Mechael (2009) described m-health as an extension of e-health and drew a distinction between voice and data. Mobile handsets can be used to support health-related hotlines and awareness campaigns (Mechael 2009: 160). In addition, mobile handsets can be coupled with location to incorporate a geographical dimension into the data collection. This would enable researchers to map the diffusion of a specific disease and implement appropriate treatment strategies. SMS can also be used to alert patients, for example by reminding them of an appointment or that medication should be taken. In South Africa, for example, when the medicine bottle is not opened at the scheduled time, patients are sent a SMS as a reminder. In Rwanda, by way of contrast, SMS has been used by nurses to send data to a centralized database so that it can be analysed on a real time basis.

Contact is one of the issues discussed by Mechael (2008). Drawing on fieldwork conducted in the Minia Governorate of Egypt, she found that benefits accrue from the increased use of mobile handsets. On the one hand, these benefits accrue from the increased use of mobile handsets by healthcare professionals to improve the care that they provide patients, while on the other hand, mobile handsets enable patients to contact their doctor or nurse. For those patients undergoing treatment, the use of a mobile handset offers the ability to instantaneously obtain an answer and reassurance regarding their treatment (Mechael 2008: 99). The use of mobile handsets by healthcare professionals and patients also saves both time and money by minimizing unnecessary travel.

However, Mechael (2008) also alluded to several problems regarding the use of mobile handsets in healthcare. One issue was that the emergency number to be called from mobile handsets routed calls to Cairo and did not provide localized information. As a consequence, mobile-handset-summoned emergency assistance occurred via a call to a friend or relative who possessed a fixed-wire line. A second issue was that most of the healthcare professionals interviewed favoured informal over formal tele-medicine (Mechael 2008: 100) – that is, the continued use of mobile handsets to coordinate doctor/nurse/patient interaction as noted above rather than tele-medicine through formal initiatives or programmes.

Conclusion

This chapter has illustrated the scope of mobile handset/device use. It is clear that mobile handsets can be used to improve the flow of information. On the one hand, this enables users to be better informed as intermediaries are bypassed, while on the other hand it enables governments to interact directly with citizens. Such interaction enables government services to be provided more effectively and efficiently than was previously the case and, through improved participation and transparency, for the corruption that is present in non-OECD countries to be tackled.

Mobile handsets also create opportunities. Reflecting the fact that mobile handsets are a business in their own right, these opportunities may take the form of selling access to others. The sale of access to others in Bangladesh through the Village Handset programme has had a far-reaching impact through economic empowerment of the female operators.

Alternatively, the widespread availability of mobile handsets has encouraged their use as a channel to deliver services such as mobile banking or healthcare. The innovative mobile banking business models that have emerged in Africa and parts of Asia reflect the considerably wider coverage of mobile compared to traditional banking networks, as well as the cost effectiveness of electronic delivery over traditional distribution channels.

This is not to suggest that the availability of mobile handsets inevitably results in their use. Notwithstanding efforts to widen access through, for example, the Village Handset programme, the cost of access may remain prohibitive for some prospective users. Moreover, while the interface on basic handsets should arguably not present a barrier to use for most users, the advent of more complicated interfaces and the widespread diffusion of smartphones are likely to engender barriers to participation – and this is before the interface and inputting requirements of new services, such as mobile banking, are taken into account. In addition, the extent to which mobile handsets can be used will also be influenced by the wider social context within which it occurs. While some individuals and societies may replace face-to-face interaction with the use of the mobile handset, others may not.

References

African Research Bulletin. 2008. Kenya: Mobile banking. 16 August–15 September, 17957.

----- 2009. Mobile handset banking: Banking the unbanked. 16 February–15 March, 18167.

Agarwal, R. Rastogi, S. and Mehrotra, A. 2009. Consumers' perspectives regarding e-banking in an emerging economy. *Journal of Retailing and Consumer Services*, 16(5), 340–51.

Alsajjan, B. and Dennis, C. 2010. Internet banking acceptance model: Cross-market examination. *Journal of Business Research*, 63(9–10), 957–63.

Al-Somali, S. Gholami, R. and Clegg, B. 2009. An investigation into the acceptance of online banking in Saudi Arabia. *Technovation*, 29(2), 130–41.

Andersen, T. 2009. E-government as an anti-corruption strategy. *Information, Economics and Policy*, 21(3), 201–10.

Appropriate Technology. 2008. Banking services for the isolated and poor. March, 35(1), 21–2

Avison, D. and Young, T. 2007. Time to rethink health care and ICT. *Communications of the ACM*, 50(6), 69–74.

Best, M. and Kumar, R. 2008. Sustainability failures of rural telecenters: Challengers from the sustainable access in rural India (SARI) project. *Information Technologies and International Development*, 4(4), 31–45.

Bezuidenhoudt, J. and Porteous, D. 2008. *Managing the Risk of Mobile Banking Technologies*. March, Bankable Frontier Associates. Available at http://www.bankablefroniter.com (accessed 17 November 2009).

Burgess, A. 2004. *Cellular Handsets, Public Fears, and a Culture of Precaution*. Cambridge, UK: Cambridge University Press.

Calisir, F. and Gumussoy, C. 2008. Internet banking versus other banking challenges: Young consumers' view. *International Journal of Information Management*, 28(3), 215–21.

Caron, A. and Caronia, L. 2007. *Moving Cultures. Mobile Communication in Everyday Life*. Montreal and Kingston, Canada: McGill-Queen's University Press.

Castells, M., Fernández-Arèdvol, M. Qiu, J. and Sey, A. 2007. *Mobile Communication and Society – A Global Perspective*. Cambridge, MA: The MIT Press.

Chandrasekhar, C. and Ghosh, J. 2001. Information and communication technologies and health in low-income countries: The potential; and the constraints. *Bulletin of the World Health Organization*, 79(9), 850–855.

Cohen, A., Lemish, D. and Schejter, A. 2008. *The Wonder Handset in the Land of Miracles. Mobile Telephony in Israel*. Cresskill, NJ: Hampton Press, Inc.

Coyle, D. 2007. Overview. In, Vodafone. *The Transformational Potential of M-Transactions*. The Policy Paper Series, No 6. Available at http://www.vodafone.com (accessed 19 October 2009).

Cukier, K. 2007. A world of connections: A special report on telecoms. *The Economist*, 28 April.

Cronin, F., Parker, E., Colleran, E. and Gold, M. 1991. Telecommunications infrastructure and economic growth: An analysis of causality. *Telecommunications Policy*, 15(6), 529–35

Davis, F. 1989. Perceived usefulness, perceived ease of use, and use acceptance of information technology. *MIS Quarterly*, 13(3), 319–40.

DiMaggio, P., Harigttai, E., Neuman, W. and Robinson, J. 2001. Social implications of the Internet. *American Review of Sociology*, 27, 307–36.

Donner, J. 2007. The use of mobile handsets by microentrepreneurs in Kigali, Rwanda: Changes to social and business networks. *Information Technologies and International Development*, 3(2), 3–19.

Donner, J. and Tellez, C. 2008. Mobile banking and economic development: Linking adoption, impact and use. *Asian Journal of Communications*, 18(4), 318–22.

Dunn, H. 2009. Teleworking the mobile Caribbean: Enabling remote work among the marginalised in Jamaica and Trinidad and Tobago. *Information Technologies and International Development*, 5(2), 52–66.

Durkin, M., Jennings, D., Mulholland, G. and Worthington, S. 2008. Key influences and inhibitors on adoption of the Internet for banking. *Journal of Retailing and Consumer Services*, 15(5), 348–57.

Falch, M. and Anyimadu, A. 2003. Tele-handsets as a way of achieving universal access – the case of Ghana. *Telecommunications Policy*, 27(1/2), 21–39.

Fink, C. and Kenny, C. 2003. W(h)ither the digital divide? *info*, 5(6), 15–24.

Futch, M. and McIntosh, C. 2009. Tracking the introduction of the Village Handset product in Rwanda. *Information Technologies and International Development*, 5(3), 54–81.

Goggin, G. 2006. *Cell Handset Culture – Mobile Technology in Everyday Life*. Abingdon, England: Routledge.

----- 2008. *Mobile Handset Cultures*. Abingdon, England: Routledge.

Goggin, G. and Hjorth, L. 2009. *Mobile Technologies. From Telecommunications to Media*. Abingdon, England: Routledge.

Guida, J. and Crow, M. 2009. E-government and e-governance. In, Unwin, T. (ed.). *ICT4D Information and Communication Technology for Development*. Cambridge, UK: Cambridge University Press.

Heeks, R. 2006. *Implementing and Managing eGovernment: An International Text*. London: SAGE Publications.

Helbig, N., Gil-Garcia, J. and Ferro, E. 2009. Understanding the complexity of electronic government: Implications from the digital divide literature. *Government Information Quarterly*, 26(1), 89–97.

Henriksen, H. and Damsgaard, J. 2007. Dawn of e-government – an institutional analysis of seven initiatives and their impact. *Journal of Information Technology*, 22(1), 13–23.

Hjorth, L. 2009. *Mobile Media in the Asia-Pacific. Gender and the Art of Being Mobile*. Abingdon, England: Routledge.

Ho, C. and Wu, D. 2009. Online banking performance evaluation using data envelopment analysis and principal component analysis. *Computers & Operations Research*, 36(6), 1835–42.

Horst, H. and Miller. D. 2006. *The Cell Handset. An Anthropology of Communications*. Oxford, England: Berg.

International Telecommunications Union (ITU). 2002. *World Telecommunications Development Report. Reinventing Telecoms, World Telecommunication Indicators*. Geneva: ITU.

----- 2009. *Measuring the Information Society: The ICT Development Index*. Geneva: ITU.

Ito, M., Okabe, D. and Matsuda, M. 2005. *Personal, Portable, Pedestrian. Mobile Handsets in Japanese Life*. Cambridge, Mass: The MIT Press.

Ivantury, G. and Pickens, M. 2006. *Mobile Handset Banking and Low-Income Customers – Evidence from South Africa*. CGAP/UN Foundation/Vodafone Group Foundation. Available at http://www.cgap.org (accessed 17 November 2009).

Jagun, A., Heeks, R. and Whalley, J. 2008. The impact of mobile telephony on developing country micro-enterprise: A Nigerian case study. *Information Technologies and International Development*, 4(4), 47–65.

Jensen, R. 2007. The digital provide: Information (technology), market performance, and welfare in the south Indian fisheries sector. *Quarterly Journal of Economics*, 122(3), 879–924.

Jipp, A. 1963. Wealth of nations and telehandset density. *Telecommunications Journal*, July, 199–201.

Juciute, R. 2009. ICT implementation in the health-care sector: Effective stakeholders' engagement as the main precondition of change sustainability. *Artificial Intelligence & Society*, 23(2), 131–37.

Kathuria, R., Uppal M. and Mamta. 2009. An economic analysis of the impact of mobile. In, Vodafone, *India: The Impact of Mobile Handsets*. The Policy Paper Series, No 9. 5 January. Available at http://www.vodafone.com (accessed 10 November 2009).

Katz, J. 2006. *Magic in the Air: Mobile Communication and the Transformation of Social Life*. New Brunswick, NJ: Transaction Publishers.

----- 2008. *Handbook of Mobile Communication Studies*. Cambridge, MA: MIT Press.

Kaul, S., Ali, F., Janakiram, S. and Wattenström, B. 2008. *Business Models for Sustainable Telecoms Growth in Developing Economies*. Chichester, England: J Wiley & Sons Ltd.

Kezang and Whalley, J. 2007. Closing the digital divide: The role of services and infrastructure in Bhutan. *Prometheus*, 25(1), 69–84.

Kwankam, S., Pablos-Mendez, A. and Kay, M. 2009. E-health: Information and communication technologies for health. In, Unwin, T. (ed.). *ICT4D Information and Communication Technology for Development*. Cambridge, England: Cambridge University Press.

Kim, G., Shin, B. and Lee, H. 2009. Understanding dynamics between initial trust and usage intentions of mobile banking, *Information Systems Journal*, 19(3), 283–31.

Knight-John, M. 2008. Making a business out of a Village Handset. In, Samarajiva, R. and Zainudeen, A. (eds). *ICT Infrastructure in Emerging Asia: Policy and Regulatory Roadblocks*. New Delhi: SAGE Publications India.

Kramer, W. and Paul, J. 2006. Mobile-enabled transactions for the base of the economic pyramid: A brief review of the 2006 'state-of-play'. In, Vodafone, *Economic Empowerment through Mobile*. The Vodafone CR Dialogues. Available at http://www.vodafone.com (accessed 12 October 2009).

Laukkanen, T. 2007. Internet vs mobile banking: Comparing customer value perceptions. *Business Process Management Journal*, 13(6), 788–97.

Lee, M-C. 2009. Factors influencing the adoption of Internet banking: An integration of TAM and TPB with perceived risk and perceived benefit. *Electronic Commerce Research and Applications*, 8(3), 130–41.

Ling, R. 2004. *The Mobile Connection – The Cell Handset's Impact on Society*. Amsterdam: Morgan Kaufman.

----- 2008. *New Tech, New Ties. How Mobile Communication is Reshaping Social Cohesion*. Cambridge, MA: The MIT Press.

Ling, R. and Pedersen, P. 2005. *Mobile Communications. Re-negotiating of the Social Sphere*. London: Springer-Verlag.

Madden, G. and Savage, S. 1998. CEE telecommunications investment and economic growth. *Information Economics and Policy*, 10(1), 173–95.

----- 2000. Telecommunications and economic growth. *International Journal of Social Economics*, 27(7/8/9/10), 893–906.

Mäenpää, K., Kale, S., Kuusela, H. and Mesiranta, N. 2008. Consumer perspectives of Internet banking in Finland: The moderating role of familiarity. *Journal of Retailing and Consumer Services*, 15(4), 266–76.

Margherio, L., Henry, D., Cooke, S. and Montes, S. 1999. *The Emerging Digital Economy*. Washington, DC: Department of Commerce.

Mechael, P. 2008. Health services and mobile: A case from Egypt. In Katz, J. (ed.). *Handbook of Mobile Communication Studies*. Cambridge, MA: MIT Press.

----- 2009. The case for mHealth in developing countries. *Innovations: Technology, Innovation, Globalisation*, Special Issue for the GSM Mobile World Congress.

Micevska, M. 2005. Telecommunications, public health and demand for health-related information and infrastructure. *Information Technologies and International Development*, 2(3), 57–72.

Miller, D. 2007. The unpredictable mobile handset. *BT Technology Journal*, 25(3/4) 321–28.

Misuraca, G. 2007. *E-governance in Africa: From Theory to Action*. Trenton, New Jersey/Ottawa, Canada: Africa World Press/International Development Resource Handset.

Molony, T. 2008. Running out of credit: The limitations of mobile telephony in a Tanzanian agricultural marketing system. *Journal of Modern African Studies*, 46(4), 637–58.

Nevens, T. 1999. The mouse that roared. *The McKinsey Quarterly*, 1, 145–48.

Ntaliani, M., Costopoulou, C. and Karetsos, S. 2008. Mobile government: A challenge for agriculture. *Government Information Quarterly*, 25(4), 699–715.

OECD. 2001. *Understanding the Digital Divide*. Paris: OECD.

----- 2009. *Communications Outlook*. Paris: OECD.

O'Hara, K. and Stevens, D. 2006. *Inequality.com Power, Poverty and the Digital Divide*. Oxford, England: Oneworld.

Peevers, G., Douglas, G. and Jack, M. 2008. A usability comparison of three alternative message formats for an SMS banking service. *International Journal of Human-Computer Studies*, 66(2), 113–23.

Porteous, D. 2006. *The Enabling Environment for Mobile Banking in Africa*. Report Commissioned by Department for International Development (DFID), Bankable Frontier Associates. Available at http://www.bankablefrontier.com (accessed 14 November 2009).

----- 2007. *Just How Transformational is M-Banking?* Report commissioned by FinMark Trust. Available at http://www.finscope.co.za (accessed 14 November 2009).

Qiu, J. 2009. *Working-Class Network Society: Communication Technology and the Information Have-Less in Urban China*. Cambridge, MA: The MIT Press.

Rajalekshmi, K. 2007. E-governance services through telecenters: The role of human intermediary and issues of trust. *Information Technologies and International Development*, 4(1), 19–35.

Ratten, V. 2008. Technological innovations in the m-commerce industry: A conceptual model of WAP banking intentions. *The Journal of High Technology Management Research*, 18(2), 111–17.

Rice, R. and Katz, J. 2003. Comparing Internet and mobile handset usage: Digital divides of usage, adoption and dropouts. *Telecommunications Policy*, 27(8/9), 567–623

Rowe, B. 2003. Rural technology deployment and access: Successes upon which to build. *Government Information Quarterly*, 20(2), 85–93.

Röller, L-H. and Waverman, L. 2001. Telecommunications infrastructure and economic development: A simultaneous approach. *The American Economic Review*, 91(4), 909–23.

Sinha. C. 2009. Effects of education and ICT use on gender relations in Bhutan. *Information Technologies and International Development*, 5(3), 21–34.

Southwood, R. 2008. *Less Walk, More Talk. How Celtel and the Mobile Handset Changed Africa.* Chichester: John Wiley & Sons Ltd.

Standage, T. 2001. A survey of the mobile Internet: The Internet, untethered. *The Economist*, 13 October.

----- 2003. Beyond the bubble: A survey of telecoms. *The Economist*, 11 October.

----- 2009. Mobile marvels – a special report on telecoms in emerging markets. *The Economist*, 26 September.

Sullivan, N. 2007. *You can Hear Me Now. How Microloans and Cell Handsets Are Connecting the World's Poor to the Global Economy*. San Francisco, CA: Jossey-Bass.

The Economist. 2005. The real digital divide. 10 March.

----- 2009. Banks and information technology: Silo but deadly. 5 December.

Thompson, S. and Dean, M. 2009. Advancing information technology in health care. *Communications of the ACM*, 52(1), 118–21.

Tookey, A., Whalley, J. and Howick, S. 2006. Broadband diffusion in remote and rural Scotland. *Telecommunications Policy*, 30(8/9), 481–95.

Trkman, P. and Turk, T. 2006. A conceptual model for the development of broadband and e-government. *Government Information Quarterly*, 26(2), 416–24.

van Dijk, J. 2005. *The Deepening Divide – Inequality in the Information Society*. London: SAGE Publications.

Veeraraghavan, R., Yasodhar, N. and Toyama, K. 2009. Warana unwired: Replacing PC with mobile handsets in a rural sugarcane cooperative. *Information Technologies and International Development*, 5(1), 81–95.

Vehovar, V., Sicherl, P., Hüsing, T. and Dolnicar, V. 2006. Methodological challenges of digital divide measurements. *The Information Society*, 22(5), 279–90.

Vodafone. 2006. *The Roles of Mobile Handsets in Increasing Accessibility and Efficiency in Healthcare*. The Vodafone Policy Paper Series, No 4. Available at http:www.vodafone.com (accessed 20 October 2009).

Waverman, L., Meschi, M. and Fuss, M. 2005. The impact of telecoms on economic growth in developing countries. In Vodafone. *Africa: The Impact of Mobile Handsets*. The Vodafone Policy Paper Series, Number 3. Available at http://www.vodafone.com (accessed 18 October 2009).

Williams, H. and Torma, M. 2007. Trust and Fidelity: From 'under the mattress' to the mobile handset. In Vodafone. *The Transformational Potential of M-Transactions*. The Policy Paper Series, No 6. Available at http://www.vodafone.com (accessed 19 October 2009).

Yakhlef, A. 2001. Does the Internet compete with or complement bricks and mortar bank branches? *International Journal of Retail and Distribution Management*, 29(6), 272–81.

Yoo, Y., Lee, J-N. and Rowley, C. 2008. *Trends in Mobile Technology and Business in the Asia-Pacific Region*. Witney, England: Chandos Publishing.

Index

Kolibri 360 GmbH
Ferdinand-Jühlke-Straße 7
99095 Erfurt, Deutschland
produktsicherheit@kolibri360.de